Psychiatry in the Scientific Image

Philosophical Psychopathology: Disorders in Mind
Owen Flanagan and George Graham, editors

Psychiatry in the Scientific Image

Dominic Murphy

The MIT Press
Cambridge, Massachusetts
London, England

First MIT Press paperback edition, 2012

© 2006 Massachusetts Institute of Technology

This book was set in Sabon by SNP Best-set Typesetter Ltd., Hong Kong.

Library of Congress Cataloging-in-Publication Data

Murphy, Dominic.
Psychiatry in the scientific image / by Dominic Murphy.
   p.   cm.—(Philosophical psychopathology. Disorders in mind)
Includes bibliographical references and index.
ISBN13: 978-0-262-13455-2 (hbk. : alk. paper) — 978-0-262-51744-7 (pb.)
1. Psychiatry—Philosophy. I. Title. II. Series.

RC437.5.M87   2006
616.89.001—dc22
                                      2006043805

Kate's

Contents

Preface

There ain't nothing more to write about, and I am rotten glad of it, because if I'd a knowed what a trouble it was to make a book I wouldn't a tackled it.
—Mark Twain, *The Adventures of Huckleberry Finn*

When, as a graduate student, I became interested in the significance of psychiatry for the philosophy of mind, I went looking for a book that laid out the various issues and tried to make the workings of modern psychiatry philosophically illuminating. I found many fascinating books and articles written by philosophers, scientists, and clinicians about particular topics within psychiatry, but I couldn't find the book I wanted. So in the end, I decided to write it myself. The result is an opinionated treatment, which tries both to introduce the issues and to persuade you to think about them in particular ways. It covers a lot of ground, for mental illness touches on most aspects of our lives and every operation of the mind. Even so, some psychiatrists and clinical psychologists will doubtless point to omissions. In response, I plead that this is chiefly a philosophical book, though I hope one not just for philosophers. It examines the chief issues raised by psychiatry for the philosophy of mind and the philosophy of the biological and cognitive sciences. For reasons I don't understand, philosophers have not paid much attention to psychiatry since (or outside) the Freudian tradition. There is a lot to be done, and there are hopeful signs that philosophy is beginning a serious engagement with psychiatry. Let's hope it will benefit both disciplines.

I have accumulated many debts along with the way, and not the least pleasant part of finishing this book is the opportunity it offers to acknowledge them. My interest in the philosophy of mind was formed as an undergraduate by the teaching and example of Bill Lyons, whose conception of the subject was marked by the belief that it should be continuous with the sciences and concern itself with all of our mental life, and not just a few fashionable topics. As an philosophy master's student, I benefited from the supervision of Mike Martin, and it was a seminar run by Mike and Gabriel Segal that first made me think about a number of the questions that crop up in this book.

A more proximate cause was a seminar on philosophical psychopathology—organized at Rutgers by Steve Stich, Terry Wilson, and Rob Woolfolk—which led me to write a distant ancestor of this book as a doctoral thesis under Steve's direction. Like everyone who has been taught by him, I am at a loss to convey just how much I owe to Steve, and certain that I can never repay him. Terry and Rob welcomed me into their classrooms in the clinical psychology program at Rutgers, and I hope that Terry still finds what I like to think of as my enlightened scientism about mental illness congenial. Rob has progressed from teacher to collaborator and friend, and much of the book bears his imprint.

A postdoctoral year at Washington University in St. Louis introduced me to a different conception of the place of philosophy of mind within the cognitive sciences. I am grateful to Bill Bechtel, Randy Buckner, Andy Clark, and Jesse Prinz for helping with that chapter of my philosophical education, and to Alison Wylie for her comments on an early version of chapter 8.

I have many debts to my Caltech colleagues Fiona Cowie, Alan Hajek, Chris Hitchcock, Steve Quartz, and Jim Woodward. They have provided a remarkably stimulating and congenial environment in which to work, and been of great help as the book dragged along. Fiona, in particular, has read many drafts and done her best to inculcate in me the virtues of crispness and clarity that mark her own work. Jim, Chris, and Alan have been helpful readers and even more helpful tutors, and much of the philosophy of science in this book is the result of conversations with them. My students at Caltech have been a delight to teach, and a spur to thinking through the issues.

A great many other people have provided encouragement and criticism along the way. I especially want to thank, in roughly chronological order, Richard Samuels, Ron Mallon, Louis Sass, Jerry Fodor, J. D. Trout, Mike Bishop, Ellen Landers, John Doris, Bill Wimsatt, Mac Pigman, Kim Sterelny, Paul Griffiths, Steve Downes, Ken Binmore, Or Neeman, Manuel Vargas, Peter Godfrey-Smith, and Jose Bermudez. An invitation from Peter Machamer and Sandy Mitchell to talk to their NEH seminar in the summer of 2003 was a great help to me in figuring out how the various pieces finally fit together. A special role was played by George Graham, who was a stimulating and sympathetic critic and the initial editor of this series, and I have more recently become grateful to his successors, Jennifer Radden and Jeff Poland. I'm also indebted to Tom Stone at The MIT Press, who put up with a lot, and bought me lunch when my wallet was stolen.

Last, I thank Kate for all her love and support, and Liam, for the periods of sleep he granted me between his birth and this book's completion. May all of us remain only theoretically acquainted with its topics.

The mind, it's said, is free
But not your minds. They, rusted stiff, admit
Only what will accuse or horrify.
Like slot machines only bent pennies fit.
—Philip Larkin, *Neurotics*

1.1 What This Book Is About

1.1.1 Insults and Obsessions

Human beings, those partially domesticated animals, can suffer in a lot of ways. Here are some people who share one of them:

The Marquise de Dampierre was from childhood afflicted with convulsive contractions of the muscles in her hands and arms that later spread to her shoulders, neck, and face. Her spasms began to affect her voice and speech, and she started to utter bizarre shouts. Then she began to curse compulsively, and hurl insults and obscenities at those around her. This rather unsettled Parisian high society.

When Tim was five he began to blink repeatedly and roll his eyes. Then he started to make unusual head movements, which led his classmates to nickname him Noddy. In his early teens he started to cough habitually even when there was nothing wrong with him. The odd "ticcing" movements persisted, preceded by feelings of tension or tickling sensations in the area of the tic. He also developed an urge to mimic other people's actions, and to lick and smell the objects around him.

As a teenager, Michael constantly blinked his eyes and would sometimes contort his body and appear to shrug his shoulders. As he grew older he became unable to prevent himself from blurting out socially inappropriate or threatening remarks. When he meets an African-American he cannot stop himself shouting racist epithets and he once told a woman "I want to rape you!" Several legal orders have been handed down barring Michael from restaurants and apartment complexes where people have been upset by his behavior. He tries to tell people who find him disturbing that he has a disease that makes him act as he does, but they seldom believe him.

These three people all suffer from Georges Gilles de la Tourette's Syndrome, or Tourette's Disorder as it is more often called.[1] Tourette's is a mental disorder. This book is a philosophical discussion of what it means to say that someone has a mental disorder, and how we might explain and classify mental disorders.

Let's look at Tourette's in more detail, starting with some of the symptoms then reflecting on questions that this brief glance at Tourette's raises: What makes Tourette's a mental disorder? How are we to distinguish it from other disorders, both mental and nonmental, that seem to share its symptoms? How do we explain Tourette's?

Tourette's is famous for turning people into compulsive swearers, although in fact only a minority of patients engages in "coprolalia," as this uncontrollable cursing is known. Nor is coprolalia necessary to receive the diagnosis of Tourette's. According to the DSM-IV-TR, the essential features of Tourette's syndrome are tics, which are sudden, rapid, recurrent, nonrhythmic, stereotyped motor movements or vocalizations (APA 2000 [DSM-IV-TR], 108). Motor tics range in complexity from blinking to complicated gyrations, and phonic tics range from clearing one's throat to simple sentences to coprolalia and to compulsive imitation of others' speech—some phonic tics are barely audible, while some are yelled (Leckman, King, and Cohen 1999). Other sorts of inappropriate verbal behavior also are found among people with Tourette's, such as involuntarily confessing to a murder

---

1. The descriptions have been adapted from cases mentioned by Gilles de la Tourette (1885), Robertson and Baron-Cohen (1998), and Kushner (1999). The disorder bears Georges Gilles de la Tourette's name and the Marquise de Dampierre (first discussed by Itard in 1825) is its emblematic patient (see, for example, Finger 1994, 233–236). Gilles de la Tourette wrote about Dampierre but never examined her personally. He affirmed that his teacher, Charcot, had examined her frequently in her old age. Kushner is skeptical (1999, 21). Dugas (1986) argues that three of Gilles de la Tourette's nine original case histories would not meet contemporary diagnostic criteria for the disease that was named after him, but Dugas nonetheless commends Gilles de la Tourette for having made sense of disparate cases and describing the course of the disease in a way that has withstood the test of time. Tourette's has been included in every edition of the *Diagnostic and Statistical Manual of Mental Disorders* (APA 1952, 1968, 1980, 1987, 1994, 2000.) Following widespread precedent, I will from now on refer to these books as DSM-I, DSM-II, DSM-III, DSM-III-R (third edition, revised), DSM-IV and DSM-IV-TR (fourth edition, revised text).

(Kurlan et al. 1996). People with Tourette's often are afflicted with obsessions and compulsions as well especially when it comes to getting things "just so." They may spend hours performing tasks in search of an elusive sense that something is right, whether it is by arranging things in their closet, parting their hair, or, in the case of a Jesuit schoolteacher, "having to recite the Ave Maria in a special way while fastening his eyes to the right on a statue of the Virgin Mary" (King et al. 1999, 51).

1.1.2 The Issues

Perhaps all of these behaviors have everyday analogues that do not strike us as pathological—any one of us might occasionally swear at people, inadvertently say the wrong thing, or engage in a ritual (think of superstitions). But when it happens on the scale of Tourette's, it stops being embarrassing but normal. It becomes a funhouse mirror distortion of normal humanity. People with Tourette's are not mad; they aren't floridly psychotic. Psychotics believe things like this: they think that someone else is speaking to them inside their head, or that insects are being spontaneously created in their vicinity because the Upper God has become less hostile to them "since the steady increase of soul voluptuousness" (Schreber 1903, 218). People with Tourette's are not insane in the way that psychotics are, but they look like they have something wrong with their minds all the same. So here's one issue—what makes certain odd behaviors symptoms of mental disorder? To make the question more complicated, many of the symptoms of Tourette's do not seem to fit the model of distorted mental life.

Consider the tics. It is the tics that really distinguish people with Tourette's, and they may just be muscular spasms without any of the social significance of cursing. In what sense are they mental? Moreover, tics are not only found in Tourette's Syndrome. DSM-IV-TR (108) recognizes four different Tic Disorders, including Tourette's. The others are Chronic Motor or Vocal Tic Disorder, Transient Tic Disorder, and Tic Disorder Not Otherwise Specified. DSM-IV-TR (111) also notes that if Tourette's is to be diagnosed, the tics must not be caused by medication or by a general medical condition such as Huntington's Disease. The manual also distinguishes the tics found in tic disorder from other sorts of abnormal movements, such as the jerky motions associated with Sydenham's Chorea. Like Tourette's, Sydenham's chorea is a disorder that strikes in childhood and manifests itself in abnormal movements. So why are the tics of Tourette's mental while those of Sydenham's chorea are part of a general, physical medical condition?

Of course, Tourette's sufferers also may have symptoms that seem more psychological, but that does not explain why DSM-IV-TR regards the tics as the business of psychiatry and not some other branch of medicine. It would be possible to parcel Tourette's symptoms out into abnormal movements and abnormal cognitive processes and regard only the latter as falling within the province of mental disease. DSM-IV-TR applies exactly this strategy to Huntington's disease. The movements of Huntington's patients are regarded not as a psychiatric matter at all but as tics due to a general medical condition, whereas dementia due to Huntington's disease is a psychiatric diagnosis (DSM-IV-TR, 165).

The strange rituals and obsessions that Tourette's patients may suffer from do seem psychological, but, like the tics, they do not afflict only patients with Tourette's. These symptoms overlap with those of Obsessive-Compulsive Disorder (OCD), and not all Tourette's patients suffer from them. So controversy exists over whether people with both tics and obsessions have Tourette's, OCD, or something distinct from both (Cath et al. 2001, King et al. 1999).

Tourette's then, is a disorder that seems to share qualities with other mental illnesses, with illnesses that are not mental, and perhaps with personality traits that are not illnesses at all. These issues raise the central question of what mental illnesses are, and more specifically what is mental about them, what makes them illnesses, and how we can explain and classify them. These are the questions addressed in this book. The endpoint is a theory of classification. It is the job of a system of psychiatric classification (a nosology, in the jargon of the trade) to help us settle these questions by identifying the mental illnesses and distinguishing them from one another so that they stand in appropriate relations. This book is an exploration of the philosophical issues that the project of nosology, broadly conceived, brings up.

One issue we have just met: the concept of mental disorder. A satisfactory nosology must at a minimum classify mental disorders and exclude other things, so error about the concept of mental disorder is an invitation to a wrongheaded taxonomy. We do not want our nosology to say that people with a belief in supply-side economics are mentally ill, nor do we want it to say that schizophrenics are not mentally ill. These are generally uncontroversial cases (although some people would disagree about schizophrenia (Laing 1965) but some controversies raise the same questions: is alcoholism a mental disorder, for example, or a physical disease, or merely a certain way of living a life that happens to be centered on strong drink the way a bibliophile's revolves around books? We can adduce factual and conceptual considerations on either side of the debate about addiction, and here as elsewhere we lack clear methodological and conceptual guidelines. And the lack is pressing, since being wrongly diagnosed as mentally ill is a nightmare, yet being wrongly diagnosed as not mentally ill is to be abandoned to forces that one often cannot master, which can cause great misery and even death.

But of course we want a nosology that's more nuanced than the well/sick dichotomy. If we classify dissimilar individuals together we will treat them in ways that ignore the important differences between them. Some researchers into Tourette's believe that the fourfold distinction among tic disorders in DSM-IV-TR lumps too many diverse cases together, and that we should recognize up to eight different types of tic disorder, most with several subtypes (Tourette Syndrome Classification Study Group 1993). To get people into the right taxa, we need to understand how their conditions are caused. So we need to know how to explain mental illness if we are to classify it properly. What should the explanations look like?

We don't know the explanation of Tourette's yet, and the same is true of most forms of psychopathology. Not that we lack theories. In fact they are abundant: some are busy and productive and at least partially successful; others are moribund, and some are stone dead, although their corpses are occasionally moved into new positions by optimistic graverobbers. Some psychiatric theories are grandiose in scope, others are very modest, but none has managed to unite the mental-health field behind it. The ensuing mix of theoretical disunity and conceptual uncertainty is bad news for classification, but its ramifications are not just technical. Errors in psychiatric classification may affect millions of people. One epidemiological survey (Kessler et al. 1994) found that 48% of Americans had suffered from a psychiatric disorder at some point in their lives. But that assumes the current taxonomy is correct, and many theorists feel that the diagnostic system that the study relied on is actually overinclusive and pathologizes some perfectly normal people.

We do not want our classification of mental disorders to depend on a false theory, and since we lack a clear understanding of how the mind works and how it breaks down, perhaps staying away from theory is a good policy. But to renounce theories altogether as a matter of principle, as DSM-III (APA 1983) officially did, deliberately cuts psychiatry off from any hope of finding out how things really are. In that case we would have no way to answer questions about what conditions, traits, or forms of life should or should not be included in a list of disorders. In the end, nosology must be based on causal explanations of what is being classified. We do not need a grand theory to achieve this, although the process might help a grand theory to emerge. In many cases, good theories of the causes of measurable abnormalities in thought and behavior can be obtained at a local level, via experiment and clinical investigation of the phenomena of interest. These inquiries go on at arm's length from large-scale theories of either the normal or abnormal mind. In looking at the philosophical issues that psychiatry raises, I tend to rely on theories of fairly restricted scope, ones that aim to understand at most several related conditions and explain their occurrence. But in this book I try to show how we can build a methodological framework that makes room for all these theories and also lets psychiatry take advantage of and contribute to developments in the cognitive sciences, broadly construed, that do show some promise. The aim of this book is to illuminate the road ahead, not to announce that everything is finished.

So there are both empirical and conceptual aspects to the three main issues before us, which are how we should conceive of, classify, and explain mental illness. The body of the book begins with a discussion of the concept of mental disorder as it appears in what I call the orthodox program of conceptual analysis. I reject a number of assumptions of the orthodox program, both in psychiatry and in the philosophical literature.[2] Some analyses are unable to deal with counterexamples, but, more substantively, I also question the whole project. The current literature on mental illness lacks a coherent concept of the mental and a satisfactory account of disorder, and it yields far too much authority to commonsense thought about the mind. I do not deny that everyday thought about the mind has a role to play, but I reject the idea that a classification of mental illness should be a regimentation of our commonsense intuitions. Some scholars think that is a worthwhile goal, but I do not.

I also dispute the desire for a purely observation-based set of diagnostic categories or some otherwise atheoretical or operationalized nosology. If we take seriously the claim that purely observation-based or operational categories are required, it becomes very hard to undertake many of the central inquiries of modern psychiatry. More importantly, we run the risk of grouping different disorders together merely because their outward manifestations are similar.

The proposal that classification should be purely operational, like the proposal that it should regiment common sense, does not help us to develop a classificatory scheme that accomplishes what we want, although both proposals have been defended artfully. I will try to extract lessons from both

2. What I call the philosophical debate here is represented by the orthodox program in chapter 2, but it does not include just philosophers. Psychiatrists, psychologists, historians and sociologists have contributed more to the philosophical debates, extensionally defined, than philosophers have, and some prominent theorists (like Jerry Wakefield) have a foot in both the philosophical and the clinical camps.

proposals that do help us toward what we want from a scheme of classification. A scheme of psychiatric classification is a set of concepts that represent particular disorders and the larger classes into which they should be aggregated. The overall conceptual family has a broader and a narrower job to do.

The broad job of a nosology is to function as any good taxonomy should: it must not be so complex as to be unlearnable and it should apply generally to its domain and guide inquiry in that domain. I also think, although some will find this an egregious addition, that it should balance with those other desiderata a desire to accurately represent a portion of nature. But as I say, this requirement must be balanced against others. Like the legendary map with the scale of one inch to one inch that was highly accurate but couldn't be unrolled for fear of upsetting the farmers, a system of classification that aims to capture nature too completely is unusable.

The second job of a nosology is to discharge its specific responsibilities to psychiatry—to help us identify, explain, and treat mental disorders. My objections to current nosological approaches are motivated by a concern that neither the broad nor the narrow job is likely to be done well if the orthodox program of conceptual analysis sets the agenda. I raise these issues and suggest an alternative agenda in chapters 2 and 3 and then work out the implications of the agenda in the remainder of the book.

Although my view that classification should be causal and tied to a theory of the hidden structure of the taxonomic domain is a very demanding one, it is exhibited in other areas of inquiry and it is a goal in still more.[3] And it may be that by now we know enough, about both normal and abnormal psychology, to make some safe bets about what the forms of explanation will look like. I argue in this book that while we do not have all the answers, we nevertheless can approach psychopathology in a way that lets us see what the eventual explanations will look like, and what discriminations among conditions they will let us make.

The picture is not overly pessimistic, though: some of the conditions I have mentioned are quite well understood by now. Can we assume that the explanations these have received will in the end be generalized to cover the con-

---

3. For example, the International Classification of Sleep Disorders was produced by deliberations of a committee who held that "organization on the basis of symptomatology was unsatisfactory." Therefore, "a structure organized pathologically and less symptomatically was sought," but too much of the relevant etiology was unknown. Nonetheless, a causal classification remains the goal (American Sleep Disorders Association 1990, 23).

ditions that are not yet understood? For example, we now understand that the physical basis of Huntington's disease is genetic: Huntington's is caused by an abnormally long CAG repeat near the tip of chromosome 4 (Ashizawa et al. 1994). Sydenham's chorea has a physical basis too: it is a complication of a streptococcal infection involving inflammation of the basal ganglia and a variety of insults to the central nervous system. This view rests on anatomical investigation conducted in the early twentieth century (Greenfield and Wolfsohn 1922) and latterly has been supported by neuroimaging studies (Giedd et al. 1995). What we know about the neurological basis of Tourette's, and its similarity to other conditions with fairly well-understood physical bases, convinces many people that Tourette's is a brain disorder. For example, obsessive-compulsive disorder appears to involve basal ganglia dysfunction just as Sydenham's chorea does, and the development in childhood of both tic disorders and OCD is associated with streptococcal infection (Swedo et al. 1998). Tourette's also seems to share some properties with "genetic expansion" diseases like Huntington's, such as having an earlier age of onset when inherited from the mother's side. Tourette's sufferers also seem to have more motor tics when they inherit the disease from their mother, but more vocal tics when they inherit it from their father. This suggests that another genetic process, known as imprinting, may be at work (Davies, Isles, and Wilkinson 2001).

Some theorists have seized on findings like these and argued that Tourette's is an organic, physical condition. But others have insisted that it must have a psychological basis because the pattern of behavior associated with the disease suggests some sort of psychological conflict (Cohen and Leckman 1999, 4–5):

Are the tics voluntary or involuntary, physical or mental? Why do they come and go? If patients don't have them when they are in the office, why do tics appear as soon as they walk down the hall to the elevator? What accounts for the specific symptoms, the virtuosity with which they are chosen and executed? What makes patients feel better when the tic is finally emitted? Are tics, then, like sneezes or masturbation? Like an itch or a scratch? Why do patients curse? How do they learn the curse words? Why are mothers so often the target of aggressive attacks—attacking words, pinching, yelling and controlling? When a child is doing something his parents and he hate, can he control it? How do you know what is intentional and what is beyond control? Is Tourette's syndrome organic or functional?

Over the years, different sorts of theorists have tried to answer these questions about Tourette's and other disorders according to the resources of their theories, with psychological explanations ranged against physical ones. At about the time that postmortem studies were establishing the pathology of Syndenham's chorea, for example, Sandor Ferenczi argued that tics were the expression of repressed masturbatory urges (Kushner 1999, 58–59).

Debates that pit physical theories of etiology against psychological ones are common in psychiatry. But they can seem frustrating and misguided even to the people involved in them. After all, everyone knows that psychological phenomena, like all of human behavior, are rooted in brain processes. I assume this unsophisticated materialism throughout this book. Working out the metaphysics in a philosophically satisfactory way is an enormous task, but it is one that we can put aside.

So it is not materialism that separates brain-based and psychological theories of mental illness. The natural way to understand debates about the causes and explanation of mental illness is in part philosophical but also a historical legacy. Historically, they reflect both a lingering everyday inability to take materialism seriously, and perhaps a long-standing reluctance on the part of brain science to look at cognitive phenomena. Philosophically, they are arguments about levels of explanation. Biological theories of psychopathology sometimes seem to deny that there are interesting or fruitful explanations to be had at psychological levels of explanation, or to assert that molecular theories are in some sense more fundamental than other kinds of explanation. A commitment to reductionism, rather than materialism, motivates the drive for biological explanations of mental illness. Also playing a role is a view of disease taken from medicine, which relies on a picture of symptoms as expressions of an underlying disruption to the normal functioning of an organ system.

The issues are partly empirical but partly conceptual. Understanding a particular mental illness like Bipolar Disorder (Manic-Depression) mostly involves sorting through a torrent of empirical findings, but it also involves more straightforwardly philosophical concerns about the nature of causal explanation and the mutual relation of theories in the sciences of the mind/brain. The wider issue of what a mental disorder is to begin with is less amenable to empirical oversight: it involves thorny conceptual puzzles.

So, at the root of this book are the questions of what we want from a system of psychiatric classification and how we can achieve it. Because we want nosology to rest on causal understanding. This issue of a correct nosology is intimately connected with the issue of how mental disorders are to be explained as well as the conceptual issue of what a mental disorder is. The three issues of conceptual understanding, classification, and explanation are intimately related. So this book discusses all three issues. In a moment I will preview the plot. But first let me give you the bottom line.

This book is deeply reactionary. It is a qualified defense of the medical model, which says psychiatry is a branch of medicine dedicated to uncovering the neurological basis of disease entities. The medical model is sometimes presented as a brave new world. In fact, it dates back to at least the third quarter of the nineteenth century: it clothes in modern dress Henry Maudsley's desire "to arrive at some definite conviction with regard to the physical conditions of mental function, and the relation of the phenomena of the sound and unsound mind" (1867, vi).

The medical model, given that we are animals with a biology, including a brain that is the foundation of our mental life, has intuitive appeal in psychiatry. I think it's especially appealing when one remembers that studying the brain draws on the cognitive and social sciences as well as just molecular biology, for reasons I will go over in the central part of the book. But there is another reason for adopting the medical model.

Psychiatry as it stands is not a particularly mature or successful enterprise. There are reasonable doubts about the success of clinical practice, including the utility of drugs, and the avowed philosophical foundations of psychiatry, I will try to show, are hopeless. But I do not want to advocate a total reform. In the face of this uneven record it might be tempting to adopt a Jacobin approach, sweep away the former regime and build a new set of structures in its place. Some people do think this is what psychiatry needs, but I do not. The situation is not one of total failure, after all, since there is plenty to admire about modern psychiatry, especially in comparison to its predecessors.

So I advocate starting from more or less where we are, excavating the promising aspects of the medical model and showing where conceptual roadblocks stop progress toward an improved nosology. This raises some problems, though. In the end, I make a number of recommendations for nosology, but given the immaturity and patchy success of psychiatry, how can we tell the difference, using existing standards, between a good taxonomy and a bad one? Without a mature theory in the field, we can't take successful practice as unreconstructed evidence. Without an archival database of follow-up studies, what evidence do we have that current practice in psychiatry produces good outcomes, outcomes that couldn't be matched by alternative nosologies that start from a completely different set of premises? I use a kind of unification approach, drawing on a number of different sciences, and I assume that it speaks well of a taxonomy if it stands up to a number of challenges to diagnosis. But to show that unification supplies evidence of theoretical success, we can't appeal just to its utility for practitioners (or some corollary notion). We have to appeal to the maturity of the theory. But that is precisely what you can't do in the case of psychiatric diagnosis.

So, my solution, in essence, is to borrow from the neighboring sciences: I see psychiatry as the study of failures of normal human nature, where what counts as normal human nature is decided by a variety of disciplines that comprise the cognitive and biological sciences. These sciences are not without their own problems, but my basic project assumes that they are in good enough shape for us to conclude that if psychiatry, which studies the same subject matter (human beings), aims for consilience with the sciences of the normal mind/brain, and adopts their best practices, it will be on a sounder footing.

I defend three theses: the current literature on the concept of mental disorder is an impediment to research; the explanation of mental illness answers to the standards of good explanatory practice in the cognitive neurosciences; and the classification of mental illness should group symptoms into conditions based on the causal structure of the abnormal mind. I am going to conclude that, while we do not have all the answers, we can nevertheless now approach psychopathology in a way that lets us see what the eventual explanations will look like, and what discriminations among conditions they will let us make. Of course, Maudsley thought the same thing, and you may conclude that my optimism is as unwarranted as his was. Anyway, here's how it goes.

1.2 The Structure of the Book

Chapter 2 distinguishes two projects. The first project is one in which conceptual analysis establishes what counts as the subject matter of an inquiry and the constraints under which the inquiry should proceed. The second project is one in which a combination of empirical and conceptual assumptions serve as a methodological framework, revisable in practice, within which research projects can develop. I try to reorient psychiatry from the first project toward the second one.

The orthodox view among psychiatrists and philosophers of psychiatry is that we can distinguish between psychological malfunction and its consequences, and that mental illness should be seen as having an objective scientific component (the malfunction) and a normative, socially negotiated component (the harmful consequences of the malfunction). Call this the two-stage picture. It has its attractions: it promises to align psychiatry more closely with the rest of medicine while making room for values and social concerns, it makes sense of the goals of much psychiatric research and it does seem to accord with many folk-psychological intuitions about mental illness. The two-stage picture, then, assumes that the concept of mental disorder must respect commonsense understanding of what a mental illness is, and do so within the disciplinary framework of psychiatry. I question both parts of this consensus in chapter 3.

I argue in chapter 3 that there is in fact no coherent commonsense concept that adequately can regiment inquiry. I further contend that the disciplinary division between psychiatry and clinical neuroscience is a historical artifact that is impeding inquiry and has no respectable theoretical basis. Therefore, arguments that a condition is not, intuitively, a mental disorder, or is not a proper psychiatric explanandum but belongs to neuroscience, have no standing as objections to merging psychiatry and clinical neuropsychology. I advocate the merger as necessary to develop the broadest and most fertile approach to understanding psychopathology. However, rejecting intuitive constraints and existing disciplinary boundaries as legitimate shapers of inquiry is not the same as denying any legitimate social interest in psychiatry. I agree that therapeutic and legal contexts are ones in which concepts of mental illness answer primarily to our values and intuitions, but I deny that these extra-scientific contexts provide or require a general concept of mental disorder that can establish what psychiatry is. Much of the spirit of the two-stage picture survives this treatment, but it is cut loose from its current self-imposed conceptual constraints and reoriented as a set of methodological assumptions.

After reorienting the conceptual foundations of the two-stage picture in chapter 3, in chapter 4 I start evaluating its scientific commitments. The two-stage picture makes a number of empirical assumptions about cognitive architecture and the relation between brain and behavior. In addition, it makes conceptual assumptions about rationality and biological function. All these assumptions, I argue, need to be heavily qualified in the light of our best understanding of how the mind works and the best explanatory practices elsewhere in biomedicine. In chapter 4 I discuss those practices in more detail. I contrast the brain-centered "medical model" with the more pluralistic "biopsychosocial" model. These two models agree that mental illness can be caused by an array of biological and cognitive factors. The dispute between their adherents turns on the belief held by proponents of the medical model that molecular neurobiological explanations are more fundamental than other explanations. However, the notion of a fundamental explanation is very hard to make precise. What counts as a fundamental explanation typically depends on what questions we ask. With a few rare exceptions, mental illnesses are the effect of interrelated causal processes at different levels of explanation. In the vast majority of cases we cannot find fundamental causal factors. However, although I reject the idea that molecular reductionist theories are more fundamental in general, I do argue that they enjoy a privileged role in some cases. In general, molecular reduction, especially as it involves animal models, gives us a special kind of understanding that is closely tied to our abilities to intervene and manipulate a system in therapy, and to test therapeutic assumptions. The result of chapter 4 is a very ecumenical scientific picture, with many explanations of different types and at different levels, all jostling together. The overall idea, though, is that the medical model is vindicated by the general endeavor of tracing abnormalities in behavior and cognition to specific causal factors that are realized in brain tissue.

In chapter 5 I turn aside from the development of the positive picture to note some obstacles. In particular, I argue that the two-stage picture's separation of scientifically determined malfunction and socially determined personal handicaps is especially difficult to sustain when personal maladjustment apparently depends on failures of rationality that are caused by malfunctions in "central" psychological systems. Here, attempts to localize symptoms in specific structures often hypothesize specialized belief-acquisition or abductive reasoning systems that are unsupported by science or depend on questionable folk-psychological assumptions about rationality. Worse, they face Fodor's arguments about the computational intractability of non-demonstrative reasoning. I conclude that many aspects of mental illnesses that involve central systems still seem ineluctably normative and that the prospects for the two-stage picture depend intimately on the resolution of difficult questions about human cognitive architecture. But, I argue, that does not mean that we cannot make any scientific headway, nor does it mean that we cannot develop a naturalistic theory of rationality and its vicissitudes. And it does not mean that we cannot explain conditions based on normative assumptions. Nor is disagreement over whether something is a mental illness an inevitable consequence of the normative aspects of its explanation.

In chapter 6 I present a simple theory of psychiatric explanation as causal validation. I criticize the existing concept of construct validity as unable to transcend its operationalist heritage and actually explain anything. The simple theory I proffer in response is that psychiatric explanation aims at articulating the causes of idealized representations (or "exemplars") of the typical course and clinical profile of a disorder. Exemplars specify a number of features or symptoms for the various disorders and will have slots left open at various levels of explanation for the causal explanation of symptoms at that level by a model of the abnormal processes, which is in turn derived from a model of how the system normally works. Thus, the exemplar for major depression might include lowered affect, negative (but complicated) self-assessment, and lethargy and lack of motivation, allowing for corresponding explanations in terms of interactions among the biology of mood, cognitive distortions that skew perceptions of reality in ways that maintain depressed mood, and failures of the behavioral activation system (which produces exploratory and reward-seeking behavior). The same process can be applied recursively to symptoms themselves within the wider structure. Thus, for instance, "hopelessness," a symptom of depression and other mood disorders, might have as stereotypical features "automatic cognitions" that lead one to make persistent, wide-ranging negative evaluations that attribute life-impairing problems to oneself ("I got a bad grade because I'm stupid; I've always been stupid and that's why I can't find a girlfriend"). Hopelessness might itself be explained by an interaction of an inherited temperament and stressful, aversive early experiences, together with a general cognitive style that leads one to overestimate one's ability to control the course of one's own life, and hence one's responsibility. I explain the exemplar approach via an extended study of recent work on schizophrenia that illustrates both the advantages of the approach and the difficulties raised in chapter 5.

In chapter 7 I discuss how sociocultural factors can be integrated into the neuropsychological approach to explanation outlined in chapter 6. This is important because aspects of mental illness, especially epidemiology and cultural variation in symptoms, are best seen as reflections of the social context. The chapter looks at recent work by Ian Hacking on the social construction of mental illness. I fault Hacking for not integrating the sociological and the neuropsychological in a way that makes apparent their causal interdependence. I offer an alternative, and very simple, explanation based on the representation of cultural information in the brain.

Chapter 8 discusses the currently fashionable view that evolutionary adaptedness might explain some mental illness. I distinguish three sorts of evolutionary explanations of mental illness: first, breakdowns in evolved computational systems; second, evolved systems performing their evolutionary function in a novel environment; third, evolved character traits designed to make people antisocial. I review these explanations with progressively less credulity in light of the empirical evidence, although, as I show, the second explanation may work when applied to phobias.

Finally, classification. Chapter 9 provides some philosophical background, and chapter 10 presents a theory of the foundations of psychiatric classification. Chapter 9 begins with a critique of the existing psychiatric classification system, enshrined in the *Diagnostic and Statistical Manual of Mental Disorders* (DSM). I argue that a lesson of recent philosophy of science is that classification should be based on the causal structure of nature. The DSM has not learned that lesson.

I then move on to the topic of natural kinds, because literature on the classification of mental illness is preoccupied with the question whether mental disorders are natural kinds. I distinguish among theories of natural kinds and disentangle the account of natural kinds I prefer from its competitors. Mental illnesses are not natural kinds according to my understanding of natural kinds, but this does not prevent us from classifying their exemplars according to the scheme I propose.

Chapter 10 offers the basis of a psychiatric classificatory scheme. I argue that psychiatric classification should be based around the exemplars that I introduce in chapter 6. The causal histories of exemplars may be very complicated, with different aspects of the disorder addressed at different levels of explanation. Nonetheless, I argue that we still can classify disorders based on their clinical presentation and the kind of causal history they exemplify, so that schizophrenia can be classified as a developmental cognitive disorder, for example, Alzheimer's as a degenerative cognitive disorder and Korsakoff's Dementia as a cognitive disorder caused by (usually alcohol-induced) damage to the Papez circuit and medialdorsal relay nuclei. I also look at the question of whether mental disorders should be seen as categorically different from normal mental functioning or situated along dimensions that are continuous with normality. I conclude that each conception seems to fit some disorders and that this issue cannot be determined in advance of further detailed inquiry.

The classificatory system I advocate is chiefly designed for researchers. However, it can be applied in clinical practice for purposes of therapy or rehabilitation. It gives therapists a set of resources that they can apply to particular patients, and causal information that helps establish where to direct therapeutic interventions.

1.3 What This Book Is Not About

This is, as far as I know, the first book-length treatment of psychiatry within analytic philosophy of science. That may suggest that there is absolutely no demand whatsoever for this volume.[4] However, philosophical interest in psychiatry has been increasing in recent years. Let me end my introductory remarks by situating what I want to do in this wider landscape.

Graham and Stephens (1994) distinguish three different ways of doing philosophical psychopathology.[5] The first draws on clinical literature to support positions in the philosophy of mind, the second studies the ethical and experiential dimensions of psychopathology, and the third, part of philosophy of science, looks at methodological issues in the science of psychopathology. Most of this book is at home in the third of these three bodies of thought. The book is a piece of philosophy of science, and it tackles three problems in the science of mental illness. One problem is that of a correct taxonomy and two problems are about how to understand the taxa. I ask how to conceive mental disorders in a methodologically fruitful way and how to explain them, as well as how we should classify them.

So I hope it is clear at the outset that I am not offering a comprehensive treatment of the many fascinating philosophical issues raised by psychopathology. For example, I offer next to no engagement of the second philosophical project that Graham and Stephens identify. I do not ask what it is like to be mentally ill. Cognitive neuroscientists have offered explanations of the problem that leads those afflicted with Cotard's Delusion to believe that they are dead and to defend their belief when challenged (e.g., Gerrans 2000; Young and Leafhead 1996). I will look at these explanations, but even if they are correct, understanding them will not allow the rest of us to experience the world the way that patients with Cotard's delusion experience it (which might be just as well, under the circumstances). Whatever the underlying cognitive explanation may be, there is surely something very strange indeed about the experience of having Cotard's delusion, and that strangeness is likely to remain elusive. I do not suppose that there is some ineffable qualitative, nonmaterial residue specific to the minds of patients with Cotard's delusion, only that a complete scientific explanation of what distinguishes them from those of who do not think we are dead—and I don't doubt that some explanation is possible—will leave us in the dark about

4. Although many of the issues have been aired by psychiatrists. See, for example, Bolton and Hill 1996.
5. See also Graham 2002 for a survey of more recent work. The trifurcation is not exhaustive, since some work fits none of the categories exactly; but it would be a shame to miss, e.g., Sorenson 2000.

what it is like to be in their position. There is a long tradition in phenomenological psychopathology, inaugurated by Karl Jaspers (Jaspers 1959)—although his achievement was great enough to cover much additional ground—that tries to understand mental illness from the inside. This project is carried on today in both explicitly phenomenological terms (e.g., Sass 1992, 1994), by more analytically minded philosophers (e.g., Stephens and Graham 2000), and by clinicians (Jamison 1995). As Goodwin and Jamison complain, (1990, 14) contemporary discussions of mental illness can look arid compared to the work of the great theorists who tried to understand insanity in all its dimensions. I will settle for aridity if I can make some progress on the issues I take up. Understanding the subjective experience of mental disorder is important, but it is not essential to psychiatric explanation or classification.[6]

Theorists who aim to understand mental illness from the inside often do so as part of a wider project that tries to settle or clarify issues in the philosophy of mind by appealing to psychopathology. This is the first prong of Graham and Stephen's trident. I think we usually lack the understanding of abnormal psychology required to illuminate the philosophy of mind, and what is presented as psychiatric warrant for a philosophical position evidence is often—though by no means always—a tendentious reading of the evidence, inspired by the very position it is supposed to support. Ian Hacking (1995b, ch. 16) makes this point vehemently about philosophical treatments of multiple personality disorder, but it applies more widely. The issues I confront belong to the philosophy of science more than the philosophy of mind. Apart from some gestures in chapter 3, I also avoid looking at issues of public policy or at the ethical dimensions of psychiatry and psychotherapy, either in the wider community or within the clinical domain, although I have some thoughts on those matters (Woolfolk and Murphy 2004). I would be proud if my conclusions helped to advance debate on these topics, but I am not going to confront them here.

6. For a dissenting view, see Mishara 1994. But Mishara runs together some topics that I prefer to keep separate. I agree with him that a full understanding of psychopathology involves knowing what it's like to suffer from it, but I am not convinced that we need to incorporate phenomenological concepts into explanation or classification. We can distinguish (i) patients get into mental states that involve a certain picture of the world (as happens when we characterize some psychotic or delusional states) from (ii) classification should be structured around patients' experience of the world. I am also very skeptical about Mishara's view that phenomenology offers a "theoretically neutral access" to experience (1994, 131).

It was Wilfred Sellars (1963a) who drew the distinction between the manifest and scientific images that my title alludes to. The manifest image is the view of the world obtained through human experience. The scientific image is what science tells us about the world. This book, then, is about the latter: it is concerned with how science understands mental illness, what the limits of that understanding might be, and how we might increase that understanding.

2 | The Concept of Mental Disorder

Anyone can make true assertions about dogs, but who can define a dog?
—William Whewell, *Philosophy of the Inductive Sciences*

2.1 Introduction

In this chapter I introduce what I call the two-stage picture of the foundations of psychiatry. Many theoretically inclined psychiatrists and clinical psychologists argue for some version of the two-stage picture, and most practicing psychiatrists take it for granted. The two-stage picture embraces an enlightened scientism, in which psychiatry works out the details of mental malfunction, and then a variety of sources assess the consequences of those malfunctions. It distinguishes sharply between (i) working out when organs in the body work improperly, a scientific enterprise that psychiatry shares with general medicine, and (ii) assessing how these findings bear on our evaluation of lives that are affected by breakdowns. The first project is what determines that someone has a frontal lobe lesion, a depressive cognition, a genetic susceptibility to anxiety or a serotonin imbalance. The second project asks if human beings can flourish if they have such physical or psychological abnormalities.

Opposition to the two-stage picture comes from those who deny that we can enforce the difference between scientific assessments of humanity and the ethical, social, or political assessments that usually are lumped together as the "normative." Social theorists and philosophers more often take this view, especially if they study psychiatry in the light of its history, which is discouraging. Opponents of the two-stage picture often believe any view about normality and abnormality in humans is irremediably normative.

This dispute usually is presented as disagreement over the concept of mental disorder. It is generally agreed—for reasons I will come to—that normative judgments play a role in assessing who is mentally ill. The dispute turns on whether these judgments of value determine or follow a prior stage at which we establish someone's psychology to be dysfunctional. The two-stage picture maintains that value judgments do not determine whether someone is dysfunctional. Judgments of value determine what we think about someone whose psychology we already know to be dysfunctional on empirical grounds. I say "subjective" since the literature tends to contrast subjective judgments of value with objective scientific findings. As Philip Kitcher (1997, 208–209) puts it:

Some scholars, *objectivists about disease*, think that there are facts about the human body on which the notion of disease is founded, and that those with a clear grasp of those facts would have no trouble drawing lines, even in the challenging cases. Their opponents, *constructivists about disease*, maintain that this is an illusion, that the disputed cases reveal how the values of different social groups conflict, rather than exposing any ignorance of facts, and that agreement is sometimes even produced because of universal acceptance of a system of values.

The acceptability of the two-stage picture, then, turns in the first instance on whether objectivism about disease can be sustained. Now, objectivism or constructivism could be revisionist or conservative. A conservative view says that our concept of mental illness is correct—it says how mental illness really is. (What do I mean by "our concept"? I'll come to that in a minute.) A revisionist view says that our concept of mental illness is false to the facts about mental illness. Revisionist objectivists say that there are indeed facts about disease, and they obtain regardless of how we think about disease—so our thinking about disease may return the wrong answers when we come to cases.

Objectivists in psychiatry tend to be conservative. They have two reasons for this. The good reason is their belief that our folk judgments about mental illness share the objectivist stress on psychological malfunction. I shall present some arguments in a moment that support this first reason for conservatism. The bad reason, which I treat at length in the next chapter, is that objectivists believe in the priority of common sense over science: they think that a folk-psychological concept sets out conditions on mental disorder and psychiatry adopts those conditions and looks for the processes or states in the world that meet them. This understanding of science's relation to common sense should be rejected. Of course there are a number of conceptual puzzles in psychiatry, and there is important foundational work to be done. Successful practices do not need foundations, but psychiatry is not all that successful: it is an immature (though not young) science with an uneven empirical record, and clinicians agree that psychiatry has conceptual problems.[1] But progress will not come from the categories and intuitions of common sense.

Constructivists, on the other hand, are more likely to be revisionists. They say the concept of mental disorder is used to medicalize behavior that is really just socially deviant. They often criticize objectivists, who regard ascriptions of mental disorder as based on facts about abnormalities in human functioning, for acting either in bad faith or out of misplaced scientism.

Both sides apply to psychiatry the regulative principle that conceptual analysis establishes what counts as the subject matter of an inquiry and the constraints under which the inquiry should proceed. I often will call the program defined by its commitment to that regulative principle the orthodox program. I promote a different project that sees a combination of empirical and conceptual assumptions as a pragmatic methodological framework, revisable in practice, within which research projects can develop.

This chapter introduces and expounds the conceptual debate. I am, with heavy qualifications, a revisionist objectivist. In this chapter, on broadly conceptual grounds, I defend objectivism against constructivism and try to motivate the two-stage picture. It is an important component of our thinking about mental illness, so before taking issue with it I want to show that it has considerable attractions: I will mention research programs that seem to fit the objectivist paradigm very well, and I will endorse criticisms that scold DSM-IV's departures from objectivism.

In the next chapter I defend my revisionism, which is based on the belief that the objectivist program only can be carried through if psychiatry and the cognitive neurosciences are synthesized. That violates some common-sense views about mental illness. Hence the revisionism. The objectivist position agrees with important central aspects of our folk theories of mental disorder. But the folk theories are an impediment all the same. We cannot provide necessary and sufficient conditions for the concept of mental disorder.

There are, of course, quite general reasons for doubting whether conceptual analyses can be successful; many philosophers of my intellectual temperament condemn the ineradicable defectiveness of conceptual analysis. But the specific reasons for my skepticism are: first, there is no stable concept

---

1. For example, Kendler (1990, 969) begins his discussion of psychiatric classification by listing some issues that he regards as "fundamentally nonempirical." See also Frances et al. 1994.

there to be analyzed, since our folk psychopathology, broadly understood, is not coherent enough to be subjected to regimentation, as the conceptual analyst requires; second, the mind is a set of capacities rather than a clearly defined entity, and as such it can become disordered in very many ways, and this heterogeneity makes it very improbable that a simple concept, rather than a loosely united family of ideas, can serve as a model for our understanding of psychopathology.

There are two ways to criticize conceptual analyses. One criticism is that they are poor analyses. Poor analyses fail to respect our intuitions, are open to counterexamples, and so on. Existing conceptual treatments of mental illness poorly respect intuitions in some respects, and I make criticisms of this first sort. But I also argue that conceptual analyses of mental illness face the second problem, empirical irrelevance. Conceptual analysis may tell us how we think about mental disorder, causation, genes, etc., but it cannot tell us what a mental disorder, a cause, a gene, etc., actually is. In advocating skepticism toward conceptual analysis and looking for a different way to think of the philosophical foundations of psychiatry, I chiefly attend to three issues. The issues are: first, the relationship between science and common sense, which I have mentioned already; second, what it means to say that a disorder is *mental*; third, the nature of psychological function and malfunction. I do not give my positive views on these issues in this chapter. I reserve that task for chapter 3.

For now, I am laying out the conceptual terrain. In this chapter I begin by distinguishing objectivism and constructivism about mental disorder in more detail. I quickly reject constructivism. It neither respects nor explains away the commonsense distinction between mental illness and nonpathological deviance. Then I motivate a broadly objectivist position with cases where objectivism justifies our feelings that something is fishy about certain diagnoses (even when those diagnoses are in the ostensibly objectivist DSM). I then look at the dominant current view of objectivism, which is based on the notion of malfunction, and I set out as an illustration of objectivism-in-action some recent work on autism that conforms to the objectivist paradigm.

2.2 Objectivism and Constructivism

2.2.1 Introduction

In this section I distinguish constructivism and objectivism about mental disorder in more detail and then attack constructivism. Kitcher introduces objectivism and constructivism[2] as analyses of disease in general, but the issues he raises apply to mental illness too. As a theory of our commonsense concept of mental disorder, or as a kind of sociology of practices surrounding the mentally ill, I think constructivism is clearly false. It does not capture the way we relate to the mentally ill. It also fails as a basis for a positive program for the science of psychopathology, in large part because it denies that there can be such a science. That is not to say that a constructivist denies that diagnoses of mental disorder are based on publicly ascertainable facts. The point of constructivism is that these facts are not facts about breakdowns in normal mental functioning. Rather, they are violations of values that are widely held within a society or within an influential segment of that society.

2.2.2   The Role of Values

If I prefer vanilla and you prefer strawberry, we have no way to adjudicate the dispute. But if I say that Adam is mentally ill and Eve is not, and you think it's the other way round, then there should be facts that we can appeal to in settling this dispute. The objectivist thinks that the facts are, as Kitcher says, facts about human nature: in principle, the judgment that Adam is ill and Eve isn't reflects some failure of Adam's workings while Eve functions normally.

A constructivist doesn't need to reply that there are no facts guiding our ascriptions of disorder, nor that there are no mental disorders (although many will say that). But constructivists dispute that attributions of mental disorder are sensitive to facts about human psychology that are true regardless of social mores. The relevant facts, for a constructivist, are not facts about how human minds or bodies work. They are social. Societies share norms, and some people transgress those norms. The transgression may be a factual matter—if there is a prohibition on masturbating in public then people who masturbate in public are as a matter of fact flouting that prohibition. Some people who violate norms are regarded as immoral, and others

---

2. Kitcher's objectivism is often called "naturalism"; however, "naturalism" vies with "externalism" for the title of most philosophically ambiguous term in the English language. Both objectivism and constructivism are compatible with philosophical naturalism in some sense. To avoid these distractions I use Kitcher's "objectivism." Also, even though one might distinguish between illness, injury, and damage, for the most part I do not. Sciences of the abnormal mind must comprehend lots of etiologies, including stroke and other causes of damage, and infection and other causes of illness. Where the distinction between illness and damage seems important, I mention it.

are regarded as mentally ill. (Others may be regarded as harmless eccentrics or seen in some other way.) A constructivist can concede that we look for distinguishing features in the biology or psychology of the deviants. But a constructivist will say that we do this only because we first decide on other grounds that these people are mentally ill and that we then cast about for something about them we can medicalize.

As I said, I am an objectivist. I agree that in psychiatry we seek knowledge about abnormal psychology and that there *are* facts about it to be discovered. That is, I think that we *can* talk about normal and abnormal psychology without just indulging our values. Hence, objective knowledge of psychiatric conditions is possible. But as soon as this is said, it becomes necessary to take some of it back. First, as many objectivists admit, our values and interests may determine what conditions we regard as pathological, even after the facts are in. We may conclude that someone is psychologically abnormal but not mentally disordered. This is not much of a qualification. It introduces norms as a source of judgments about mental illness, but only after we have made prior judgments on grounds of abnormalities in human nature. The second qualification is that the scientific assessment of abnormality is not always possible, and that when it is possible it varies across contexts of inquiry. That will take some elaboration, which I provide in chapters 4 and 5. For now, I want to continue with the defense of the possibility in principle of separating scientific and social assessments of human psychology. It is this possibility that is the premise of the two-stage picture: first, we agree on facts about the failure of someone's psychology to function properly. When we have decided that, there is still a question about how to think of the person who is malfunctioning. The second stage, of normative judgment, is universally recognized to be necessary, even by objectivists. Let's call *simple objectivism* the view that all there is to disorder is the failure of someone's psychology or physiology to work normally. A simple objectivist about mental disorder just identifies Adam or Eve as mentally healthy or disordered depending on the facts about their mental functioning relative to our best current theory of what mental functioning should be. There may actually only ever have been one simple objectivist about mental disorder, but the exception is illuminating: it's Thomas Szasz.

Szasz (1987) claims that mental disorder is a mythic notion that is used as a pretext for repressive acts and institutions. This is usually read us a constructivist view because it denies the reality of mental illness. But in fact Szasz's concept of disorder is quite different from that of the constructivists. Szasz has a very strict objectivist concept of mental disorder, but unlike other objectivists he argues that nothing in fact falls under that concept. Szasz thinks that the only respectable notion of disease is that of damage to bodily structures, and putative mental disorders are not the result of tissue damage. In reply, psychiatrists often agree with the first point but insist that mental disorders will eventually receive an explanation in terms of brain abnormalities. Szasz's response is that if something is a brain disease then it's physical, not mental: he appears, in fact, to be an eliminativist about the mental.[3]

Normative judgments cannot be all there is to the concept of mental disorder, but they cannot be neglected. They do inform our conclusions about whether it is bad to have an abnormal psychology, whether it makes no difference, or even whether it is desirable. A recent stir was made by evidence of a specific kind of brain lesion which turns the patient into a gourmet, fascinated by the intricacies of fine dining (Regard and Landis 1997: they call the result a "gourmand syndrome").

Someone who comes down with a "gourmet lesion" is not clearly any worse off, and may even see her quality of life improve. Here we seem to have an example of something—a brain trauma—that causes damage but can have beneficial effects, as in the gag about the hurricane that hit New Jersey and did a billion dollars worth of improvements. It is hard to see why an interest in fine dining should count as a disorder if it enhances one's life. (Britain could well do with an epidemic of this lesion.) We can imagine other cases—job performance improved by a malfunction that increases attention span, or an abnormal psychological trait that makes for greater popularity, charisma, or sexual attractiveness. The straightforward identification of dysfunctions with mental disorders is untenable, and simple objectivism is false. So theorists have looked for a more nuanced objectivism about mental disorders.

---

3. Szasz is also a libertarian about the political, inclined to think of all judgments of deviance as attempts to limit individual freedom. His antipsychiatry stance has had a bad effect on libertarian theorists. Sullum (2003) for example, has a perfectly defensible Millian view about the state's having no business regulating private drug use, and he correctly points out that plenty of frequent drug users experience little lasting harm. But he also defends his policy proposals with the absurd assertion that psychiatric models of chemical addiction as a neurological disorder have no scientific standing. The two-stage view allows us to ask, first, about the effects of drugs on the brain and, second, whether those effects, conjoined with our values, warrant our seeing drug use as a medical condition, a criminal act, or just a hobby. I think this prospect shows the attractions of the two-stage view.

In his influential writings on disease concepts, for example, Christopher Boorse (1975, 1976a) departed from simple objectivism by distinguishing "disease" from "illness." Boorse understood illness to depend on value judgments about suffering or deviance in addition to the presence of disease, which he saw as a perfectly objective matter, decided in the objectivist manner by whether someone's innards are working normally. He argued that a disease only counts as an illness if it is undesirable, entitles one to special treatment, or excuses bad behavior. Boorse represents the way objectivism acknowledges a normative dimension. The result is the two-stage picture: scientific judgments that a destructive or abnormal bodily or psychological process has occurred, plus, at the second stage, normative judgments about the extent and manner of the process's impact on a person's life. The constructivist denies that the first stage is anything but an ex post facto rationalization of our decision to call someone insane. The next section criticizes constructivism.

2.2.3 Constructivism Cannot Explain Our Judgments

Constructivism as a view about mental illness usually says that judgments of mental disorder rest on the norms of the surrounding society rather than empirical psychological abnormalities (e.g., Kennedy 1983). This view faces the problem of distinguishing the pathological from the merely disapproved of, a distinction we clearly make. We regard psychological and behavioral phenomena like racism, boorishness, cowardice, and hypocrisy in a bad light, but we do not think they are mental disorders. Therefore, theorists like Thomas Scheff (1999), who regards mental illness as a form of "residual deviance," face the task of saying why we pathologize selectively, singling out only some forms of residual deviance as mental disorders.

Scheff claims that whereas some forms of deviance are forbidden by law or other sorts of codification explicitly, other forms rely for their existence on unstated social norms that rarely are acknowledged but nonetheless guide our everyday actions—the idea that men should not wear skirts in Western societies is a standard example. When someone breaks one of these residual rules, claims Scheff, he or she may or may not be labeled "mentally ill" and that when we do apply that label our subsequent behavior will tend to preserve that individual's status by providing feedback that will produce more of the same behavior. But Scheff does not provide any details about why some people rather than others are labeled in this punitive way when they break the residual rules, nor why only some rules count, and he admits that after years of work the theory lacks any empirical support (although he seems unsure whether that is important). So here is one hard problem for the constructivist: why do we not regard all deviants as mentally ill (for example, we don't assume a man in a skirt is mentally ill), if mental illness is just a matter of violating norms? The objectivist answer is straightforward—some people have something wrong with them that explains their departure from normal human functioning.

Furthermore, avoiding relativism with respect to mental disorder is difficult if one is a constructivist. Even if one cites medical opinion as constituting the norms that are violated by the mentally ill, unacceptable relativism still ensues. In part, this is because scientific opinion changes. More troubling for constructivists is that relativism arrives if we do not pay attention to underlying causal processes, since then we are left with the view that any behavior that society chooses to pathologize is a mental disorder.

2.2.4 Constructivism Cannot Avoid Relativism

Constructivists cannot distinguish moral and political disputes from psychiatric ones. But this is a distinction that we do indeed make, and we can even criticize putative diagnoses on the grounds that they are politically motivated and hence fraudulent. So our thinking about mental illness seems alert to a distinction that constructivism elides. Consider some famous diagnoses that look politically motivated. Samuel Cartwright argued in 1843 that American slaves who tried to escape were afflicted with "drapetomania" or the compulsion to flee; slaves also were found uniquely prone to "dyesthaesia Aethiopica," which made them neglect the property rights of their masters (Brown 1990). More recently, Soviet psychiatrists argued (or were told to argue) that political dissidents suffered from "sluggish schizophrenia." Doubtless plenty of American doctors used to believe quite earnestly that many slaves were happy, and that those who tried to escape were miserable and abnormal—and indeed they probably were abnormal, statistically speaking, since not that many slaves made a run for it. Moreover, "drapetomanic" behavior undoubtedly was disvalued in the society that produced the diagnosis. Yet arguments over what makes something a mental illness are not just political: drapetomania was not a worthwhile diagnosis before slavery was overthrown. It was rubbish all along, and scientific rubbish to boot, and we need a view that lets us say so.[4]

---

4. Drapetomania and masturbation are the classic examples of now discredited psychiatric diagnoses which some take to support relativism about disease, on the grounds that they once were considered diseases and now they aren't. I share Boorse's incomprehension at this position (1997, 72–78). As he points out, practically everything doctors used to believe about masturbation was always false. It does not

More recent Western psychiatry has seen several putative disorders contested on the grounds that they are the result of politics and media influence rather than scientific discovery. (See, for example, Hacking 1995a, 1998 on dissociative disorders and Young 1995 on post-traumatic stress disorder.) Whereas psychiatry often has appeared to serve the ends of reactionary politics—Ingleby 1980 comes close to suggesting that the concept of mental illness is part of a capitalist plot—more recent disputes often have been over politically progressive diagnoses. There was some debate about whether "racist personality disorder" should be included in DSM-IV, and the claimed epidemic of multiple personality disorder was bound up with the rise of feminism and the attention paid to child abuse in consequence (Hacking 1986, 1995a, 1995b; Showalter 1997).

Constructivism ignores an essential fact about our everyday thought on mental illness, which is the importance of appropriate causal explanation. Our everyday beliefs about human nature explain why we do not count drapetomania as a disorder. We regard escape attempts as normal human responses to enslavement. In the same way, we regard grief as a normal response to bereavement even though people who grieve may feel or behave like people who, in other circumstances, we would think of as mentally disturbed (Engel 1961; Wilkinson 2000). Beliefs about how our minds normally operate influence our judgments that behavior is caused abnormally. Or, since people seldom make explicit judgments that other people are normal,

cause epilepsy, blindness, acne, or rickets. It neither shrinks the male genitals nor enlarges the female ones, nor is it the result of an excess of nervous energy. Nor is it found only among a minority of individuals. Likewise, any putative justifications of slaves' compulsions to flee or thieve in terms of abnormal causal mechanisms were bogus. Some people said so at the time: Thomas Jefferson, for example, pointed out that theft by slaves was explicable in terms of their situation, rather than a "depravity of the moral sense" (quoted in Cohen 1997, 110). I am not saying that nineteenth-century psychiatrists did not think they had scientific support for their views on drapetomania, pro or con: the point is that a constructivist thinks that values, not science, drive our judgments of pathology, and the psychological facts are merely marshaled to support prior value judgments. So drapetomania's status as a disease depended on who won the battle over whether the behavior was deviant. Objectivists can say quite straightforwardly that it was never a disease, regardless of what anyone believed, because it was not caused by malfunctions according to any even moderately correct psychological theory. Constructivists either must weaken their constructivism to take that into account or simply say that whether or not drapetomania counts as a disease depends on who imposes values on whom.

a better way of putting the point is this: we need some reason to regard the processes underlying deviant behavior as themselves abnormal if the behavior is to strike us as symptomatic of mental illness. We tend not to think that normally operating psychological processes in a standard environment give rise to disorders; a deviant causal history is important. Not just any sort of story about the causes of abnormal behavior will do, of course: if we discovered that someone was acting strangely due to hypnosis, or because his body was under the control of malicious extraterrestrial scientists, we would not consider him mentally ill. In order to be properly applied, the concept of mental disorder seems to require that a condition have the right sort of causal history. This is obviously a dismayingly vague condition as stated, and in section 2.3. I will discuss some theories that cash it out in different ways. But the causal constraint, despite its vagueness, means that objectivism beats constructivism. Objectivism embodies the important insight that we do not regard disease judgments as unconstrained by the biological or psychological facts. We do in fact think disease judgments depend on appropriate causal explanations. So far, at least, I am siding with common sense.

2.2.5 The Moral
The problem for constructivism is the thought that only our prior judgments of norms being broken, and not a constraint on causes, determine our opinions about whether behavior or psychology is abnormal. My objections to objectivism do not dispute this, but concern whether objectivism concedes too much to common sense and whether its concept of the mental means that it can ground inquiry fruitfully. So although objectivism does beat constructivism as a specification of pre-theoretic thought about mental illness, it is not a good approach to mental disorder from the standpoint of scientific investigation.

2.3 Objectivism

2.3.1 Introduction
I gave constructivism short shrift on two grounds. It does not respect the distinction between mental illness and other sorts of deviance, and it is impossibly relativistic. I commended instead a generic objectivism that says judgments about mental illness are sensitive to the causal antecedents of the right sort. In this section I try to flesh out this minimal objectivism and begin to ask what the right causal antecedents are. Clearly, if the arguments in the

last section were sufficient, mere social disapproval is not a causal antecedent of the right sort, and we also can rule out some philosophical fantasies; you are not mentally ill if you behave like a schizophrenic because you have been hypnotized or your mind is controlled by malevolent aliens or mad scientists. But what positive conception of conceptually respectable causal process can we come up with?

To answer that question, I start with a position that is a little more specific than the minimal objectivism introduced above. In the remainder of this section, I introduce and criticize the DSM conception of mental illness. DSM's concept of mental illness admits too many causes that are not, in fact, sources of mental illness. My criticisms are not original. They are borrowed from other authors whose conceptual work on mental illness starts from the DSM. Having endorsed their criticisms of the DSM's failure to accord with our intuitions, I will discuss their own analyses. I argue that Culver and Gert's attempts to make objectivism more precise through the notion of a "nondistinct sustaining cause" are still too vague to be a useful source of constraints. Then I discuss the work of Wakefield and Boorse, who require that mental illness is necessarily the product of a dysfunction. This is a much stronger constraint. I criticize it in chapter 3, but I end this chapter by noting its attractions, including its excellent fit with some recent cognitive neuropsychiatry.

2.3.2 The DSM Conception of Mental Illness

DSM defines mental disorder in objectivist terms, as the result of dysfunctions. Officially, DSM-III and its successors exclude specific theories of the causes of disorders from the taxonomy. This is not always true—some causal information does sneak in, and some disorders actually have a causal condition built into their names. The concept of mental disorder in DSM-IV-TR is taken over wholesale from DSM-IV (APA 1994, xxi–xxii):

In DSM-IV, each of the mental disorders is conceptualized as a clinically significant behavioral or psychological syndrome or pattern that occurs in an individual and is associated with present distress. . . . Whatever its original cause, it must currently be considered a manifestation of a behavioral, psychological or biological dysfunction in the individual.

This passage, note, mentions not just malfunctioning mental or biological mechanisms but also "behavioral dysfunction."[5] Wakefield 1997a argues

5. The other international psychiatric classification, ICD-10, also admits behavioral disorders. It defines disorders as "a clinically recognizable set of symptoms or behav-

very persuasively that some of the more counterintuitive diagnoses in DSM-IV are the result of a failure to take *underlying* malfunction into account. This permits exclusively behavioral criteria for diagnosis, leading to a failure in many cases to distinguish mental disorders from "problems in living"—fully comprehensible, indeed wholly expected, reactions to stressful circumstances. DSM-IV, Wakefield charges, sometimes attributes disorder when the behavior that is supposed to be pathological is actually a perfectly normal response to perturbing or disruptive events in one's life. These "problems in living" lead people to seek treatment, but that does not make them mental disorders: we should preserve a distinction between the mentally ill and the "worried well." Dissidents living under a totalitarian government may be miserable, isolated, and even irrational in thinking they can improve the government. But although their wish for a better polity is not psychologically beneficial, they are not insane.

Wakefield notes a number of examples of DSM-IV criteria that result in diagnostic over-inclusiveness, arbitrariness, and a flouting of our common-sense intuitions. DSM-IV's section on Child and Adolescent Disorders appears to go quite a long way toward making almost any normally rebellious teenager into a mental patient—Huck Finn and Tom Sawyer were suffering from Conduct Disorder, apparently. Perhaps most spectacularly of all, Substance Abuse may be diagnosed solely on the basis of social or interpersonal difficulties attendant on substance use, such as arguments with family members; this means that if you like a drink, "your spouse can now give you a mental disorder simply by arguing with you about it, and can cure you by becoming tolerant" (Wakefield 1997a, 640).

DSM-IV-TR does make some patchy concessions to our views about human nature. What would otherwise be symptoms of depression are excluded if they are culturally sanctioned or expected, so that grief

ior associated in most cases with distress and with interference with personal function" (World Health Organization 1992b, 5). I should note that DSM employs "multiaxial" assessment. Nearly everything I have to say applies to Axis I (Clinical Disorders, and Other Conditions That May Be a Focus of Clinical Attention. The latter includes the V codes, which are problems with relationships). I will mention the personality disorders occasionally, and in chapter 10 I discuss mental retardation. Those are the Axis II conditions. Axis III covers general medical conditions and Axis IV covers psychosocial and environmental problems, like being unemployed, which may be relevant to the patient's overall condition. Axis V rates the patient's overall level of functioning out of 100. A comprehensive assessment usually will include ratings for at least I and V, and possibly all.

following the death of a spouse does not count towards diagnosis of Major Depressive Disorder—but the bereaved person gets only two months to recover, after which time a diagnosis of clinical depression may be made. Not even that qualification, however, is permitted if one shows depressive symptoms after an unhappy love affair, loss of a job, or the receipt of a terminal medical diagnosis.

Wakefield complains that in countenancing behavioral dysfunction DSM-IV is licensing periodic departures from objectivism in its pages. He plausibly contends that diagnoses of major depression after being jilted or Substance Abuse based just on family troubles are counterintuitive because they attribute disorder on the basis of behavior alone. The diagnoses do not hypothesize underlying psychological malfunctions that cause the behavior. Disagreement with these diagnoses, then, supports the view that our intuitions about mental disorder require psychological malfunctions as causes of abnormal behavior.

The problems are most pronounced in the sexual disorders or "paraphilias," where unthinking constructivism appears to be the order of the day. The paraphilias are conceptualized in terms of the unusual objects or fantasies required for sexual satisfaction: no attempt is made to represent these proclivities as due to any underlying abnormalities. Although, after a bitter fight, the American Psychiatric Association decided in the early 1970s that homosexuality is not a mental illness, it appears that traces remain of the attitudes that led to the medicalization of homosexuality. As Culver and Gert (1982, 104–107) pointed out with respect to DSM-III, the DSM conception of sexual health seems to be that you ought to be conventionally heterosexual, and if you are not, you are sick.

It is unclear that we should regard any of the Paraphilias as psychological abnormalities in anything other than an irrelevant statistical sense. They are a mixture of conditions like Pedophilia, which we have compelling reasons on moral grounds for disliking, and other conditions, like Tranvestism, which are harmless. The DSM discussion of paraphilia captures the sense of deviance but fails, because of its constructivist bent, to respect our intuitions about the difference between kinds of deviance. The Welsh novelist Alice Thomas Ellis once said that the English don't have enthusiasms, they merely have tastes "for things like porcelain and flagellation." DSM distinguishes the latter taste from the former on medical grounds. But then it must be caused by medically significant factors. The attacks on DSM I have just noted deny that the factors exist in the instances at issue. But what sort of factors are the right ones?

2.3.3 Specifying the Causes of Disorder

I have argued that objectivism requires appropriate causal histories. But so far I have not supplied any details about what sorts of causal histories count. I now turn to that issue. I begin by discussing a very general proposal, Culver and Gert's requirement of a "nondistinct sustaining cause." I fault this on the grounds that it is too general, permitting conditions to count as disorders that are not, intuitively, disorders at all. I should admit up front that in the next chapter I will end up with a position that has some of the generality and lack of specificity of Culver and Gert's proposal, but defended on different grounds. The present point is that their proposal cannot be defended as conceptual analysis.

After doubting the conceptual warrant of Culver and Gert's view I move on to the work of Boorse and Wakefield. They argue that disorder necessarily rests on malfunctions in evolved systems that make up a species-specific design. This "harmful dysfunction" view, as Wakefield calls it, does prima facie a better job of fitting our commonsense thinking. But it makes a number of empirical assumptions that, I argue in chapter 3, only makes sense within a synthesis of psychiatry and cognitive neuroscience. Ironically enough, once this is done the view loses its appeal as conceptual analysis because the conceptual foundations of psychiatry and cognitive neuroscience only partly overlap. My reaction is not to abandon the synthesis, but to abandon the defense of it on conceptual grounds and defend it on pragmatic, methodological grounds. I develop it in more detail in chapter 4, where I supply a more defensible set of empirical commitments.

So, first, Culver and Gert's (1982, 72) view: they derive their concept of mental disorder from the more general concept of a *malady*, which involves suffering evil, or an increased risk of evil, that depends for its presence on "a condition not sustained by something distinct" from oneself. A sustaining cause is one with effects that come and go almost simultaneously with the presence or absence of the cause itself. A wrestler's hammerlock is a sustaining cause, but a person trapped in one does not have a malady because the cause (the wrestler) is distinct from the sufferer. If the cause is within the person it is a nondistinct sustaining cause if it is either biologically integrated (like a retrovirus) or if it cannot be removed without difficulty (their example is a clamp mistakenly left in the body after surgery).

Culver and Gert say one has a mental malady if one suffers from evils in the absence of a distinct sustaining cause. To count as mental, the disabilities that cause suffering must be cognitive or volitional. The cause can be physical or mental, as long as it is a sustaining cause that is not distinct from the sufferer (1982, 88); for example, the cause could be an inherited nervous system abnormality or a bad childhood (1982, 87).

Culver and Gert recognize the problem of specifying the acceptable causal constraints on mental illness. They try to solve the problem by the principle of nondistinctness. Anything that is not a distinct cause and that produces suffering or a risk of suffering counts as an acceptable cause. When it comes to mental maladies, though, the principle of the absence of a distinct cause is too inclusive. Someone who is unable to get a job because he is blacklisted could count as mentally ill, if the blacklisting were due to the right factors. Here's why.

Loss of freedom, opportunity, or pleasure count as evils, according to Culver and Gert (1982, 71): prolonged unemployment would seem to count as an evil then, since it brings with it losses in opportunity, if nothing else. Hence, if Boris is unemployable due to facts about the sort of person he is, he is suffering due to a sustaining cause that is nondistinct. It is nondistinct because it is part of Boris's nature to be that way.

Now, some aspects of Boris's nature do not count as causes of his suffering qua malady because they reflect Boris's rational beliefs and desires. Culver and Gert do not regard rational beliefs and desires as causes of maladies. So if Boris is blacklisted because he believes correctly that the government is unjust and the government controls the labor market, Boris is not mentally ill, just a victim of persecution. But if Boris can't get a job because he has been repeatedly sacked for sexual harassment in the past, his misfortunes are not caused by his rational beliefs and desires, but by his being a pig around women. Worse still, suppose Boris is a domestic servant who is sacked because he has too much self-respect and independence of spirit for his employers' taste, and then he can't find another job because they refuse him a reference. In this case, unlike the sexual harassment case, Boris suffers because of admirable aspects of his psychology, like an unwillingness to grovel. Nonetheless, they are *aspects of his psychology*, and hence nondistinct sustaining mental causes of his unemployability, which causes evils.

Boris shows that the principle of nondistinctness is not enough, because people can have nonpathological psychological properties that nonetheless are bad for them. There may not be analogous physical properties, although ugliness might be one. Although nondistinctness rules out bereavement or marital strife as causes of mental suffering, it does not rule out aspects of one's psychology, the proviso about rationality notwithstanding. In the next section, I discuss Boorse and Wakefield, who attempt to distinguish pathological and nonpathological psychology by appeal to dysfunction.

2.4 The Harmful Dysfunction Analysis

2.4.1 Introduction
Culver and Gert's analysis is too inclusive. In this section I look at the less inclusive harmful dysfunction account of mental disorder, which descends from contemporary objectivism's most influential theorist, Christopher Boorse (1975, 1976a, 1976b, 1977, 1997). The harmful dysfunction view has achieved a dominant position largely thanks to the work of Jerome Wakefield, who defends a Boorsian view. It is close to the official statement of the two-stage view that is, I earlier insisted, the consensus in contemporary psychiatry. I next try to explain why the view has such wide appeal, and I end the chapter with examples of explanations of psychopathology that might vindicate the Boorse/Wakefield analysis if the empirical bets they make are correct. The next chapter argues that, in fact, there are problems besetting the objectivist views of Boorse and similar scholars, making their work deficient as conceptual analysis. However, what is valuable about the objectivist take on mental illness can be preserved if we think of it not as capturing something in common sense but as mapping out a direction in which psychiatry might develop.

2.4.2 Boorse
Boorse's views have changed over the years. I mention some of the changes, but I want to stress his key objectivist contention that "at least at the theoretical foundation of Western medicine, health and disease are value-free concepts. . . . [T]he classification of human states as healthy or diseased is an objective matter, to be read off the biological facts of nature without need of value judgments" (Boorse 1997, 4). Boorse has long argued that a disease is any state that fails to conform to the "species-typical design" of humans, where that is understood as a specification, at various levels of analysis, of the component parts of the body and the functions they perform, function being understood in evolutionary terms as failure to contribute to survival and reproduction (1976a, 62–63).

Disease is the failure of a component part to function as it should, Boorse said, because disease is absence of health, which is the functioning of an organism according to its species-typical design. Boorse took departure from species-typical functional design to define mental disease as well as physical disease, since "there is such a thing as mental health if there are mental functions" (1976a, 63). This is a stricter condition than Culver and Gert's requirement of a nondistinct sustaining cause. It does not count a propensity to sexual harassment nor nobility of character as underlying causes of mental illness, no matter how much trouble they cause. Boorse's view, then, distinguishes between psychological sources of suffering and dysfunctional psychology. It respects the intuition that neither sexual-harasser Boris nor non-groveling Boris is mentally ill, but it also says that we would change our minds if we discovered that sexual-harasser Boris has some pathology that causes him, for example, to think that women are robots plotting his downfall. If Boris's misogyny were caused by such a delusion, I think we would agree that it was symptomatic of a disorder.

In this view, as I noted in 2.3.1, "disease" is defined as failure of species-typical function, but "illness" is defined normatively. One is ill when one's disease is undesirable, entitles one to special treatment, or excuses for one's behavior. Subsequently, Boorse (1997) has argued that both illness and disease can be understood objectively via his "biostatistical theory" (BST). The difference between disease and illness is no longer a matter of judgments of suffering but is revealed objectively, says Boorse, by the fact that in everyday speech "illness" refers to a systemic affliction that affects the whole organism, whereas diseases affect only components of the design-plan.

These revisions may or may not be right about our disease vocabulary. They do not touch the influential central insights of Boorse's original theory. These were, to recap, that we can objectively judge when a component of our species-typical psychological endowment is malfunctioning, and that mental disorders are (i) psychological malfunctions that (ii) cause people to suffer or justify their receiving special treatment. Most objectivist analyses, like Boorse's and Culver and Gert's, depart from simple objectivism by insisting on this combination of causal explanation and normative judgment.

Spitzer and Endicott (1978, 18) make the point by calling disease categories "calls to action"; they see them as contentions that something has gone wrong with a human organism that has led to negative consequences. (See also Margolis 1980 and Papineau 1994.) To analyze the concept of mental disorder, Spitzer and Endicott conjoin "organismic dysfunction" with "distress, disability or disadvantage" where these are psychological or require explanation using psychological concepts (1978, 18). The qualification of objectivism with avowedly normative criteria of disability or distress is a dominant one in the literature on mental disorder, represented more recently by the work of Jerome Wakefield (1992, 1993, 1996, 1997a, 1997b, 1999a, 2000).

Unlike Boorse, Wakefield has kept up a running dispute with his critics about conceptual matters. The evolving fight lets us to see how the orthodox program of conceptual analysis pans out in detail. And Wakefield, as we saw earlier, has fought on another front, using his conceptual analysis to attack the way DSM thinks about mental disorder.

2.4.3 Wakefield

Wakefield says that our concepts of both mental and physical disorders involve two individually necessary and jointly sufficient components. These are a judgment that an internal mechanism is malfunctioning,[6] and a judgment made by the surrounding society that the malfunction is harmful (cf. Kovacs 1998). So Wakefield would count as disordered a woman who cannot bear children if her peers regard her as harmed thereby, even if she doesn't want children and would have a tubal ligation if she were fertile. Wakefield's account, therefore, can deal with the gourmet lesion, which fits only the "dysfunction" part of his analysis, since the gourmet lesion may very well be neither undesirable, nor disvalued.

Wakefield follows Boorse in arguing that our psychology consists of evolved functioning components and that breakdowns in these systems are a necessary condition of mental disorder. This is the standard objectivist elaboration of what we earlier found to be the intuitive, if somewhat elusive, requirement that mental disorder depends on an appropriate etiology. Boorse and Wakefield interpret the causal requirement very narrowly, so that only failures of evolved function count.

The harmful-dysfunction analysis avoids some of the conceptual problems that Culver and Gert's view faced. The notion of psychological dysfunction fleshes out the causal requirement more strictly. In addition, the analysis makes room for the idea that conditions like the gourmet lesion raise both a scientific question about normal mental function and a nonscientific question about whether it is a good or a bad thing to have abnormal mental functions. However, the satisfying features of the harmful-dysfunction analysis come at a price. Boorse's objectivism about mental disorders requires that human psychology has component parts whose failure to work normally we can specify. The natural way to carry out an objective determination of mental structure and function is to look to the sciences of the mind.

---

6. Like Boorse, Wakefield understands dysfunction in evolutionary terms as a "failure of a mechanism in the person to perform a natural function for which the mechanism was designed by natural selection" (Wakefield 1993, 165). Unfortunately, this won't do, since other concepts of function exist in the life sciences, let alone in common sense. (This reading of Wakefield, along with many more detailed criticisms, is developed at length in Murphy and Woolfolk 2000a and 2000b. See also Lilienfeld and Marino 1995.) I take up the issue of function in chapter 4.

The harmful-dysfunction analysis makes empirical bets about the nature of our psychology and the ways in which we can specify its breakdowns. Before we attend to the caveats and qualifications, it might be helpful to have an example in play that shows in some detail the empirical assumptions that Boorse and Wakefield are making. The explanation of autism as a theory-of-mind disorder is a good example. The theory is not without support, but it remains controversial. I am not advancing it because I am sure it is true, but because it is a perfect match for Boorse's analysis. It shows us what psychiatry would look like if Boorse is right about its conceptual basis, because it is premised on Boorse's assumption that mental illness is due to malfunctioning mental organs. Section 2.5 outlines the theory that autism involves, as a core symptom, a theory-of-mind deficit. Then it looks at the empirical commitments the theory makes, the commitments that result from Boorse's model. I conclude that something has gone badly wrong with the whole project of conceptual analysis of mental disorder because of where it leads us. The harmful-dysfunction analysis cannot be accepted because it commits a priori to an implausibly modular psychology.

I conclude that the empirical commitments seem out of place in a conceptual analysis, but they can be weakened to fit naturally with a conception of objectivism as a programmatic orientation on psychopathology. I then argue in chapter 3 that this reorientation of the objectivist program is strengthened by other objections to the orthodox program of supplying a conceptual basis for psychiatry.

2.5 Autism

2.5.1 Autism and the Theory of Mind

Boorse and Wakefield claim our concept of mental disorder is a concept of a failure of a component part of our psychology to do its job. A number of theorists have approached autism in just this way: as an example of a breakdown in a specialized "theory of mind module" (Baron-Cohen 1995; Baron-Cohen, Leslie, and Frith 1985, 1986; Frith 2003; Leslie 1987, 1992, 1994; Leslie and Thaiss 1992.). They assume that in nonautistic individuals there exists a specialized mental system with a proprietary body of knowledge that has the job of attributing intentional states like beliefs and desires to other people and explaining their behavior in intentional terms. (Phillips et al. 1998 showed that autistic children also have difficulty monitoring their own intentions.) Some other theorists who are less enamored of modularity also think that autistic children suffer from a deficit in a distinctive theory of mind; that, for example, they "may lack an initial theory of persons"

(Gopnik, Capps, and Meltzoff 2000, 50). That's the lack of common "folk-psychological" knowledge, rather than a special-purpose, or "domain-specific," computer. I will put aside these differences for now. The idea we are exploring is that autism is a failure to develop that part of normal psychology that enables us to have a theory of mind.

Possession of a theory-of-mind has been operationalized in cognitive psychology as the ability to pass a first-order false-belief test. *First-order tasks* involve making inferences about a second person's intentional states, whereas *second-order tasks* require one to infer what a second person thinks about a third person. One standard version of the first-order task is the "Sally-Ann" test. The subject watches Sally put a piece of chocolate in location A and later, while Sally is absent, the subject watches Ann move the chocolate to location B. The subject is then asked where Sally will look for her chocolate when she returns. To answer the question correctly the subject needs to appreciate that, since Sally was absent when her chocolate was moved from A to B, she will have the false belief that the chocolate is still at location A.

Normal four-year-olds pass the first-order task, but autistic children fail it. Indeed, they do much worse than Down's Syndrome children of the same mental age, even though the autistic children surpass the Down's syndrome children in general causal understanding. Some adults with Asperger's Syndrome (high-functioning autists with IQs in the normal range) have offered quite moving accounts of their puzzlement at the realization that the people around them seemed to know what other people were thinking. Oliver Sacks, for instance, made famous Temple Grandin's comment that her adolescence was like being an anthropologist on Mars. (Frith 2003 contains several such stories. Grandin's experiences are related in Grandin and Scariano 1986 and Sachs 1995.)

These experimental results lead investigators to agree that there is a theory-of-mind deficiency in autism, even though they may differ as to the precise details of the underlying mental structure. The harmful dysfunction analysis of mental disorder rests on the expectation that this strategy can be carried out for all disorders. In each case, we expect to explain disorder by showing how symptoms arise from a malfunction in a cognitive system.

To be told that autism is, or involves, a theory-of-mind deficit gives us a handle on what the problem is. But this demystification, though welcome, is in fact only an explanation of why autism has its characteristic profile. It is not an explanation of why autism occurs. We have an explanation for the pattern of deficits characteristic of autistic individuals, but not an explanation for why those deficits arise—it's one thing to be told that Temple

Grandin lacks a theory of mind, but why does she lack a theory of mind?

2.5.2 Explaining the Absence of Theory of Mind

To explain autism, Baron-Cohen assumes that the mind is organized into modules. Now, given that modules are supposed to be distinct and specialized, the truth of that assumption would be a great present for Mother Nature to have given psychiatry (Murphy and Stich 2000 exploited the assumption for all it was worth). Boorse's analysis of disease concepts in general rests on the premise that a disease is a problem afflicting an organ, and that medicine in general is the study of things going wrong with organs. Advocates of a heavily modular approach to the mind are prone to explicitly making the analogy between modules and organs (Pinker 1997; Tooby and Cosmides 1992), and if we do have modular minds, the prospects for establishing psychiatry as the science of mental organ failure are greatly improved.

Many proponents of a modular view claim that modules are innate, but that further assumption is not required. One could, for example, see normal development as a process of modularization (Karmiloff-Smith 1992), and Baron-Cohen's view involves the idea that a module depends for normal development on first receiving information from an ontogenetically prior module. In Baron-Cohen's account, (Baron-Cohen 1995; Baron-Cohen and Swettenham 1996) normal development of the theory-of-mind module (ToMM) requires that three upstream systems be in place: an Intentionality Detector (ID), an Eye Direction Detector (EDD), and a Shared-Attention Mechanism (SAM). In autism, a variety of biological hazards, most notably perinatal problems, damage the SAM. As a result, the downstream ToMM is deprived of the input it needs to develop properly.

The Shared-Attention Mechanism has the function of forming "triadic" representations, which specify the relations between an Agent, the Self, and an Object (which might be another agent). The triadic representation is supposed to be formed by taking two dyadic representations. The first is a representation of the perceptual relation between the self and an object and the second is a representation of the perceptual state of another. One is then embedded within another. Baron-Cohen gives this example (1995, 45):

Mummy-sees-(I-see-the-bus).

Baron-Cohen's contention is that, in development, ID, EDD, and SAM between them get us to the point where we are able to read behavior in terms of volitional mental states (desires and goal representations) and to interpret eye direction as a clue to others' mental states. They also let us verify that different people can be experiencing these particular mental states about the same object or event (Baron-Cohen 1995, 51).

Baron-Cohen argues that to add a theory-of-mind module to this base, one needs two further capacities: the representation of epistemic mental-states and a way of "tying together" mental-state concepts. His fundamental idea for explaining this addition is that a triadic representation produced by SAM has a slot for a dyadic relation that can take an M-representation (Leslie 1994): an M-representation is an ascription of a relation to a proposition rather than to an object. In most autistic children, SAM appears to be seriously impaired. Baron-Cohen alleges that this impairment means that ToMM cannot receive inputs from ID and EDD via SAM, as would be the case in normal children, and without this characteristic input it is impossible for ToMM to develop normally.

2.5.3 Assumptions of the Theory

Let's ask what the mind needs to be like, if the methods and assumptions behind Baron-Cohen's picture of autism are to be generalized. Baron-Cohen himself does not quite go so far as to generalize it to all mental illness, but he does endorse a highly modular picture of the mind, familiar from evolutionary psychology, with each module having a discrete function bestowed on it through natural selection. So we have a picture of the mind as composed of mental organs with identifiable jobs. This is exactly the picture that Boorse said his analysis of mental illness required. There are mental functions, so there can be mental illnesses, with the distinct, organ-specific causal explanations that Boorse requires.

Note too the isomorphism in this picture between personal and subpersonal psychology. We can distinguish the kinds of behavior or psychological capacities that are distinctive of our species—humans possess emotions and a theory of mind, understand language, track others' gazes, remember facts and skills, and so on. These are personal traits, or properties of human animals. Unconscious subpersonal processes inside us enable us to do all these things. To understand language, for example, we have learning mechanisms, parsing mechanisms, knowledge of syllabic stress, and so on. The isomorphism assumption is that there is a correspondence between personal traits and subpersonal mechanisms so that for each job that we do we have an internal structure that is dedicated to that job.

The assumption that our brain has computational systems that correspond to different mental capacities in a one-to-one way is stronger than anything Boorse or Wakefield need, although close to some things they say, and it is probably stronger than anything Baron-Cohen needs. But all three of them do share the view that there are mental functions, and that our mind/brain contain systems that carry out these functions. It is not the case that the brain is one big information processor, rather it is a collection of them.

Now, although Wakefield often seems tempted by it (Murphy and Wool-folk 2000a) there is little to recommend the "gratuitous and indefensible" (Glymour 2001, 126) view that a deficit, such as theory-of-mind deficit, implies the existence of a component of the mind that has the job of producing the normal form of the deficient behavior. But the more general idea that there are components or "cognitive parts" (Glymour 1992, 2001) in the mind/brain does not imply a one-to-one relationship with particular behaviors or cognitive capacities.

Cognitive parts are brain components that are causally interconnected and physically localized (although not necessarily anatomically all in one place, rather than connected via neuronal projections across the brain). The joint contribution of the causal relationships these parts bear to each other is the function of the overall system. The idea of parts is typically combined with the view that each part has a particular job, as a part in a machine does: but, as I said, the job of each part does not need to correspond one-to-one with behaviors, normal or abnormal. So the idea of parts with corresponding jobs is defensible even if the jobs are not discoverable without looking inside the system. Add to that picture the idea that the jobs are varieties of information processing and you get the dominant picture of reductive explanation in contemporary neuroscience (and, minus the information processing, other areas of the life sciences). I shall adopt Bechtel's terminology and call this explanatory strategy *decomposition and localization* (Bechtel 2002; Bechtel and Richardson 1993; Zawidski and Bechtel 1996; cf. Wimsatt 1976, 1986).

In chapter 4 I say more about how this functional decomposition and localization works in cognitive neuroscience, in chapter 5 I wonder about its limits, and in chapter 6 I put it to work in a theory of psychiatric explanation. My present point is that the idea of mental functions has a natural affinity with the idea of a functional decomposition of the mind/brain into different systems. That claim, which is weaker than the isomorphism assumption, is the conventional wisdom in cognitive neuroscience. It is also a plausible gloss on the idea that we have mental functions, since the cognitive parts uncovered by this method have their functions to perform within the overall functional capacity of the brain; but, as I argue in chapter 4, the necessary construal of "function" cannot be evolutionary, which is contrary to the conclusions of Boorse and Wakefield.

So the assumptions of the view of abnormal psychology I have reviewed are: (i) the failure of a normal mind to function as it should is a necessary condition of mental illness; (ii) normal functioning is a product of natural selection of the components of the mind; and (iii) the components of the mind depend on functionally distinguishable components of the brain. There is one last assumption to discuss in this quick tour of the empirical commitments of the harmful-dysfunction analysis. The last assumption is of a common human nature—normal psychology and the brain mechanisms that maintain it are common to all human beings. Indeed, failure to exhibit one of these capacities is the result of a failure to develop the relevant mechanism, i.e., to have a mental disorder. So we add (iv) normal psychology is common to all *homo sapiens*. I will review these assumptions presently, and deny or qualify all four. For the moment, I am exploring the relationship of these assumptions to the orthodox program of conceptual analysis.

The four assumptions reflect the application of the medical model to psychiatry. Together, they make up a picture of the mind that, if correct, vindicates Boorse's and Wakefield's conceptual assumptions about what mental illness is by making their analysis fit existing explanatory models. This is a crucial point: if the mind turns out not to be organized like this, we will—if Boorse and Wakefield are correct about our intuitions—be unable to tell the correct causal stories and will in fact discover that many of what we thought were mental illnesses are not mental illnesses at all, since they do not fit the analysis. The first assumption, of mental illness as dysfunction, is not empirical. However, if Boorse is right to say that there is such a thing as mental illness if there are mental functions, it is a conceptual truth that if there are no mental functions, there can be no mental illnesses. This is alleged to follow from both common sense and a conception of disease that originates in western medicine (the two have influenced each other).

The claim of conceptual truth for the idea that if there are mental disorders there are also mental functions forces us, on conceptual grounds, to deny that some conditions are mental disorders at all if they turn out not to rest on an identifiable malfunction. Perhaps conditions that receive such treatment might deserve another status, such as moral failures or as annoying personality traits. This position does, though, imply that if we discovered that schizophrenia or depression were not due to malfunctions in identifiable mental organs, but had some other cause, then we would have to conclude that they were not mental disorders after all. In other words, we could not discover causes of mental disorder that have not been recognized by the analysis—the analysis fixes the nature of mental disorder in advance, and the job of science is to find out what things in human nature meet the analysis. So, for Boorse, we might find out that something we think is a mental disorder is not actually a mental disorder after all, because it is ruled out by the superordinate concept of mental disorders.

Wakefield has a similar conception of the role of common sense but is more bullish about the capacity of commonsense analysis to pick out a realm of mental disorders that corresponds fairly closely to what we already think is there. In Wakefield's view, our current ignorance of human psychology is beside the point, because "we do not have to know the details of evolution or of internal mechanisms" to make judgments that some inner mechanism is malfunctioning (1997b, 256). Rather, he thinks we all have an intuitive folk theory of human design that means it is "obvious from surface features" when underlying mechanisms are functional or dysfunctional (Wakefield 1997b, 256). So if someone is depressed but not in a situation that is normally depression-provoking, Wakefield (1997a) says we can infer that his internal sadness-generating or loss-response mechanisms are malfunctioning. Horwitz (2002, 98), following Wakefield, argues that symptoms of depression are sometimes *appropriate* responses to stressful events, and hence not evidence of mental illness. But Horwitz doesn't say what "appropriate" means, beyond pointing out that it is normal to get depressed in these cases. We do indeed have expectations about the psychological aftermath of bereavement. But we also think it's normal to get blisters after ingesting mustard gas. That typical response scarcely shows that nothing is wrong internally. There is a regular pattern of depression after the death of a spouse, but there also is a regular pattern of brain damage after repeated kicks to the head—we expect environmental stressors to have psychological effects, but that does not mean that those psychological effects do not result from stressors having made inner mechanisms dysfunctional.

Because some stressors are normal parts of life, our commonsense theory of human function makes room for our reactions to them, and it has been proposed that normal reactions to psychosocial stress simply should be ruled out as symptoms of mental illness (Spitzer and Wakefield 1999). But given our ignorance of human psychology it should be an open question whether the depression that sets in after traumatic experiences is a consequence of the disruptive effect those experiences have on inner mechanisms, or whether, as Wakefield and his follower Horwitz maintain, no dysfunction is present at all. Wakefield has in mind an everyday body of belief about human nature, similar to the much-discussed "folk psychology," which is the philosopher's term for theory of mind (Baron-Cohen 1995; Churchland 1981; Fodor 1987; Sellars 1963b).

Philosophers tend to see folk psychology as a theory about beliefs and desires as causes of behavior, but Wakefield appeals to a much more encompassing set of theses about normal human nature. And, obviously, commonsense views about human nature do advert to character traits, drives, emotions, and other aspects of our psychology, and not just the propositional attitudes. But Wakefield and, following him, Horwitz (2002) assume that our folk theory *rules out* some behaviors in some contexts as indicators of mental illness. This is equivalent to saying that it's a priori whether somebody's psychology is functioning normally. But we can't tell when someone's psychology is or is not functioning as designed just based on commonsense evaluation of surface features—whether someone's psychological systems are intact certainly looks like a substantive empirical question. Wakefield's denial that it is empirical at all reflects, first, the belief that commonsense intuitions are the ultimate court of appeal when we are trying to decide what counts as mental disorder, and second, the way our expectations about human functioning inform our scientific theories of human nature. Boorse, as I noted above, may share these assumptions.

We should reject these assumptions. It is not a priori that all causes of mental disorder are failures to perform an evolved function, nor that we can figure this out via knowledge of folk psychology. Although objectivists have embraced the view that all the causes of mental disorder are evolutionary failures, it hardly seems required to vindicate our everyday sense that mental disorder must have an appropriate etiology. I shall discuss many different causal hypotheses in this book, and the details can wait. The present point is that we can take from objectivism the requirement of causal explanation of mental disorder without endorsing any particular type of explanation in advance of looking at the details of psychopathology. But then how will we know whether Boris is guilty of sexual harassment because of mental illness or because of simple boorishness? I am afraid that I am skeptical about our ability to know this a priori, but I do not deny that conceptual issues have some role to play in deciding it. So I turn now to the issue raised by Wakefield's faith in the efficacy of our folk psychology as a guide to mental disorder. What is the relationship between folk psychology and the scientific picture of humanity, and how does it matter to psychiatry?

This issue is one of three that I take up in chapter 3. The second issue I discuss there also is illustrated by Baron-Cohen's theory of autism. Baron-Cohen's theory is part of cognitive neuropsychology, not psychiatry, at least as those disciplines traditionally are understood. What is their relation, especially if you are an objectivist who believes that mental illness is a matter of mind/brain malfunction?

A psychiatrist who reviewed an early draft of this book said the conceptual debates were beside the point since everyone now acknowledges that mental illnesses are brain diseases. That same reviewer went on to say that I should not discuss agnosias as examples of mental illness, because they are neurological disorders, not psychiatric ones. The tension involved in reconciling these two views is the second issue I take up in chapter 3, where I admit that there are counterintuitive consequences involved in synthesizing psychiatry and the cognitive sciences. But we should not care about them unless intuitions have authority over science, which they do not. Even though judgments like my reviewer's are rooted in current scientific demarcations, those demarcations reflect accidents of history, not respectable theoretical motivations.

The third issue I take up in chapter 3 is whether there is any interesting difference between mental and physical disorders. What's mental about mental disorder? This is another question that we cannot answer by relying on conceptual analysis. We have no consistent intuitions about the mental that can apply across the disciplines, let alone apply to a conjunction of folk psychology and science.

2.6   Conclusion

Objectivism is better than constructivism at vindicating our intuitions about mental disorder, and the harmful-dysfunction analysis is an improvement over more inclusive statements of the requirement that mental illnesses have the right type of causal history. I also have pointed out that at least one research program makes empirical assumptions that suit the harmful-dysfunction analysis. The harmful-dysfunction analysis also reflects the two-stage picture of psychiatric objectivity and normative debate as distinct enterprises that psychiatry has adopted. So the harmful dysfunction analysis is prima facie an attractive analysis of the concept of mental illness. However, as I argue in chapter 3, the harmful-dysfunction analysis is not defensible as a conceptual analysis. But this is not a problem, since conceptual constraints need not be accepted by psychiatry. Why this is so, and what relationship folk psychology should bear to psychiatry, is explained in chapter 3.

3 Psychiatry and Folk Psychology

O the mind, mind has mountains; cliffs of fall
Frightful, sheer, no-man-fathomed.
—Gerard Manley Hopkins, *No Worst, There is None*

3.1 Introduction

In this chapter I first distill a methodological program for psychiatry from the foundational program of conceptual analysis, then I articulate a way of thinking about the two-stage picture, with its separation of science from questions of value. The methodological program involves a take on mental illness that is counterintuitive, both in commonsense terms and within theoretical psychiatry, so it needs some defense. I provide the defense by denigrating the intuitions that tell against me. Despite that, I admit that there are important roles for nonscientific thinking about the methods of psychiatry, and especially its social and political role. There are hard issues to be addressed in psychiatry that are not purely empirical, and I want to separate worthwhile conceptual questions from the scientifically irrelevant preoccupations of common sense. The result orients psychiatry around a set of empirically informed answers to some puzzles in the philosophy of mind and philosophy of science.

In advocating skepticism about conceptual analysis and a different view of the philosophical foundations of psychiatry, I chiefly attend to three issues. They are: first, the relationship between science and common sense, which I have already mentioned as the orthodox view among conceptual analysts; second, what it means to say that a disorder is *mental*, where our intuitions pull in different directions; third, the nature of psychological function and malfunction, which is not, despite what Boorse and Wakefield seem to think, a matter of evolutionary dysfunction. This chapter introduces these issues and shows why they are important. I give different views of the relationship between science and common sense, the nature of the mental, and function and malfunction in this chapter. I finish by revisiting, in light of my views, the core commitment of the two-stage picture: that scientific theories of mental breakdown can be developed apart from moral or political views about what human beings ought to be like.

I have mentioned three issues, then, that I use in this chapter to cause trouble for question the project of conceptual analysis, which I reconstructed in chapter 2: the relation of science to common sense about mental illness, the nature of the mental, and the nature of psychological function and malfunction. As far as the relation of common sense and science goes, conceptual analysts seldom acknowledge a commitment to the priority of common sense over science, but their commitment is apparent in what they write. Science must be much more autonomous from common sense, but that does not mean intuitions have no role to play in our thinking about psychopathology. It is one thing to say that our intuitions are relevant, and another to insist that they are hegemonic, and still another—as I show—to say that they are consistent.

The question of what it means to say that a disorder is mental, leaves objectivist conceptual analyses open to counterexamples that cannot be finessed. While we regard as mental some human capacities and traits, such as the workings of the perceptual system, we deny that disorders afflicting these phenomena are mental disorders.

Last, objectivist analyses say that a mental disorder is a malfunction in a component of our mind. But psychiatry is not so closely tied to functional ascription, and our concept of function is not inevitably evolutionary. My view, finally, is that objectivism does for psychiatry what adaptationism does for evolutionary biology: objectivist assumptions cannot explain everything, and they sometimes fail to apply, but they are a good heuristic for approaching a problem. We can assume that mental disorders are malfunctions in our psychological mechanisms as a starting point, but there are cases where this assumption is false and cases where we cannot know whether it is true or false, and this does not impugn the diagnosis.

My pragmatic position depends on a series of empirical bets that can be made once psychiatry and cognitive neuroscience are synthesized. This chapter defends that pragmatism against the foundationalist project of the conceptual analyst. Chapter 4 elaborates by looking at the empirical and philosophical assumptions made by the psychiatric "medical model," then I move on to positive theories of explanation and classification.

The conceptual position adopted in this chapter helps to explain and classify mental illness, and it is in step with, and inspired by, promising scientific approaches. On the other hand, it ends up committed to the position that blindness and diabetic comas, among other unlikely candidates, are mental disorders. The utility of that position will, I hope, outweigh its counterintuitive nature. But its utility is scientific: it clearly is not an analysis of the commonsense concept of mental illness. I also acknowledge that the picture of psychiatry I give is one according to which mental illness is not a natural kind and psychiatry lacks a clearly delimited subject matter of its own. That is not a problem, since many sciences share these features, and do so in response to a variety of epistemic and nonepistemic pressures. Psychiatry must answer to both kinds: it must be shaped by epistemic demands in such a way that, we may hope, it will one day be a reliable and fruitful source of knowledge. It also must be able to deliver this knowledge in ways that a variety of nonscientific projects can use.

I end this chapter with a quick advertisement for one way of viewing the interaction of psychiatry and our ethical, legal, and political values. Unlike my views on explanation and classification, this proposal, after being sketched in this chapter, is not discussed in the rest of the book. I just hope to calm any worries that I might be adopting an implausible scientism. We want the resources of science to be available to help in the understanding and treatment of the problems that beset the mind/brain. But this understanding and treatment should be shaped by the goals of medicine and the wider practical and moral interests to which those goals answer. In the last part of this chapter I defend the idea that the science of the abnormal mind will produce knowledge that can be used by several different projects that will require proprietary concepts of mental illness that respond primarily to nonscientific agendas. I defend this view against the idea that it insists on an unsustainable and undesirable separation of fact and value. I concede that in the end the separation may be unsustainable in some cases, but I argue that it is not undesirable.

So this chapter is rather diffuse, with several different discussions that take place around the conceptual foundations of psychiatry. The first is the relationship between our commonsense concepts of mental illness and the science that investigates their empirical content.

3.2   Science and Folk Theory

Wakefield, you will remember from section 2.3.2, uses his conceptual analysis of mental disorder as harmful dysfunction to evaluate the DSM concept. He argues that a number of DSM diagnoses should be rejected as anathema to our intuitions. Wakefield's contention that some DSM diagnoses are counterintuitive is persuasive. The important question is whether they show anything scientifically interesting. A detailed look at Wakefield's arguments shows that he regards intuitions as scientifically decisive, that he has an unduly restrictive notion of the causes of psychopathology, and that his analysis is open to counterexamples that rely on different conceptions of the mental. Moreover, these faults are shared widely by the orthodox program, devoted as it is to conceptual analysis.

3.2.1 Traditional Conceptual Analysis

Wakefield claims to be analyzing the scientific concept of "mental disorder." However, if his analysis fails to accord with scientific usage, which he claims to be investigating, Wakefield does not concede that the term "mental disorder" is used in psychiatry or psychology in ways that go beyond his analysis. Rather, he insists that the scientific, clinical concept in those instances is just wrong (1997a). But why challenge a scientific concept solely on conceptual grounds (as opposed, say, to challenging it on the grounds that it does not apply to anything in nature)? Wakefield believes that some scientists are misapplying a folk concept, rooted in our everyday understanding of human nature, that other scientists share and to which psychiatry should remain faithful.

This idea, that science discovers the empirical application of our pretheoretical folk concepts, is an old corollary of conceptual analysis. It stems from the time when analytic philosophers agreed that their subject was just the analysis and clarification of our fundamental concepts. The evidence used for such conceptual analysis is just our everyday intuitions, which means that its proponents often hold empirical findings to be irrelevant. Snowden (1990, 122), for instance, insists on the following principle:

It is wrong to incorporate as an element in the analysis of a concept C any condition F which can be revealed as necessary and essential condition for the correct application of C only by arguments relying on what are, broadly, empirical considerations.

I'm assuming that this principle must be restricted. It applies to concepts in commonsense contexts, since it surely cannot apply to concepts that one acquires only through an education in some theory, like *isotope, gamete* or *double dactyl*. Even so, Snowdon's principle is very strong, and it leaves us unclear regarding why we should care about conceptual analysis at all. All we are doing, it seems, is inquiring from our armchair about what other people think.

The reasoning behind Snowdon's principle is that it makes no sense to imagine finding out empirically, for example, that you can know there is beer in your fridge even if you do not believe there is beer in your fridge. Our concept of knowledge just rules out p in the absence of belief that p. Now, for some concepts this is entirely plausible, despite my skepticism. For instance, it is unlikely that empirical inquiry could convince us, against our intuitions to the contrary, that betrayal is good.

But when Snowdon's approach is applied to psychological concepts, it implies that our folk psychology defines what mental states are. Hence, psychological states that do not meet the definition simply do not (indeed, could not) exist. In his influential work on emotion, for example, Anthony Kenny argued (1963, 51) that although investigating human physiology could "lead to results of the highest interest," it nonetheless "cannot have the status of an experimental analysis of the nature of the emotions." Kenny's view is that emotions are defined by our everyday thought, and hence scientific theories are just irrelevant to emotion concepts. Emotions are whatever our lay concept of emotions defines them as, irrespective of what science says. The objection to the approach (put forcefully by Griffiths 1997) is that it tries to give the meanings of terms without investigating what in the world those terms actually refer to. Instead, conceptual analysis merely tells us what people believe about emotion or mental illness. That is moderately interesting, but how does it help us discover what emotions or mental disorders *are*?

This approach does not always write off science. It can champion science at the expense of common sense, as in Churchland's (1981) thesis that neuroscience has not discovered anything that satisfies the definition of our commonsense concept of belief, and therefore beliefs do not exist. Wakefield's periodic appeals to empirical findings may seem to show that he is not doing this very traditional conceptual analysis but something with a closer tie to science. However, what he is doing is a version of traditional conceptual analysis that has developed in recent years.

3.2.2 Newfangled Conceptual Analysis

These days, instead of offering necessary and sufficient conditions, conceptual analysts are likely to think of themselves as constructing accounts of folk theories that "functionally define" our concepts by showing how they figure in those folk theories. The analysis must satisfy certain key "platitudes" about the concept that we all share, which constrain any analysis of the concept, and which tell empirical inquiry what to look for.[1] Wakefield

---

1. Menzies (1996) offers a very clear picture of the idea of platitudes as constraints on the analysis of causation. Lewis (1972) inspired this approach, which has roots in Ramsey and Carnap and was originally intended to introduce theoretical terms in the

seems to have something similar in mind when he faults the DSM analysis of mental disorders for flouting our intuitions. More generally, it seems that both sides of Kitcher's objectivist/constructivist divide assume that there is a lay concept of disorder that should constrain the scientific understanding of what is or is not a disorder. How does the authority of folk theory feature in Wakefield's work, and who else endorses the hegemony of intuition?

We already have come across the first piece of evidence that Wakefield privileges folk thought over science. I refer to his contention that the DSM is over-inclusive relative to common sense. Wakefield thinks some DSM-IV diagnoses outrage our intuitions by attributing disorder on the basis of behavior alone without looking for malfunctioning mental mechanisms. But whose intuitions are outraged? Not, presumably, those of the psychiatrists and psychologists who wrote the DSM chapter on substance abuse, since they are content with the idea that "mental disorder" can be a merely behavioral dysfunction. Wakefield is *criticizing* the scientific, theoretical picture of mental disorder by an appeal to intuitions. This theme comes up time and again in Wakefield's writings, as when he says that the evidence for his analysis is not DSM criteria but our intuitions about disorder (Wakefield 2000). He is searching for necessary and sufficient conditions for the folk concept of mental disorder and assuming that science should search for the psychological processes that fit the concept thus defined.

Folk psychology's authority over the psychiatric concept of mental disorder is widely accepted by conceptual analysts. For example, Brian Grant complains that sleep disorders seem "obviously misplaced" (1999, 65) in DSM-IV. And Denis Dutton denounces a definition of mental illness on the grounds that it "would lump Alzheimer's sufferers and brain-damaged accident victims with delusional psychotics" (1996, 561). Like Wakefield and Grant, Dutton assumes that a clinical concept of mental illness ought to capture a set of everyday intuitions to do with insanity, and if it does not it is illegitimate. Boorse is less restrictive than Grant or Dutton in his conception of what the mental disorders should be, but he too is quite happy to adduce everyday linguistic usage and commonsense intuitions as evidence, even though he claims to be discussing the clinical concepts of health and disease.

sciences. Jackson (1998) defends the idea at length. Peacocke (1994) employs the idea of "primitively compelling" inferences to develop a different, though similar, project of conceptual analysis, one explicitly aimed at establishing the conditions of inquiry in the cognitive sciences.

But why should the study of psychopathology simply obey our pre-theoretic views? Well, it might seem that when we are dealing with an aspect of folk thought, we have some sort of privileged vantage point. After all, it is ourselves that we are talking about, and surely our folk theories might be informative.

The point is not that our folk theories might be informative, or a good place to start, perhaps by roughly pointing out the paradigmatic instances of the sort of thing we should investigate. That is unobjectionable. The problem is the orthodox program's assumption that folk psychology is in charge of science. That view erects a barrier to inquiry. Scientific practice, in order to establish the correct taxa and the most useful generalizations, may need to range over as many conditions as possible, covering all the ways in which the mind/brain can break down. If we discovered that some process that does not conform to commonsense thinking is the cause of Schizophrenia, we would not conclude that schizophrenia is not a disorder. Mental disorders may form a very heterogeneous category, including few conditions (possibly only the major psychoses) that correspond to our ideas about insanity or madness, and also including some conditions that are not currently thought of as psychiatric. So, there may be good reasons for rejecting the view that the job of psychiatry is to understand what in nature corresponds to the everyday sense of "mental disorder." However, the rejection of common sense can take several forms, depending on whether one seeks to simply ignore pre-theoretic and unscientific intuitions, or to preserve some role for them in the conceptual structure that replaces the orthodox program. I will end this chapter by facing these issues, but for now, I want to continue my discussion of the three main issues. The first issue was the relation between science and common sense: I said that privileging common sense, as opposed to admitting some role for it, hampers empirical inquiry. I now move on to my second issue: what is mental about mental disorder? This issue does not just touch on the relationship between psychiatry and folk psychology; it also touches on the relationship between psychiatry and the other sciences of the mind.

3.3 Psychiatry and the Nature of the Mental

3.3.1 Introduction
Rejecting the hegemony of folk theory allows us to discuss the nature of explanation in psychiatry without one irrelevant constraint. But another constraint derived from folk thought is still with us. Psychiatry is supposed to be about mental illness. What is *mental* illness?

The limits of the mental establish the scope of the abnormalities psychiatry is concerned with, but the question of scope hardly ever is discussed.[2] However, it is particularly pressing now, as the science of the abnormal mind develops in ways that bring together traditions of inquiry that view the mental in different ways (or have not been concerned with it at all). The limning of the mental is also important for me, since the revisionism I advocate takes a broader view of the mental than is customary in psychiatry and also flouts commonsense intuitions by, for example, concluding that blindness and comas are mental illnesses. So I will argue that there is no coherent conception of the scope of the mental in psychiatry, or in common sense, that one is bound to respect. A historical story explains in outline why that is.

3.3.2 What's Mental about Mental Disorder?
There is an absence of clear discussion of the nature of the mental within the orthodox program, except for a few statements on the mind-body problem (e.g. Kendler 2001). A form of eliminativist materialism that denies that there are any minds at all may motivate members of the antipsychiatry movement (Szasz 1987), but while I agree there are no such *things* as minds, we do not have to give up on mental talk in consequence. Philosophy is full of nonreductionist and functionalist materialisms about the mind. These discussions show that one can be a materialist while still thinking of the mental as autonomous, and thus preserving a role for distinctively mental disorders (Papineau 1994). I will not delve into the metaphysics: a generic materialism suffices for psychiatry.

What I want to discuss is what we might call the scope of the mental. Which phenomena are the mental ones? Despite optimists like Fulford (2002, 351) who think that saying what "mental" means is much easier than saying what "disorder" means, we lack a clear general conception of the mental. And there are two reasons why it matters. First, it is another reason to deny that folk theory should steer psychiatry. There is little prospect of reaching general agreement on the nature of the mental in a way that lets the various sciences of the mind fall into line with folk psychology. Second, lack of clarity about the nature of the mental has implications for how psychiatry relates to other sciences of the mind, including inquiries that deal with abnormal

2. Culver and Gert (1982) and Margolis (1980) both raised some of the points I am about to make, noting some difficulties with the psychiatric concept of the mental. But they seem to think that some defensible folk concept of the mental could be found. I am less sanguine.

mental phenomena but are not thought of as part of contemporary psychiatry. If psychiatry is to unite with these other sciences, these issues must be resolved. I propose a way of resolving them and a diagnosis of why our thinking about the nature of the mental is so inchoate. My approach also provides reasons for abandoning the conceptual constraints imposed on psychiatry by the two-stage conception in its conservative, foundational garb.

Scholars in the orthodox program do not concern themselves with whether a given human capacity is mental or not. In so far as there is agreement about what the realm of mental disorder amounts to, it is a tacit endorsement of the opinions of psychiatry—the discussion concerns the commonsense propriety and scientific objectivity of existing diagnoses. The orthodox program assumes that commonsense "folk psychology" and the scientific study of psychopathology share a conception of the mind and that the live issues concern the application of that shared conception to questions of disorder. It is in the application of concepts within this shared picture of the mind that the orthodox program dictates to science.

What does an objectivist conceptual analysis implies about the "mental" part of "mental disorder"? On an objectivist reading, the sciences of the mind have the role of searching for psychological dysfunctions. So a natural way to understand Boorse and Wakefield is that they construe the mental as the province of those sciences that discover what mental mechanisms are. But, by that test, perception is an exemplary mental process, yet nobody I have asked thinks blindness is a mental disorder.[3] However, generating the intuition that vision is a mental phenomenon is easy, which implies that blindness is a mental disorder.

Vision often counts among the paradigmatic examples of mental phenomena. For example, when philosophers worry about the mind/body

3. With one (partial) exception: true to his background as a neuroscientist treating traumatic brain injury, Warren Lux (in conversation) insists that *of course* Cortical Blindness (unlike other forms of blindness) is a mental disorder. Everyone else I have spoken to agrees that our concept of mental disorder has no place for blindness in any form, so I feel justified in assuming that I have the intuitions on my side. If some clinicians disagree, though, it does strengthen my case that scientific assumptions about mental illness do not cohere with each other or with the lay concept, despite the orthodox program's assumptions to the contrary. The extent to which intuitions can be taken be for granted is of course a vexed one. It seems possible to study the general opinion as to what conditions count as mental illnesses empirically. We might also look to historians to see how concepts of the mental have changed over time.

problem they fret about consciousness, especially visual experience. "How," wonders Colin McGinn, "can technicolour phenomenology arise from soggy grey matter?" (1989, 349). Responses to McGinn's pessimism about our ability to answer that question do not deny that visual experience is a legitimate example of a mental phenomenon. After all, according to the sciences of the mind, vision *is* a quintessential mental phenomenon. Perception is a central issue in psychology and cognitive neuroscience, so surely its claim to be mental is beyond dispute. So blindness should be a mental disorder, but it is not.

So it appears that we have a counterexample to objectivism as defined by Wakefield and other objectivists like Horwitz (2002) and Papineau (1994). Their analysis makes a disease a mental one if it results from the failure of a *psychological* mechanism to do the job it evolved to do. So blindness is a mental disorder, since it is both mental and a disorder.

The example of blindness shows that uncertainty about the nature of the mental gets the orthodox program, *qua* foundational conceptual analysis, into trouble. Disruptions to the normal mind/brain cause parts of our perceptual systems to fail. The ensuing conditions therefore meet Wakefield's first criterion for being a mental disorder, and the requirement of harmfulness also is met, since they are undesirable. But these undesirable consequences of a failure of our mental apparatus do not count as mental disorders in folk thought or appear in the DSM. Hence we have a counterexample to Wakefield's analysis.

Vision is not the only example of problems that some bodies of thought view as mental and others do not. Papineau's (1994) example of a mental state is the state of being in pain, often discussed by philosophers as an example of qualia, but more bound up with physical medicine than psychiatry. Examples like pain, perception, and tics (are they physical or mental?) suggest there is no coherent concept of the mental extending across the folk theories of the mind. What is mental in one context becomes physical in another.

Confusion can result from not noticing that we lack a coherent general concept of the mental. Horwitz (2002, 22) includes perception among the functions of human psychological systems, but almost immediately goes on to say that the eyes are studied by general medicine rather than psychiatry because they are part of a physical system (2002, 24). (So, of course, are the systems that psychiatry studies.) Horwitz says that the function of this physical system is to convey visual information, which seems like a psychological claim. But according to objectivism a mental disorder is one with an etiology or symptomatology involving those properties or phenomena falling into the scope of the sciences of the mind, for how else can the objective identification of psychological malfunction proceed? And the scope of the sciences of the mind surely includes vision. Still, Horwitz says that blindness is a physical disorder (232, n. 7). When he discusses the mind in general terms he relies on a generous conception of the mental derived from the cognitive and brain sciences, but when he considers psychiatry his conception of the mental narrows to fit an intuitive understanding of the scope of that discipline. It thereby excludes some of the phenomena that he has already placed in the realm of the mental in the wider context. The general position Horwitz has embraced has revisionist consequences for folk psychiatry, but he fails to notice them because he is trying to preserve a commonsense demarcation of mental and physical ailments, which does not treat perception in the way that the study of the normal mind treats it.

Now, as I pointed out, Wakefield's charges of overinclusiveness and counterintuitiveness are an embarrassment to the DSM if the concept of mental disorder contained in the DSM is supposed to be our folk concept.[4] So Wakefield's objections might seem easy to defeat if we drop that requirement. However, although I think that dropping it would be a wise move, it seems clear that the DSM as written does in fact have aspirations to respect ordinary thinking, and thus is unable wholly to rebut Wakefield's counterexamples, and it too faces the problem of blindness. I do agree with Wakefield that the DSM's neglect of causal histories is problematic. Many of the arguments in the next few chapters are devoted to making that last charge stick, so the full extent of the difficulties will not become apparent until much later. But here's a preliminary.

In section 3.3.3 I say why the concept of the mental in psychiatry is just as incoherent as the folk view. In section 3.4, I assert that mental illness is not always evolutionary dysfunction. In section 3.5, having rejected the current state of affairs, I offer a positive proposal about the scope and nature of psychiatry that draws on objectivism but strips it of its present conceptual foundations. I also suggest an explanation for the present state of affairs

4. Another example: despite Wakefield's strictures, it is not at all implausible to talk about behavioral function and dysfunction in scientific contexts (cognitive ethologists do it all the time). Of course, Wakefield may be right about the stupidity of certain applications of the idea of behavioral dysfunction, even if science can recognize behavioral dysfunction as a basis for mental disorder attribution in principle. There seems no reason, for instance, why arguments about drinking should be singled out as behavioral dysfunctions: if arguments are indicators of dysfunction, then why not arguments about debts, epistemology, or cheese?

that portrays it as a theoretically unwarranted historical accident, which supports the positive program by making the existing picture look even more unattractive. In section 3.6, I pile on the justification by looking at some empirical work that shows the benefits of ignoring disciplinary constraints or constraints due to folk conceptions of the mental. And in the last sections of the chapter, I look at the two-stage picture again in the light of this new conception of its supporting science.

3.3.3 Psychiatry and the Concept of the Mental

Wakefield and Horwitz chastise the DSM-IV for neglecting commonsense intuitions. If the DSM-IV rejects common sense as a constraint, though, their criticisms are ineffective unless there are compelling grounds for accepting constraints from common sense, which there are not. However, it does seem that the architects of the DSM suppose its concept of mental disorder to be constrained by folk thought to some degree, although the case is not open-and-shut. Psychiatrists and psychologists clearly have not been afraid to flout common sense sometimes, and they have discovered many surprising facts about the mind. But it appears that psychiatric investigation takes place within a view of the mental that is a product of folk psychology, interacting, as I shall suggest, with factors more internal to the history of psychiatry. The lack of clarity is the result of the inexplicitness of the DSM's conceptual framework. And it does seem that the DSM aims for a scientific concept that nonetheless respects the cultures and beliefs of the people to whom it is supposed to apply (DSM-IV-TR, xxx–xxxiv), which makes room for everyday thought as a constraint on psychiatry. As I showed in the discussion of depression in the last chapter, some concessions are made for circumstances in which being miserable is to be expected. So perhaps the DSM adopts not the subordination to folk thought that the orthodox program concedes, but only the view that our everyday intuitions are to be respected, but not treated as definitive. Nevertheless, the DSM definitely suffers from conceptual confusions about the mental.

Modern psychiatry harbors a body of intuitions about the nature of the mental that are hard to render coherent. As I said, the visual deficits are mental disorders according to some understandings of the mental, but are excluded from psychiatry, which suggests a view of the mental similar to that of folk conceptions of psychiatry.

As psychiatry thinks of itself more and more as a clinical brain science, the exclusion of some conditions becomes harder and harder to understand. Nancy Andreasen, in the course of a manifesto for biological psychiatry, says simply that "mental illnesses are diseases that affect the brain" (1984, 8) and she also says that the retina is part of the brain—"an extension of the brain outside the skull" (1984, 85). It seems then that biological psychiatrists should see blindness as a mental disorder even if it results from retinal damage. But psychiatrists do not think in this way, and Andreasen does not want to reform them, even though she thinks psychiatry deals with brain disease and retinas are parts of the brain.

Psychiatrists who reason like Andreasen fall short of calling blindness a mental disorder because it sounds so counterintuitive, despite their apparent commitment to doing so on wider grounds. Vision is part of psychology. Many important works in cognitive psychology have treated both normal and abnormal vision. Brain lesions or other psychological disruptions can cause a failure to process perceptual information normally, leading to, among other problems, visual agnosias. Prosopagnosics, for instance, suffer from no loss of vision and can identify inanimate objects. But they cannot recognize faces, even faces of family members (Palmer 1999). These failures of the visual system clearly are departures of the mind from its proper functioning. Are they also mental disorders? Psychiatrists do not treat agnosia, and they are not mentioned in DSM-IV-TR. If you agree with that exclusion, then you presumably agree that at least some problems identified by psychological and neurological theory are not mental disorders. Suppose that we decide, however, that the agnosias are in fact, mental disorders. Intuitions about that are not clear-cut (among the people I have asked, anyway). But what about plain old blindness? Nobody thinks that general blindness is a mental disorder, which means that as damage to one's vision gets worse, the intuitions that one is suffering from a mental illness become weaker even though they originally were based on facts about loss of vision.

However, although psychiatrists shy away from calling blindness a mental illness, just as folk psychology does, their conception of the mental does not accord with our wider intuitions. If there is a conception of the mental guiding the DSM, it is a very strange one. According to the DSM-IV-TR, perception is not mental, but sleep and literacy are. The DSM-IV-TR includes sexual dysfunctions, which seem dubiously mental: the problem in Male Erectile Disorder (impotence) seems to be the flow of blood rather than the stream of consciousness. Sleep disorders clearly do fall into the provinces of sciences of the mind/brain, but I am unclear whether commonsense has a view as to whether sleep is a mental phenomenon: as we saw, Grant (1999) finds sleep disorders out of place in a compendium of psychiatric disorders. Along with sleep disorders like Insomnia, there are other puzzle cases in the DSM-IV-TR, like Eating Disorders and Caffeine Intoxication. It is far from clear whether these diagnoses conform to any commonsense understanding of what a mental disorder is.

The folk conception of mental illness does not cover these cases, so Wakefield's complaints about the DSM-IV's overinclusiveness may well accord with our intuitions about mental illness. The issue is whether the intuitions Wakefield appeals to should constrain the development of a theoretically motivated taxonomy of the way normal minds can become abnormal. I argue that we must dispense with them if the scientific case is strong enough.

The scientific case turns on the utility of looking for generalizations that cut across folk, psychiatric, and other scientific approaches. Achieving a synthesis of the clinical mind/brain sciences requires scrutiny of the relations between psychiatry and the neurosciences as well as the relations between psychiatry and common sense. This is true even for phenomena that everyone thinks of as mental, such as cognitive deficits. The cognitive disorders section in the DSM-IV-TR maps poorly onto the conditions that cognitive neuropsychology recognizes. In part, this is because the heterogeneity of DSM conditions contrasts sharply with the fineness of grain that neuropsychology often adopts. Cognitive neuropsychology typically recognizes very fine-grained disorders based on specific symptoms that arise from breakdowns in a psychological mechanism. The DSM-IV-TR groups cognitive symptoms into very broad classes without any precise specification: all we are told about delirium is that its essential feature is "a disturbance of consciousness that is accompanied by a change in cognition that cannot be better accounted for by a preexisting or evolving dementia" (DSM-IV-TR, 136). A Dementia is characterized by "multiple cognitive defects that include memory impairment and at least one of the following cognitive disturbances: aphasia, apraxia, agnosia, or a disturbance in executive functioning" (DSM-IV-TR, 147).

The DSM is interested in the underlying brain or (other organic) diseases and uses these, rather than functional impairment, to subtype cognitive disorders: the DSM-IV distinguishes dementia on the basis of an underlying disease or injury—Alzheimer's, Huntington's, Pick's Disease, AIDS, head trauma—and not on the basis of a particular pattern of cognitive malfunction. This illustrates the systematic neglect in the DSM-IV of any level of explanation between the brain and "clinical phenomenology," which is most naturally attributed to the positivism and operationalism of the architects of the DSM-III. A cognitive approach, on the other hand, looks for a functional specification at the intervening level of cognitive architecture. The neglect of this level may be what causes the DSM to omit altogether well-entrenched cognitive neurological diagnoses like Capgras's Delusion, Hemineglect, or

Simultanagnosia. Any general account of psychopathology, however, should surely address these phenomena.

Our confusions regarding cognition and perception (and movement, as we saw when talking about Tourette's in the introduction) shows the puzzles that arise when we push at the concept of the mental in different contexts. The confusions show that we cannot find in the DSM a concept of mental disorder rooted in contemporary science. Psychiatry contains no principled understanding of the mental. When we appreciate that we can look in new ways at many issues in the philosophy of psychiatry. A general theory of psychopathology must discuss conditions that are currently divided along disciplinary lines and it must reject commonsense intuitions as a source of authority.

3.4 Objectivism as Revisionism

3.4.1 Introduction

We have arrived at the following position: The orthodox program errs in thinking that a clear concept of mental illness exists in folk psychology and can provide guidance for students of mental illness. And no objectivist account of mental disorder that accepts common sense and works within existing disciplinary conventions can preserve both our wider thinking about the mental and our narrower thinking about the mental within folk psychiatry.

Hence, our current demarcation of mental disorder is *scientifically* uninteresting (which is not to say that it is not interesting at all). We have very diverse conditions lumped together as disorders, and it is no surprise that a general analysis fails to account for the heterogeneity we see. As section 3.4.3, argues the diversity results from developments in folk and scientific thinking that have left us with no clearly demarcated overall concept of mental disorder. I do not say that individual concepts or diagnoses are always messy and confused, and I am not arguing that scientific and clinical research always has been infected and undermined by the problems I am raising. Many questions about mental illness have been answered scientifically via high-quality studies. But the superordinate concept of mental disorder, as it stands, does not pick out a nonarbitrary class. And as well as suggesting that the conceptual issues need to be reframed, this arbitrariness has scientific consequences. It can obstruct our ability to generalize across related conditions, mislead us into thinking that disciplinary boundaries correspond to an interesting break in nature, and stymie the linkage of psychiatric research with research in other disciplines, notably the cognitive sciences.

My alternative account takes over aspects of objectivism. The revisionist consequences for common sense do not concern me much. The important conceptual question from my point of view is how we should approach the concept of mental disorder from the standpoint of scientific investigation into human psychology. That is the important question because it guides the first of the two inquiries—the objective, scientific one—that, according to the two-stage picture, should inform our thinking about mental disorder. The second inquiry, what our judgments should be about malfunctions in human psychology, requires a different answer, one that appeals to the cultural understanding of well-being and normality. Whether this separation can be vindicated depends on a number of issues that I take up in the next chapter. At the moment I am exploring the scientific basis of the two-stage picture, and I have suggested that it should reject both common sense and disciplinary constraints.

This double rejection involves biting a large bullet. Normal psychological function can be disrupted in many ways. Common sense denies that you have a mental disorder if you are in a diabetic coma, or have no cortical activity because you are on a life-support machine and with only a brainstem intact. Yet mental functioning is massively impaired in these cases, so I have to call them instances of mental disorder.

Taking all these conditions on board certainly distances me from current opinion in folk thought and in psychiatry. I do not deny the counterintuitiveness of lumping blindness or comas in with depression and schizophrenia. But there are substantial benefits to expanding the objectivist remit across the whole of psychology, and we should not lose them just because the expansion does not agree with common sense. This desire stems from the respects in which my position is conservative. Our investigation of mental illness should respect certain core examples: if schizophrenia and major depression are not mental illnesses, then it is hard to know what a mental illness could be. So I imagine psychiatry proceeding via ostension, in the sense that we start with core examples, try to explain them and explain things like them, then extend the explanatory structure that the ostensive approach generates as far as it should go. However, a project that proceeds in that way must be revisionist in this sense: as it grows away from the core it may need to reject folk beliefs and accept revisionist positions, in order to extend the principles that explain the core into new areas, in search of maximum power, generality, and progress.

So, we want to explain, taxonomize, and conceptualize mental illnesses without being inhibited by folk categories if they impede the search for power, generality, and progress. For example, the same symptom or process may show up in cases that common sense sees as mental disorders and cases that common sense (or psychiatry) says are not mental disorders. We might find the motor or cognitive deficits of Tourette's, Huntington's, Sydenham's chorea, and even schizophrenia to be manifestations of the same underlying disruptions, or different disruptions to the same system. If they share a causal basis in this way, the intellectual landscape should not get in the way of our searching for it.

The positive conception of the mind that I recommend, and its consequences for the philosophy of science, are examined in detail as the book proceeds, but the argument is easier to follow with an introduction here, accompanied by an example of the sort of inquiry that accepts the revisionist consequences I have urged. The next section briefly sketches the positive conception of the mental that falls out of the revisionist project, discusses the history that has left us where we are, and then provides an example of revisionism in action.

3.4.2 So What Is a Mind, Then?

The view of the mind I rely on says that when we talk about the mental we are referring to capacities that enable us to perceive, understand, and act on our environments, among other things. So "the mental" covers states and processes that play a very direct role in intelligent action, including processes such as perceiving, remembering, inferring, and a wide variety of motivational states. I think of the mental as functionally defined rather than picked out by a concept in our vernacular metaphysics.

Think, by analogy, of how we might try to understand causation. Rather than trying to define causation "all at once" as it were, so that all causal relations are comprehended by our definition, we might try a different tack. We might look at a variety of prima facie causal relations in an attempt to understand what we can about causation (Hitchcock 2003). Of course this requires some intuitions about the subject matter, but the intuitions need not be any more uniform than the projects in which causal talk features, and although intuitions play a role in getting inquiry started, they can be overridden as it develops.

So, by analogy, as we proceed, I look at a wide variety of putative mental disorders, with the boundaries set by projects that try to explain mental disorders, not by folk thought. I also connect these projects in abnormal psychology to projects that try to understand the normal mind. As I've said, the picture of the mind that the behavioral sciences produce has important implications for how far we can understand mental illness. This reliance on science has another consequence that many will find odd, which is that the nature of the mental in part depends on what the sciences come up with. Some pictures of the mind stress embodiment very heavily (e.g., Clark 1997; Hurley 2000, 2001). Others prescind from details of our embodiment to stress a more purely computational theory of the mental (e.g., Fodor 1987; Pylyshyn 1984). On my view, what counts as mental depends in part on who is right in these debates. If we need to include facts about our embodied nature to explain the functions that allow us to behave intelligently, then the scope of the mental probably will contain more in the way of movement and physical action than it would if more narrowly computational treatments do the best explanatory job.

So my understanding of the mental is loose and contains some vagueness and indeterminacy, but I am unmoved by that: minds are not objects whose boundaries are apparent. On the other hand, we are not without clues to the nature of minds, courtesy of the contemporary biological, cognitive, and social sciences. The scientific picture I entangle with psychiatry traces the exercise of various capacities to the way the brain works in socially situated humans, assuming a basic evolved cognitive toolkit as the foundation with the superstructure provided by culture. No single science or model of the mind is privileged, because we can ask many different questions about the mind, and also because there is no reason to suppose that one science or one theory within a science can explain the diversity of mental phenomena. The mind is not an object with a well-defined essence; it is a loosely related set of capacities, and no theory in philosophy or science explains everything about it. So the structure, for example, will not bet on the general rightness of the model that Baron-Cohen develops to explain autism, the model that falls out of Boorse's objectivism.

The promiscuous explanatory structure I recommend understands the mind in a variety of ways, borrowing from many cognitive, neurological, and social sciences. If we think of the mind as these sciences recommend, many departures from normal functioning will not accord with our pretheoretic thinking about mental disorders.[5] But that is consistent with my view that although our interests set the terms of the project, they do not determine it.

Section 3.5 closes this discussion with an example of the benefits to be gained by riding roughshod over the traditional boundaries between psychiatry and other disciplines. It looks at some recent attempts to understand a variety of behavioral and affective abnormalities in terms of breakdowns

---

5. This shows, incidentally, that the indeterminacy of Culver and Gert's nondistinctness principle is not the problem. The problem is their abstractness about etiology.

in the control of purposeful action and our conscious awareness of it. The discussion's intent is to give positive reasons for rejecting the boundaries of the mental in psychiatry. But first, I offer a negative argument, providing some history with a view to undermining contemporary psychiatry's self-demarcation on other grounds.

In virtue of what facts *do* psychiatrists judge a condition to be a mental disorder? Even if the demarcation is a mess theoretically, it must have some explanation. And categories "are consequential. Accordingly there is important work—philosophical? archeological? historical?—to be done in reconstructing the ways in which our most influential divisions were constructed and how they have left their mark on the world we inherit" (Kitcher 2001, 53). The way to understand the psychiatric concept of mental disorder is to see it as a blend of folk and scientific thinking that has left an important mark in a number of contexts. The species of inquiry Kitcher has in mind is most strongly associated with Michel Foucault. Like Kitcher himself and some other scholars (Gutting 1989; Hacking 1979; Lilla 2001), I think it is wrong to see Foucault as some sort of epistemological nihilist, and instead read his work as a blend of history and philosophy that investigates the process by which certain pictures of the world came to seem inevitable. If we conceive of this excavation of concepts more broadly than Foucault did,[6] we can see that it is starting to emerge in interesting ways in contemporary philosophy. The tracing of the history of a concept through its theoretical transformations is a staple of historians and philosophers of science. Paul Griffiths (1999) recommends seeing concepts as epistemic projects, but perhaps a better perspective would see them as components of various projects, including but not limited to epistemic ones. This sort of conceptual examination, with its stress on historical influences on thought, represents a welcome way of understanding the concepts that shape our thought about the world without trying to perform mere conceptual analysis on them, although it preserves a role for the sort of nuanced scrutiny of concepts that many philosophers of the analytic tradition excel at (for more discussion see

6. But also more narrowly. More broadly in the sense that it draws on methods and addresses issues that Foucault never looked at, in particular analytic philosophy's concern for commonsense intuition, not as a source of definition but as a guide to commonsense epistemic and normative concerns. More narrowly in that the history may be entirely devoid of the governing assumptions of much work influenced by Foucault. In particular, it may not take up what Stefan Collini calls the guiding suspicion of cultural studies: "Somebody Is Trying to Screw Somebody Else" (Collini 1999, 263).

Davidson 2001 and Geuss 2001). In that spirit, let me attempt a preliminary outline of some of the projects and developments that have left us with the ideas about mental disorder that we currently possess.

I will not discuss many of the most frequently voiced concerns about the development of DSM psychiatry. For instance, it has been argued very convincingly that a variety of political pressures played a role in the development of the DSM-III. One example will suffice here: some DSM-III categories appear to have been negotiated by American psychiatry to satisfy the demands of the U.S health insurance industry (Cooper 2005). A thorough discussion of the DSM requires a full-blown "externalist" history. Such a discussion is not my goal, but I do want to note some of the processes that produced the present confusion about mental disorder. Apart from their intrinsic interest, I hope that alluding to these processes further serves to undermine the idea that a clear concept of mental disorder exists and can be analyzed, either inside the disciplines or outside them. People who are already convinced, or who have no appetite for the history, should skip ahead to section 3.5.

3.4.3 Where Do Our Psychiatric Categories Come From?

To begin, folk thought about mental illness is bound up with the development of attitudes towards madness, rationality, emotion, and the division between mind and body. Standard outlines of the history of at least Western conceptions of madness tend to see them as arising in classical Greece or earlier and persisting, bound up with notions of perversion, holiness, disease, and innocence, until the movement toward institutionalized care for the insane began around the time of the Enlightenment (e.g., Ellenberger 1970; Porter 1987; Shorter 1997). It seems that when the science of the abnormal mind emerged in the nineteenth century, it was born within a set of pre-existing folk conceptions about its proper province, which it has never shaken off. No matter how sophisticated nineteenth-century cognitive neuroscience or psychiatry became, the study of blindness never would qualify as a branch of psychopathology. Existing conceptions of disorder, mind, and body would not allow it.

In some cases forms of behavior did accord with folk intuitions about mental illness, but even these folk intuitions and categories were challenged as the scientific study of the mind began. The classic example of such a challenge is the one Charcot and Freud undertook to convince the medical and lay communities that hysteria is a disorder that might affect men as well as women (Ellenberger 1970; Hacking 1998). Hysteria was regarded as a quintessentially feminine disorder, chiefly because of its historical association with the uterus: the Egyptians and Greeks believed that the uterus was given to wandering about the body, causing behavioral symptoms as a consequence. If the uterus traveled as far as the throat, for example, it was thought to produce the characteristic choking sensation that later became known as *globus hystericus*.[7] Charcot and Freud triumphed, unlike Thomas Beddoes, the great English physician who had argued the same case during the Napoleonic Wars. The paradigmatic old-time hysteric now probably would be diagnosed with Somatization Disorder (still occasionally known by another old name, "Briquet's Syndrome"), although only if the symptoms first occur before age 30 (DSM-IV-TR, 486).

These days the history of Hysteria is such a big business that it has its own history (Micale 1995). The story of hysteria, though, can tell us something about the development of the science of psychopathology. For example, we might want to know why hysteria survived as somatization disorder, but the Aphasias are excluded from the DSM-IV (although some forms of developmental language disorders are in the DSM-IV, including stuttering). The explanation for the DSM's inclusion of the descendants of hysteria but exclusion of aphasia, as far as I can see, is that the first two editions of the DSM had a heavily psychoanalytic orientation, and there simply were no psychoanalytic explanations of aphasia at the time. This contrasts with, for example, autism. Autism is a disorder for which psychodynamic explanations have long existed, courtesy of Bruno Bettelheim, who passed the notoriously annihilating judgment that blamed the condition on "refrigerator mothers." When the theoretical foundations of the DSM changed to become at once both semi-operationalized and neo-Kraepelinian, a number of disorders were inserted, and most of the DSM-II disorders were just grandfathered in.[8] It appears, though, that no systematic effort was made to include a number of cognitive disorders that previous editions had omitted. This is

7. To remove the uterus from the throat, insert rat droppings up a woman's nose. Micale (1995) contains a short history of premodern views on hysteria. A hair-raising treatment of Egyptian gynecology can be found in Nunn (1996), a book that should reassure anyone who doubts that humanity makes progress.

8. After I had written this, I was delighted to find that Allen Horwitz (2002) had done the historical and sociological work required to turn my indictment into a conviction, and I refer readers to his book for more details. But Horwitz takes dynamic psychiatry as his starting point and in so doing ignores the earlier split of neuropsychology and psychoanalysis. Mace (2002, 66–68) appeals to that split to support a concept-centred evolutionary view of scientific change, about which I am more skeptical.

partly because the reformers of the DSM-III Task Force worked with disorders that were already paradigms of psychiatry. The cognitive disorders had been the province of brain scientists rather than psychoanalysis, and hence were seen as something that psychiatrists, at least in America, did not treat or study but regarded as general medical (Axis-III) conditions.

However, the fact that neuroscience and psychoanalysis divided up the field of psychopathology differently was itself a historical accident (Martin 2002). Freud's early career was as one of a group of neuroanatomically inclined investigators of both normal and abnormal psychology, a group that included his teachers Brücke and Charcot (Glymour 1992, Sulloway 1979), and indeed Freud's early publications included studies of aphasia and infantile cerebral paralysis. It was not until his middle age that Freud's theorizing lost its affiliation with neuroscience and evolutionary biology, and it was in this form that his theories swept American psychiatry (although their American reception was heavily influenced by the work of Adolf Meyer, whose distinctive vocabularies also found their way into early editions of the DSM). Psychoanalytic approaches dominated American psychiatry for much of the twentieth century.[9]

By the time psychoanalysis began its rise, a distinction had hardened between psychiatry and neuropathology, with different professional associations and separate meetings at large medical conventions (Shephard 2001). One important impetus behind the replacement of psychoanalysis by neo-Kraepelinian psychiatry was the rising success of drug therapy, applied to patients within the existing psychiatric population. Hence, the cognitive and

9. Enoch and Trethowan (1979) discuss in Freudian idiom a number of unusual conditions that originally and subsequently received putative explanations drawn from cognitive neurospychology, and they supply historically oriented literature reviews that allow readers to trace the rise of psychoanalysis in the interim. I restrict my remarks to American psychiatry, but much of what is told here is true of psychiatry in other countries. France, where psychoanalysis remains influential, is one obvious exception, and phenomenological psychiatry is an important force in many parts of Europe. British psychiatry, on the other hand, often has stressed cognitive methods. However, the medical model is gaining ground all over the world as the DSM becomes the standard diagnostic manual internationally. A standard British textbook adopted DSM-III criteria some time ago (Gelder, Gath, and Mayou 1989), and Dugas (1986) uses it in his retrospective of Tourette's. One member of a DSM-IV working party told me that his committee's deliberations did not touch on ICD-10 (World Health Organization 1992b) because "we knew that DSM was going to take over the world." As the DSM's influence expands, the ethos of American psychiatry will spread in its wake.

neurological disorders were largely unaffected by this development. An even earlier factor was the nineteenth-century distinction between mental diseases and nervous disorders. This distinction was hard to justify theoretically, but it allowed "nerve doctors" to make a lucrative living treating wealthy families who were afraid of the stigma of mental illness (Shorter 1997).

These accidents of scientific history are not relevant to just disciplinary boundaries and conceptual distinctions in contemporary science. They also played a part in informing the recent evolution of folk thought about mental disorders. Immense lobbying is undertaken with the aim of having a disorder added to the DSM because its adherents are convinced of its real, pathological nature. For example, Allan Young (1995) says that, from the point of view of clinicians convinced of the reality of Post-Traumatic Stress Disorder (PTSD), the lobbying has been essential. They say that PTSD has always existed, but was taken seriously only after being included in the DSM-III, "following a political struggle waged by psychiatric workers and activists on behalf of the large number of Vietnam War veterans who were then suffering the undiagnosed psychological effects of war-related trauma" (1995, 5). This was not just a matter of clinical politics. Many PTSD activists were not trained in psychiatry or psychology at all, especially those who first demanded recognition from the medical community as bearers of genuine symptoms. These ordinary veterans interpreted their psychological state in terms of psychiatric concepts that had penetrated the wider culture. In turn, they affected mainstream psychiatry in a form that was partly modified by folk thought.

The example of PTSD shows how concepts of mental disorder appeal to people who want their suffering to be taken seriously. On the other hand, other lobbying movements that undertake to have putative conditions such as Chronic Fatigue and Gulf War Syndrome recognized think that this means treating them not as psychological disorders but as straightforward physical pathologies—a category that in their thinking includes brain disorders (Showalter 1997). What motivates such projects is often a belief that *mental* disorders carry a stigma.

I repeat that these disputes are not just a matter of DSM politics unrelated to the development of folk thought. Many laypeople object to the idea that a certain condition could be a mental disorder, whatever the experts say, on the grounds that it is just the wrong sort of problem. Sometimes the experts listen. Furthermore, the theories of Freud and his followers, in however ill-digested a form, clearly have had an impact on our concepts of mental disorder and normal functioning, so the assertion that developments in psychopathology can influence folk psychology should not be too surprising.

One result of the history is that whether something counts as a physical disorder or a mental one is partly a matter of the niche it finds in the medical community. That cannot be the whole story, though, since Huntington's disease, a clear candidate for a straightforward organic brain syndrome, is also a DSM-IV-TR diagnosis (as a source of dementia). Other forms of neurologically based disorder were removed from psychiatry once their physical basis was found, for example Neurosyphilis and Epilepsy. The case of neurosyphilis is especially interesting, since in the nineteenth century it was a paradigmatic mental illness, much like schizophrenia today, and the discovery of its etiology was regarded as a psychiatric triumph. But the victory was purchased so dearly as to be indistinguishable from defeat, since the disorder immediately shifted niches to become the property of physicians. A subset of patients who were demented and paraplegic, it turned out, could be distinguished from their fellow sufferers by a propensity to keel over if they were asked to stand up straight with their eyes closed. If they shared this propensity, they probably suffered from tertiary syphilis, not the hitherto diagnosed General Paralysis of the Insane (GPI). The switch from a functional explanation to one couched in terms of detectable organic lesions led to GPI's annexation by the wider medical community (Shorter 1997, 55–58, 129).

Disorders are still assigned different niches depending on who is defining them. A clear case is narcolepsy, which is classified by the World Health Organization as a neurological disorder but is included in the sleep disorders section of the DSM-IV-TR. Some disorders generally assigned to neurologists are not in the DSM-IV-TR, although some of their characteristic symptoms are. But other disorders are in the DSM-IV-TR even though they may also be the property of cognitive (and noncognitive) neuroscientists. Aphasias are not in the DSM-IV-TR (although some forms of developmental language disorders are, including Stuttering), but Autism, with its characteristic language impairments, is included. The ability to do math certainly seems like a paradigmatic exploitation of cognitive faculties (see Shallice 1988 on acalculia), so why does the DSM-IV-TR include Mathematics Disorder? No clear rationale exists for the current demarcation of psychopathology between psychiatry, as enshrined in the DSM, and the cognitive sciences.

So, in conclusion, we may suspect that whether a condition counts as a mental disorder is bound up with a long series of transformations of folk and scientific psychology. (Although, it is not part of the clinical or folk concept of mental disorder that if something is a mental disorder it must have the right sort of institutional history.) This history manifests itself today in complicated and inconsistent ways: in judgments of value, personal worth, and personal responsibility; in folk concepts of the mind; in social phenomena like expectations about who treats what kind of disorder; as well as in scientific disputes and conflicts about the authenticity of dissociation or post-traumatic stress. Probably all these factors play a role in how we think about some disorders, and some may dominate in particular cases. But they do not neatly distinguish mental disorders from other things. Moreover, they seem to be inconsistently applied. Almost any phenomenon that in some contexts counts as evidence for a condition's being a mental disorder—such as brain damage, or addiction—may be cited in other contexts as evidence that the same condition, or another one, is not a mental disorder.

We are left with a mess. Philosophers and mental health professionals have gone astray in accepting a commonsense conception of their domain without stopping to ask if the folk concept of the mental is a reasonable, or even coherent, constraint. Judging by the practice of the sciences of the normal mind, there are phenomena such as vision whose pathologies appear to have perfectly defensible claims to be mental illnesses (blindness is, after all, both mental—since vision is a mental phenomenon—and an illness) yet are excluded from psychiatry without any defense of that exclusion. If the defense is to be mounted on conceptual grounds, then we may ask why folk psychology should dictate a subject matter to psychiatry since folk psychology itself appears to be no better off. This is an even more urgent question now than in the past, as neurocognitive explorations of mental illness become more common and psychiatry moves closer to the cognitive sciences. Bringing these disparate inquiries together is essential to understanding mental illness, but it involves some violence to common sense and to the current self-understanding of psychiatry. This is fine, since, to remind you, the self-understanding that psychiatry currently possesses includes a conceptual framework that, at important points, is a historical accident.

Section 3.5 turns from the science of psychiatry back to its alleged conceptual foundations, and deals with the issue of dysfunction. It has two questions to answer: first, is mental illness dysfunction? Second, when it is dysfunction, is it evolutionary dysfunction?

3.5 Function and Dysfunction

3.5.1 Introduction
Chapter 2 introduced the harmful-dysfunction analysis of the concept of mental disorder and showed how well it fits the empirical assumptions of

Baron-Cohen's theory of autism. And it is very close to the official analysis adopted in the DSM-IV. According to the DSM-IV, dysfunction is a necessary condition of something's being a mental disorder. The harmful-dysfunction analysis construes dysfunction narrowly as the failure of some component part of the mind/brain to perform its evolved function, and does not regard behavioral abnormalities as dysfunctions, unlike the DSM. In section 3.5.4, I explain how function is to be understood, but for now I want to look at some challenges to this view based on the distinction often drawn in contemporary psychiatry between what McHugh and Slavney (1998) call the concept of diseases and the concept of behaviors.

A number of thinkers (for example, Beck 1996 and Bandura 1986) have suggested that much psychopathology arises, at least in part, as a result of problematic social learning. These theorists conceive of faulty social learning as the acquisition of irrational, distorted, or negatively valued information, as opposed to the breakdown of information-processing mechanisms in the brain. In the language sometimes adopted by computational psychologists, the problem is with the input, not with the hardware or the software. In section 3.5.2 I discuss this idea and suggest that the apparent gulf that McHugh and Slavney introduce between diseases and behaviors can be brought into the same explanatory structure by acknowledging a common level of computational explanation and by looking more closely at the relations between the brain and the social environment. In section 3.5.3, I provide an illustrative example, which shows that both cases of faulty social learning and neurological dysfunction can be seen as precipitating causes acting on our cognitive mechanisms.

3.5.2 What People Have and What They Do

The dysfunction assumption is often equivalent to the claim that mental illness inevitably depends on abnormalities in brain structures. McHugh and Slavney (1998) argue that this picture needs to be complemented by other views of mental illness, of which the most important is the concept of behavioral problems. Patients with behavioral problems, they argue, do not have broken brains, but suffer from a combination of excessive or misdirected physiological drives, operant conditioning, and bad social learning. Their view is widely shared.

The distinction between diseases and behaviors does remind us that not all mental illness rests on brain lesions, but it also obscures the common basis that they may have, which is a distinctively psychological level of brain function. The distinction between diseases and behavioral disorders blurs once we pay attention to the cognitive level. Although the distinction between dysfunction and structures of motivation or learning remains, the disease perspective and the behavioral perspective can be brought together in the same theoretical structure once we recognize that neuroscience is not just molecular, it is a cognitive (and indeed a social) science too.

McHugh and Slavney assert that many deviant behaviors occur as a result of unfavorable cognitive development, poor inhibition of drives, and poor social learning. They do not insist that deviant behavior is just a matter of conditioning, and they are happy to posit inner processes, but they do enforce a sharp distinction between odd psychology (which causes and is manifested in behaviors) and damage to brain tissue (which causes and is manifested in disease systems). Their sharp separation reflects their belief that the big philosophical issue facing psychiatry is the mind-body problem.

Without worrying about the metaphysics, we can just take on board the lessons of cognitive neuroscience and unite the two perspectives of disease and behavior: the results of brain damage can be seen as functional problems, and so can the results of poor learning or inhibition. In both cases, neurocomputational processes give us a way of understanding the basis of behavior, and McHugh and Slavney's separation of causes into biological and psychological turn out to reflect different ways that neurocomputational cognitive parts can be affected so as to produce deviant behavior.

The neurocomputational perspective does not vindicate the dysfunction assumption, though. It is consistent with the view that neurocomputational systems underlie mental illness generally to argue that those systems cause deviant behavior not because they are lesioned. Instead they might be fed, as input, pathogenic information that the subject acquired through idiosyncratic learning. The dysfunction assumption rules out a priori an account of psychopathology acquired through normally operating mechanisms of learning and conditioning. Such a view threatens the basis of much cognitive research and therapy. Re-educating the subject, not assuming that some internal mechanism is broken, is the right reaction to such behavior. There is a clear and consequential distinction between a broken mechanism and a functioning mechanism operating over deviant input. The dysfunction assumption obscures this distinction.

The provision of a common neurocomputational level between tissue and behavior merely shows that the perspectives of brain abnormality and faulty learning can be brought under a common theoretical structure. They both change behavior through neurocomputational pathways. The following section introduces an example that makes the point, then notes the conceptual issue that is raised: who decides that a behavior that does not depend on dysfunction is pathological? The resolution of that issue depends on our theory of rational behavior.

3.5.3 Dysthymia

McHugh and Slavney seem to understand dysfunction narrowly in terms of tissue damage. On that understanding, much of mental disorder may result not from dysfunction but from drives, conditioning, and learning that are normal parts of our psychology. But we can understand dysfunction more widely once we recognize that the disease process and the learning process operate through neurocomputational structures. It could apply to computational capacities that go wrong in the absence of detectable tissue damage. That is a theory of computational, rather than physical dysfunction. On this view, faulty social learning could be faulty because of computational abnormalities. Even on this wider understanding of dysfunction, however, there still may be mental illnesses that result from normal systems operating in an environment for which they are ill-suited.

I turn now to consider the explanation of Dysthymia given by McGuire and his colleagues in order to illustrate the second and third possibilities: the idea of neurocomputational dysfunction and the idea of normal performance in an unfavorable environment. Depending on how we read their theory, McGuire and Troisi offer both options when they explain dysthymia as an incapacity to engage in reciprocal exchanges. The idea of such a capacity has played an important role in evolutionary studies of cognitive behavior (Trivers 1971), but I am putting the evolutionary considerations aside and just asking what is required for understanding social reciprocation.

Dysthymia is an affective disorder characterized by depressed mood for most of the day, for most days, extending over two years, but without major depressive or manic episodes. McGuire and his colleagues (1998) found that dysthymic patients had a notable deficit in their ability to achieve social goals and carry out simple social tasks. They tended to blame others for their dissatisfactions, rather than considering their own behavior (as did a matched control group.) Dysthymic patients were also less likely than controls to interact socially with others. Perhaps the most striking finding of the study was that dysthymic subjects "believed that they helped others significantly more than they were helped by others. Thus, by their own reckoning, they were cooperators." However, "a detailed analysis of their social interactions, which involved collecting data from siblings or friends, strongly suggested otherwise." Subjects with dysthymic disorder "not only tended to exaggerate their helpfulness to others, but they also downplayed the value of others' help. . . . In addition, they were skeptical of others' intentions to help as well as to reciprocate helping that [they] might provide. For the majority of [dysthymic subjects], these views began prior to adolescence. . . ." (McGuire 1998, 317)

Dysthymics seem to overestimate the value or importance of their own contribution in a reciprocal relationship and/or regularly underestimate the value or importance of the other party's contribution. From the point of view of the dysthymic person (though not from the point of view of those she interacts with), she is regularly exploited or cheated in social exchanges, and this might well lead to lessened social interaction and depressed mood for extended periods.

McGuire and Troisi offer a dysfunction-based explanation of their findings, relying on the notion of an impaired reciprocal altruism system that normally computes what is expected in reciprocal relations. But this is not the only sort of defect that might lead dysthymic persons to exaggerate their own helpfulness and downplay the helpfulness of others. The information such a system needs could not possibly be innate: even if the basic principles of reciprocal exchange are human universals, the value of specific acts varies enormously from culture to culture. In our culture, giving your neighbor a hot tip on a stock counts as a valuable favor, while paying a shaman to chant prayers does not. In other cultures this pattern is reversed. Therefore, the learning process needed to master cultural information introduces a source of potential error. A person who had failed to master the local culture's value system might well end up thinking that she helped others vastly more than they helped her. Some people diagnosed as dysthymic may, as McGuire and Troisi suggest, have defective reciprocal altruism systems, while others simply fail to master the prevailing principles of social exchange. If so, the latter group, but not the former, might be treatable via cognitive psychotherapy that seeks to inculcate the social codes they have failed to internalize.

McGuire and Troisi reject an explanation for dysthymia in terms of adverse environments. However, they have in mind "subjects who had experienced repeated emotional abuse during childhood and whose symptoms had begun in close association with the abuse" (McGuire and Troisi 1998, 247). But someone could receive deviant information about cultural expectations without being emotionally abused. And when McGuire and Troisi consider normality in combination with adverse circumstances, they only raise the possibility of a normal genetic makeup. They consider this possibility implausible, and hold out for "infrastructural suboptimality." It could be that the cognitive mechanisms are impaired, in which case no information can be used normally: this is what McGuire and Troisi seem to have in mind when they blame dysthymia on "suboptimal algorithms." It might be, though, that the information-processing mechanisms are working as designed, and the suboptimality of the architecture is due to the information flowing through the system, which represents social norms out of kilter with those prevailing in the culture.

Wakefield's (2000) solution to the difficulty that learning-based accounts of mental illness raise for the dysfunction assumption is to hold on to the assumption by claiming that in cases of social learning the dysfunction is a belief. His only suggestion for identifying dysfunctional belief is that the belief is "outside the expected range of input" of an evolved mechanism (Wakefield 2000, 264). What is that? If the expected range of input is defined in evolutionary terms then most of our beliefs are outside it, since most of our beliefs are about things that did not exist when our minds evolved—we cannot be adapted to expect beliefs about basketball or compound interest as input. An evolutionarily novel belief is not the same as a dysfunction.

Wakefield argues that the self-esteem system might dysfunction as a result of parental ridicule, producing punitive self-evaluation as input to the system and thus ruining self-esteem. But if there is a self-esteem system then it must make allowance for the odd bit of ridicule. If it did not, one weary parental joke might doom a child. And how can punitive self-evaluation be outside the input range of a self-esteem system? We do sometimes err grossly and blame ourselves harshly, after all.

There is not an objectively defined class of dysfunctional beliefs but a list of beliefs-in-context that can harm people's mental health. Causal explanation in psychopathology sometimes proceeds by citing beliefs and desires as pathogenic. But when we do this we are not necessarily pointing to dysfunctional mechanisms. Belief acquisition and revision may be working well within normal range, but in an environment that produces what we regard as irrational behavior.

The question now arises whether, if all that is wrong with, say, dysthymics is their lack of appropriate social information, we should call them mentally ill, their chronic low-level depression notwithstanding. Plenty of people lead worse lives than they would if their beliefs or habits were revised; which of them need their beliefs and habits revised because they are pathological? This is a general issue affecting all learning- or conditioning-based theories. It is especially pressing if one is sympathetic to the dysfunction assumption, since dysfunction is often seen as an objective finding on which to anchor a diagnosis, whereas claims about deviant behavior and poor social learning might seem to apply to problems in living rather than mental illnesses.

It will be easier to answer this question once we have in hand an understanding of how far we can give human psychology a satisfactory mechanistic explanation—an explanation in terms of interacting components in systems that cause behavior. If we can specify normal neurocomputational structures and the developmental processes that give rise to them, we may be clearer about the prospects for calling some learning process, physiological drives, or forms of conditioning abnormal or normal. Such a resolution is unlikely in all cases, however: in many cases we will be unsure whether abnormal behavior is a sign of mental illness or a problem in living. My point here is that we cannot just endorse a priori the claim that mental illness is inevitably due to dysfunctioning neurocomputational structures.

One other lesson to draw from the explanation of dysthymia in terms of normal faculties in an abnormal environment is the important point that the relation between a disturbed person and his environment needs to be considered. The idea that we can explain mental illness just by looking inside the skull is bound to lead us astray. I pick up this theme again when I sketch a theory of psychiatric explanation. At the moment, I am still trying to articulate the framework within which psychiatric explanation works. I have said already that there is no reason to assume that mental disorders are invariably the result of dysfunction. The next section looks more closely at dysfunction.

3.5.4 Functional Analysis in Psychiatry

Psychopathology that is not caused by a malfunction is possible, so the harmful-dysfunction analysis of mental disorder looks dubious. I turn now to the second question raised by the reliance of the harmful-dysfunction model on an evolutionary concept of function. Is it correct that the relevant concept of malfunction is an adaptive one, and that dysfunction is a failure of a system to fulfill its adaptive function? In this section, I introduce psychiatry to the contemporary philosophical literature on functions (for a fuller treatment, see Woolfolk 1999). The Boorse-Wakefield picture, and evolutionarily inspired treatments of mental illness such as Baron-Cohen's, do not take the lessons of this literature into account. They assume a misleading picture of how the biomedical sciences think of functions.

When one attends to the details of scientific practice, a picture often emerges of the same term's being used in different ways. The concept of a gene, for example, plays different roles in different areas of biology, and what counts, physically, as a gene varies across these projects (Kitcher 1984, Moss 2003). Arguing about what a gene *really* is seems beside the point, since the concept refers to different entities in different inquiries. I argue at the end of this chapter that the same is true of the concept *mental disorder*. The term is used differently in diverse theoretical, therapeutic, and judicial settings, and no generally correct analysis covers all cases. Another concept with different meanings is *function*. Functional ascriptions in biology and the biomedical sciences are not invariably ascriptions of teleological or evolutionary function.

The modern literature on function stems from two seminal papers. Wright (1973) argued that ascriptions of function to a structure were causal-historical, explaining the current existence of the structure in terms of its past contribution to its own selection. Wright's analysis applied to any structure that participates in a selection process and thus was not explicitly evolutionary; Millikan (1989) and Neander (1991) narrowed the analysis to evolutionary contexts. The other key paper was Cummins's (1975) analysis of function as the causal contribution a structure makes to the overall operation of the system that includes it. Cummins's concept of function is not a historical or evolutionary concept. According to Cummins, a component may have a function even it was not "designed," and, therefore, parts with no selection history can be ascribed a function. In this sense of *function*, Harvey understood the function of the heart two centuries before Darwin.

Wakefield is certain, quite without argument, that functional ascription is teleological throughout biology. Boorse, as we saw, thinks that there is mental illness if there are mental functions, and endorses an evolutionary view of psychological function. However, many theorists have argued that there are biological sciences for which Cummins's ahistorical analysis is broadly correct (Godfrey-Smith 1998, Kitcher 1998). Allen and Bekoff, for example, distinguish cognitive ethology from molecular and cellular biology on the grounds that whereas molecular and cellular biology talk about how organs do what they do, ethology talks about what organs do and why they do it (1997, 92): function concepts vary as the life sciences attend to different explanatory projects. *Function* in biochemistry and physiology simply refers to the contribution a structure makes to the overall organismic system containing it and therefore differs from talk of function in evolutionary disciplines; thus, "philosophical analyses reveal unresolved ambiguities in biological practice" (Kitcher 1998, 266), and philosophical analysis should respect, rather than try to reform, the differing scientific concepts: "let no philosopher join together what science has put asunder" (Godfrey-Smith 1998, 291).

So we need to see how functional analysis and ascription work in psychiatry. Psychiatry, I have said, is a branch of medicine. And Schaffner (1993) has argued very convincingly that although medicine might use teleological talk in its attempts to develop a mechanistic picture of how humans work, the teleology is just heuristic. It can be completely dispensed with when the mechanistic explanation of a given organ or process is complete. Schaffner argues that ascriptions of evolutionary function, which he calls the "primary sense" of *function*, are parasitic on the "secondary sense" of *function*, in which "some organ, mechanism, or process is 'useful' to the organism in the sense that it keeps it alive and/or allows it to thrive" (1993, 389). Schaffner thinks that as we learn more about the causal role a structure plays in the overall functioning of the organism, the need for functional talk of any kind drops out and is superseded by the vocabulary of mechanistic explanation, and that evolutionary functional ascriptions are "necessary, though empirically weak to the point of becoming almost metaphysical (1993, 389–390)." For Schaffner, teleological functional ascription is merely heuristic; it focuses our attention on "entities that satisfy the secondary sense of function and that it is important for us to know more about" (1993, 390).

Schaffner imagines a mutation that gives an individual an altered but functionally enhanced biological structure while at the same time rendering him sterile. Schaffner asserts that we would ascribe some function to the new structure even though it was not selected for and could not enhance fitness. So biological structures can have a job to do irrespective of evolutionary analyses of biological jobs. Godfrey-Smith and others are right that in some areas of biology functional ascription is teleological. Schaffner, though, is correct about biomedical contexts: biomedical ascriptions of function to an organ or structure do not make assertions about adaptedness. Even though theorists in the relevant sciences do believe that evolution is responsible for the structures, functional analysis does not require an evolutionary explanation: it requires only that the role played by the structures in the overall functioning of the living organism can be identified. The evolutionary concept of function is not what drives biomedical functional analysis. What drives it are the demands of theory-building in biomedicine. Biomedicine requires systems to be analyzed in terms of their contribution to the overall maintenance of the organism as a living system. In section 3.5.5, I go into this in more detail and argue that function concepts vary depending on what the theory is trying to explain. When we understand the explanatory point of psychiatry, we will see that it requires a nonhistorical function concept that is at home in mechanistic explanation.

3.5.5 Function Concepts Vary with Theory Construction

Psychiatry is a branch of medicine, which suggests that its proprietary function concept ought to be continuous with that of biomedicine. There are reasons, too, for preferring a nonhistorical account of function in the cognitive neurosciences. As I show in chapter 4, a complicated interplay of psychological and physiological investigation is needed to arrive at a functional decomposition of the brain, and evolutionary considerations seem to play only the limited heuristic role in this process that Schaffner suggests. And yet, if we agree that adaptationist accounts of function play a role in behavioral ecology, should we not think that the behavioral sciences are cognate with behavioral ecology in the study of humans, and hence embrace an evolutionary account of function as correct for human psychology? This is a reasonable question for at least human behavior at a fairly coarse-grained level of description: it is hopeless at a fine grain, since many human traits have a function that reflects culture rather than the selection history of the species. A shake of the head, for example, indicates agreement in India but disagreement in America, and I would not want defend the view that this difference arises from distinct natural selection.

Still, there seem to be considerations pulling us both toward and away from evolutionary construals of function, depending on the theories psychology aims to build. At an abstract level, we may be able to identify capacities that any cognitive organism would need to survive on our planet, and derive a basic cognitive toolkit from them (Godfrey-Smith 1996, Sterelny 2003). At slightly less abstract levels, we might be able to use evolutionary considerations to suggest the existence of more precise descriptions of things that an organism might do based on our general knowledge of the evolutionary demands facing organisms with a certain size, lifespan, or social arrangement. But these general specifications of cognitive tasks tell us next to nothing about the organism's innards, where the capacities are actually realized biologically (Murphy 2003). A general sense of what cognition is good for, combined with some knowledge of how cognitively sophisticated organisms work, provides us with plausible and potentially fruitful evolutionary analyses of the function of various cognitive capacities. But when we ask about the details of implementation in the brain, our thinking is constrained much less by general beliefs about the evolutionary process, and much more by the demands of fitting a specification of system into the overall functioning of the higher-level system of which it is a part.

Whether or not a biological enterprise uses a historical/teleological concept of function or an ahistorical one, then, depends on the constraints that a theory in that area meets. And that is largely determined by the purpose of the theory. An ethological theory is interested in accounting for an organism's behavior, typically in terms of optimal strategies for maximizing fitness. It is constrained by evolutionary models and employs an evolutionary concept of fitness. A theory in cognitive neuroscience, on the other hands, employs the sorts of descriptions of cognitive behavior that we see in ethology only as heuristics that suggest some capacity must be realized somehow in the brain. When cognitive neuroscientists look to see how the brain implements these capacities, they must do so in a way that makes the contribution of a system to overall brain function perspicuous, so the constraints are different: they derive not from the nature of the evolutionary process but from the ahistorical systematic nature of brain function, and the requirements of mechanistic explanation. So the relevant concept of function is the mechanistic one of the contribution a component part makes to overall functioning.

Explanation in psychiatry, if it is to be part of the cognitive neurosciences, will proceed by showing how a system fails to make its customary contribution to overall functioning. The failure, as I suggested in section 3.5.2, may not be a matter of dysfunction at all, since the failure may result from normal function over pathogenic information or in an adverse environment. But in that case we still need to understand the functional role of a system to see what its overall contribution is. So the relevant notion of function is the mechanistic one, and so is the relevant notion of dysfunction. Explanation in psychiatry does not explain how fitness is lowered by failure to reach optimal behavior: it shows how bits of your brain fail to do what they usually do.

In the next section I make these ideas clearer by looking at the way explanation in the cognitive neurosciences works. Before I do that, I want to point to some of the benefits of understanding psychiatry's functional ascriptions in nonevolutionary terms. A commitment to evolutionary concepts of function has landed exponents of the harmful-dysfunction concept of mental illness with horribly implausible commitments and started an irrelevant debate about whether functional ascription is normative. Both these difficulties are evaded if we move to a mechanistic-explanatory concept of function.

3.5.6   How Not to Talk about Function in Psychiatry

In the human sciences, function must be assessed in different ways and at different levels: psychological traits can be assigned an evolutionary or cultural function, but they are realized in physical structures that have functions in virtue of their contributions to overall systemic organization. Specifically, in psychiatry we want to understand behavioral neuroanatomy, and in that endeavor we need a function concept that fits the constraints of mechanistic, rather than evolutionary, explanation. I say more about how mechanistic explanation works in neuroscience in chapters 4 and 6, but first let me provide some extra support for the position by pointing out that if one ties mental illness conceptually to malfunction, and understands malfunction in evolutionary terms, one faces three problems. The first is the necessity of showing that mental illnesses are evolutionary fitness-lowerers. This is very hard to do: so hard, in fact, that nobody has ever tried to do it. But if you take the evolutionary construal of function seriously, it looks as though malfunction should reduce fitness, and so the theory should explain how it happens. In general, discovering the historical evolutionary function of a structure is much harder than discovering its present contribution to systematic function. The discovery of evolutionary function would be less hard if every complex structure had an evolutionary function, but current evolutionary theory recognizes that many may not. That brings me to the second problem for evolutionary analyses of psychological dysfunction.

The second problem that one faces if one ties mental illness to evolutionary dysfunction is that if a mental illness depends on psychological structures that have no evolved function, it cannot really be mental illness. This appears to be a consequence of the claim that we saw Boorse make, that disease is a matter of failing to carry out evolutionary function. And in discussing the diagnosis of Inhibited Female Orgasm, Wakefield applauds the DSM-III-R for discriminating between cases in which women fail to have organisms from inadequate stimulation and those involving true dysfunctions. Wakefield assumes that the diagnostic criteria have specified "the external conditions under which a woman would experience an orgasm if the functioning of her *internal orgasmic mechanisms* is normal" (1992b, 244, emphasis added). Now Wakefield's positing of internal orgasmic mechanisms here seems more than the assignment of a mere metaphor; rather, it seems an attempt to do evolutionary psychology based on psychiatric data. This choice of example turns out to be especially unfortunate given the very controversial evolutionary status of the female orgasm. A number of authorities contend that the morphology underlying the female capacity for orgasm is not an adaptation at all and has no evolutionary function, being a byproduct of the embryological development of the structures underlying male sexual response (Symons 1979; Lloyd 2005).

More generally, if any elements of our psychology are not the direct result of natural selection but arise from other biological causes, they cannot malfunction in the sense of failing to perform their evolutionary job. On a strictly evolutionary analysis of psychiatric dysfunction, if a psychological mechanism has evolved not by natural selection and it is involved in the etiology of a syndrome, we have to conclude on purely conceptual grounds that the syndrome is not a mental disorder. But one should not discount dyslexia as a mental disorder just because written-word comprehension is not an adaptation but a byproduct of other adaptations. We think that illiteracy is a misfortune whereas dyslexia is a disorder, which is why some freshmen receive extra time to sit exams and others take remedial English. On the two-step picture this different treatment reflects our focus on the structures that underlie failures of language comprehension or production, and not just the deficits in behavior. Illiteracy is a "problem in living" whereas dyslexia indicates malfunction. But it does not indicate malfunction on an evolutionary account unless it can be shown that the cause of dyslexia is the failure of an adaptive mechanism to do its adaptive job. It is not enough to show that there is some failure that causes dyslexia if you take an evolutionary account of function: to be a malfunction, the failure must be adaptively significant. And there cannot be an adaptation for reading and writing given how recent literacy is.

The conceptual affiliation of mental illness and evolutionary dysfunction has unhappy consequences, then: first, it is hard to find out what the adaptive function of a complex structure is and whether changes in its performance lead to lowered fitness; second, there are no a priori reasons to regard something as psychologically normal regardless of the toll it takes on well-being, just because it involves nonadaptive complex structures.

I turn now to the third problem for evolutionary analyses of function and malfunction in psychiatry. It is widely believed in the literature that the evolutionary function concept is normative by virtue of being teleological and thus imports normative considerations into the scientific foundations of the two-stage view. Whether or not this is correct is irrelevant if the mechanistic conception of function is not normative. I argue that, *in principle*, it is not inherently normative, but that in crucial parts of psychology it may be.

We have already looked at the constructivist view that judgments of disorder are merely about desirability. Against that claim, I said that our concept of mental disorder insists on appropriate causal histories. The issue we need to address now is whether the appropriate causal histories can be identified without importing normative claims. However we understand function talk in the sciences, the claim that mental dysfunctions are objectively identifiable in principle must be defended if objectivism is to be accepted. Constructivists of various stripes deny that objectivists like Boorse can identify mental dysfunctions without importing value judgments.

It is important to distinguish two objections here. One objection seems to be that any causal explanation in psychiatry is bound to be normative because it attempts to explain something that we (negatively) value. This is a terrible objection. Of course it is true that, as Sedgwick says, a fractured femur "has, within the world of nature, no more significance than the snapping of an autumn leaf from its twig" (Sedgwick 1982, 30). But independent of human interests, nothing has any significance at all—significance isn't a property of the universe, and Sedgwick's point holds equally well for any phenomenon in any science. Sadler and Agich (1995) say that we want to survive and so we are interested in discovering the mechanisms of survival and adaptation. They conclude that "the normative goal of these mechanisms is assumed at the outset" (1995, 225). But this simply confuses the idea that we value something with the idea that our explanation of how it works is evaluative.

Sadler and Agich do suggest a better defense of the idea that dysfunction is intrinsically normative (cf. Bermudez 1998, 450); this second claim is not that our values make something salient, but that evolution is teleological and hence evolutionary function concepts are normative. This objection stems from an unduly literal interpretation of the biological metaphor of design, but that is irrelevant here. The function concept in psychiatry is one that fits mechanistic, not teleological, explanation. Mechanistic functional ascription, I argue, is not *necessarily* normative.

Let me begin with an analogy to Peter Railton's "naturalized norms." Railton (1986, 185–186) imagines explaining why a roof collapsed as follows:

". . . For a house that gets the sort of snow loads that one did, the rafters ought to have been 2 x 8's at least, not 2 x 6's." This explanation is quite acceptable, as far as it goes, yet it contains an "ought." Of course, we can remove this "ought" as follows: "If a roof of that design is to withstand the snow load that one bore, then it must be framed with rafters at least 2 x 8 in cross-section." An architectural "ought" is replaced by an engineering "if . . . then . . ." This is possible because the "ought" clearly is hypothetical, reflecting the universal architectural goal of making roofs strong enough not to collapse. Because the goal is conceptually fixed, and because there are more less definite answers to the question of how to meet it, and moreover because the explanandum phenomenon is the result of a process that selects against instances that do not attain the goal, the "ought"-containing account conveys explanatory information.

Railton calls this *criterial* explanation, because it explains why something happened by reference to a criterion that, although defined naturalistically, can regulate a human practice, like house building. The fact that the criteria govern a human practice is what makes the explanation normative; the fact that explanations employing the criteria can be stated in "if . . . then" terms is what naturalizes the norms.

Railton's picture, applied to the explanation of why someone exhibits symptoms of mental illness, says that the patient behaves in an abnormal manner because of some fact about her psychology or physiology—she ought to have different habits of self-evaluation or more blood flow in the entorhinal cortex. And this, in the spirit of Railton, can be reformulated as the claim that "if this part of her brain (or psychological property) is to play its customary role in maintaining the overall system, it must have more projections to the hippocampus (or different computational properties, or a smaller legacy of her troubled relationship with her mother, or whatever)." The reason an exponent of the two-stage picture would not regard this as a normative explanation is that, on the two-stage view, the criteria for assessing adequate performance are supplied by nature rather than by a human practice. The criteria are not supplied, as in Railton's example, by regulative criteria derived from human goals. They are discovered in nature.

The view that the functional analysis of human psychology can be read off nature in this way is very strong. It is not the view that relative to human goals and interests, we can establish what psychological systems should be like and how they should be arranged to meet those goals and further those interests. Rather, it is the view that psychological normality imposes non-human, natural functional standards. Those standards exist independently of what people think they should be. Some people will say that since even this view licenses statements about what some biological system ought to be like, it is in fact normative in a fairly weak sense (this seems to be Bermudez's view). All of medicine is normative in this sense—the problem is whether any science is not, though, because all sciences license expectations about what ought to happen in a normal system: stars, for example follow a reliable progression through developmental stages, so we can predict what ought to happen to them. I will not push this point, because the more important issue is whether or not human interests establish the relevant norms, if there are norms. As I said, the view that they can be read off nature is very strong. My view takes time to develop, but in a nutshell it is this: norm-free mechanistic explanation is possible in some areas of our psychology, such as the visual system. Brian Keeley (1999) has argued convincingly that we can establish a watertight understanding of the function of a biological system by a consilience of different lines of biological inquiry—not just some adaptationist hypotheses, but ethology and various forms of neuroscience too. His example is the discovery of an electrical sense modality in some species of marine vertebrates. In cases where such convincing ascription of function to a mechanism is possible it does not seem at all odd to suppose that the standards of good performance are to be found in nature (although that does not mean it will be easy to establish what the norms are). The mark of such natural norms, I think, is that a theory can make a generally correct prediction that is nonetheless false in a particular instance, since in that instance the prediction comes out false because of a deficit in the biological system, not a shortcoming in the theory. This is not a general property of scientific theories, and it does show something distinctive about functional systems. But it does not show that the functional standards are not natural. If this is normativity, it is not the kind of normativity that critics usually raise against psychiatry. The standards guiding judgments of malfunction are natural, rather than social, and that is what the objectivist needs.

We can study a system and assess its performance relative to a picture of functioning that, making allowance for normal variation, rests on models of the capacity that relate it to natural standards of performance. However, although this holds in principle, I worry that it is not possible in areas of our psychology that implicate judgments of, for example, rationality because in those cases standards do seem to be imposed by human deliberation, rather than by biological nature. (From now on, when I talk about norms, it is human rational or moral norms that I have in mind.) The extent to which explanation that does not invoke norms is possible, then, depends on how much of our psychology is like the visual system—i.e., decomposable into structures to which we can ascribe a natural function.

Reflecting on some disputes illustrates the difference between vision and reason. Many advocates for the deaf deny that being deaf should be considered a handicap, for example, to be rectified by cochlear implants. But they don't deny that their constituents are deaf: that is, they admit that their auditory system is abnormal in the sense that ordinary explanations of its function are not true of it, but they deny the implications that are usually taken to follow from that. This is just the sort of situation that the two-stage picture is supposed to reflect: a theory of malfunction and a dispute over how we should regard that malfunction. In contrast, debates over addiction or other putative irrationalities build the departure from human norms into the characterization of the disorder. To diagnose addiction, on some views, is just to say that people are failing to flourish—they are irrational, and under the influence of cravings they cannot control. To get around this, we need a mechanistic explanation of addiction that provides a body of fact that can be separated from the question of the norms governing the realm of human behavior in which the facts are found.

In chapter 5 I link this issue with whether we can get a mechanistic explanation of "central systems": the parts of our psychology that appear to resist modular decomposition. Decomposition is the first step in a mechanistic explanation, and these aspects of our nature may not be mechanistically tractable in the relevant sense: we can, perhaps, model our understanding of them, but we cannot explain them in a way that tells us whether our understanding of them can be separated from how they really are.

Unfortunately, these are just the areas in which we most want norm-free behavioral science, and hence it may be unattainable where we need it most. So some normative disputes are inevitable in psychiatry. Also inevitable is some normative agreement. Some conditions, like addiction, will continue to involve substantive value judgments that people will disagree over; others, like psychoses, will continue to involve substantive value judgments that command widespread assent. That is, even when reason cannot be mechanized, it may be *naturalized* well enough to permit agreement. We can justify some normative ascriptions of pathology on grounds that reflect our understanding of reason as part of our animal nature.

So the final theory is messy. But so is human nature. And the position needs more preliminaries before I can defend it. In chapter 4 I review mechanistic explanation in more detail and also discuss the decomposition and the universality assumptions. More immediately, in the next section I defend the idea that the study of what goes on outside central systems is legitimately part of psychiatry. The moral of the current section is partly provisional: I have noted that mental illness is not always a matter of dysfunction, and that the relevant function concept in psychiatry, seen as cognitive neuroscience, is not evolutionary. It may or may not be normative in the sense of having normal performance defined by human, rather than natural, standards. This conclusion puts more strain on the framework we inherited from chapter 2.

To further motivate the dissolution of that framework, I now provide a concrete example of scientific gains accruing from abandoning the distinctions among minds, bodies, and kinds of disorder that the orthodox program makes. The example encompasses work on schizophrenia, a paradigmatic mental disorder, and some other work on conventionally nonpsychiatric disorders. It also involves work on movement disorders, and thereby blurs the line between the physical and the mental. Having gone over the case in some detail, I now step back to ask about the proper role of everyday thought in the study of mental illness. But first, as promised, I introduce a case where science benefits from ignoring existing distinctions.

3.6 Explaining Abnormalities of Motor Control

3.6.1 The Theory
Frith and his associates (Frith et al. 2000a, 2000b; Blakemore et al. 2002) offer an account of movement disorders that addresses both symptoms of

schizophrenia and symptoms traditionally regarded as part of neuropsychology. They argue that several abnormalities in the control and awareness of action can be explained in terms of one model of the underlying processes, even though these disorders traditionally have been seen as very different. The pathologies they try to explain include: Optic Ataxia (in which patients have trouble grasping objects that they have a clear sight of); "the alien hand" (or "anarchic hand," in which patients feel that the movements of their hand are outside their control); contextually inappropriate, stereotypical "utilization behavior"; the subjective experience of sensations in an amputated "phantom" limb; the lack of awareness of an existing limb when it is not visible; the delusions of alien control of one's limbs that is sometimes reported by schizophrenics.[10] This is a rag-bag of condition, in two respects: it contains conditions that by present lights are both psychiatric and nonpsychiatric, and it mixes up, again by present lights, the clearly mental, the not clearly mental, and the probably not mental. That is, it has just the irreverent attitude I recommend to conventional conceptual and disciplinary boundaries.

Frith, Blakemore, and Wolpert (2000a, b) explain the movement disorders via a model of the control of action derived from contemporary cognitive neuroscience. On this account, the motor system employs a "forward model" and an "inverse model." The forward model makes two predictions: first, before an action is underway, it compares the outcome of a motor command (such as the command to your arm and hand to pick up the bottle of Scotch in front of you), and it compares the predicted outcome to the desired outcome. Second, after the action has been initiated, the forward model predicts the sensory consequences of the movement and compares them with the actual sensory feedback (the results, in our example, of the arm reaching for the bottle).

The inverse model, on the other hand, provides the motor command necessary to carry out the action, fine-tuning the movement of the limb toward the goal and the motions needed to, for example, pick up the bottle and pour its contents on to your cornflakes. Frith et al. (200a, b) argue that we are normally unaware of the fine adjustments to the inverse model. Individuals

10. For details of these conditions, see the following: Harvey (1995) (a translation of Balint's original 1909 paper) and Perenin and Vighetto (1988) on optic ataxia (or Balint's syndrome); Brion and Jedynak (1972), Marchetti and Della Salla (1998), and Parkin (1996) on the alien hand; Lhermitte (1983) on utilization behavior; Ramachandran and Hirstein (1998) on phantom limbs; Fish (1962), Mellor (1970), and McKenna (1994) on passive motor experiences in schizophrenia.

with form agnosia can make appropriate reaching and grasping movements despite being unconscious of the position or shape of the object they are reaching for (Milner 1998). But we do have some conscious access to the forward model, notably its prediction of the sensory consequences of movement. Frith and his colleagues attribute to our conscious access to those predictions our ability to distinguish changed sensory input resulting from our movement from sensory changes caused by the movement of the world relative to us.

Frith and his colleagues explain the different abnormalities of movement I listed in terms of damage to different components of the system. For example, they explain the alien hand (they call it the "anarchic" hand) as follows: (1) the subject unconsciously forms representations of nearby objects in terms of the movements that would be appropriate for grasping them; (2) the subject is unable to inhibit action based on these representations of nearby objects. A lesion undermines the normal connection between intended actions and sensory input, so that the representations of nearby objects suffice to launch the action, which is carried out in a stereotyped fashion. The alienness of the experience is accounted for by the subject's awareness of the intentions she has actually formed, which tells her that the hand is operating independently of those intentions.

This is consistent with the observation that alien hand behavior occurs only in some circumstances. For example, Parkin (1996, 181) reported that his patient MP only exhibited the abnormality in two situations: either situations in which only one hand could be employed or complicated ensembles of actions that were not overlearned to the point of becoming automatic, but required some thought or monitoring, such as cooking a meal. MP's left hand would lob all manner of unnecessary objects into the frying pan while the right hand was trying to cook.

Parkin suggests that the circumscribed nature of the alien activities implies that MP's left hand is not inhibited from making a contribution to simple activities but rather engages in actions that would normally have been disregarded as a means. The left hand is trying to do its bit to help but the useful action is already taken. But because it is not inhibited from engaging in other stereotyped contributions that would normally be rejected as possible actions in the context, the left hand is free to engage in them.

Patients with the alien hand may deny that the hand is theirs but usually they do not; in either case, they experience its movements as uncontrolled, by themselves or by anyone else. In contrast, schizophrenics experience their limbs as moving under the control of other agents, either people or other forces like "cosmic strings. When the strings are pulled my body moves and I cannot prevent it" (Mellor 1970, 18).

Frith and his associates explain the differences in alien hand and schizophrenia by positing a deficit in the part of the motor system that generates the forward model and represents the predicted state of the system. With no awareness of the predicted state one's actions appear to be carried out without any volitions on one's own part. According to this view, schizophrenics suffering from delusions of external control have a goal and move their limbs to attain it but are unaware of initiating the action—in addition, "the patient's belief system is faulty so that he interprets this abnormal sensation in an irrational way" (Blakemore et al. 2002, 240). They argue (Frith et al. 2000b) that this experience is like seeing a slide projector moving on by itself just as one is about to press the forward button—the device is doing what you want, but not because of you, so you assume it is under some external control.

I will not discuss all the conditions that Frith's team examines. I hope that the examples indicate the potential that is possessed by the search for large scale explanatory generalizations based on models that ignore current conceptual constraints. We must look to our best current understanding of the mind/brain to unite inquiry into what are currently thought of as diverse phenomena. I now move on to the morals that the story raises.

3.6.2 The Morals

Frith et al. disregard existing boundaries in search of explanatory unification. Their discussion raises three points. First, recall that Blakemore et al. (2002, 240) explain delusions of control via a general problem with the patient's beliefs as well as a deficit in the motor system. This is a point that I examine at length in chapter 5. Many conditions involve failures of rationality as well as problems in particular systems. These rational shortcomings are likely to be especially difficult to explain, because we have neither a generally accepted theory of rationality nor any clue about the computational systems that underwrite our inferential capacities. The concept of "executive function" is just a placeholder for a general theory of thinking and deciding.

The second point to extract from Frith and his colleagues' treatment of movement disorders is that in so far as we can explain psychological abnormalities in terms of underlying systems, we may need to ignore traditional disciplinary boundaries. The movement disorders Frith et al. consider include symptoms of schizophrenia and of traditionally neurological conditions. The model does not respect the difference between psychiatry and neurology and does not worry about the place of reaching and grasping among our mental states.

Third, I am not insinuating that explanations of psychiatric phenomena will always take the form of cognitive models. Let me rebut one bad objection to the idea that cognitive models will become the dominant form of psychiatric explanation and then connect the issue of cognitive models with the discussion so far and the discussion left to come in the remainder of the book.

The bad objection sometimes raised against the expectation that cognitive models will dominate in psychiatry is that the whole point of the cognitive neurosciences, and indeed cognitive science in general, is precisely that it is cognitive, whereas most of what afflicts people with mental disorders is not cognitive but a matter of mood, affect, habit, and the other aspects of mental life about which the cognitive sciences have little to say. My quick response is that we should not take the components of our psychological architecture to be purely cognitive even when they are studied by cognitive science: among our psychological mechanisms, for example, some of the best candidates for psychological modularity include the basic emotions (Griffiths 1997). Furthermore, it is not just cognitive psychology that I am looking to, but the whole panoply of cognitive neuroscience, and neuroscience does have a great deal to say about the affective areas of our mental life. An account that explains pathology in terms of embodied cognitive architecture is not barred from appealing to cognitive or chemical processes underlying affective states, all the way down to molecular levels. Of course, doing this will not tell us what it feels like to be depressed, or why such qualitative states issue from our brains, but those are other problems, and noncognitive aspects of neuroscience do not solve them either. It may be a general problem for materialism, but in fact dualists cannot explain what being depressed is like either; you would just have to experience depression.

That concludes my discussion of the immediate significance of the work of Frith et al. on motor disorders. Many scientists are working along these lines, and I could have picked other bodies of research, so scientists, I think, will be unmoved at the practical level. They will agree that of course you should just go ahead and do good work without worrying about intellectual demarcation, so what is the point of making a big philosophical deal of saying so?

Well, the big deal has two parts. The first concerns the philosophical issues that arise when we wonder what psychiatry would be like in the light of revised conceptual foundations. If we can make these foundations explicit, we can ask questions about both the internal structure of the science and the relation of psychiatry to the wider world. The internal structure of psychiatry raises issues about explanation, reduction, realism, operationalism, and the nature of classification. They are addressed in the coming chapters. One main theme is that the promise of synthesizing psychiatry and cognitive neuroscience is being stifled because of mistaken philosophical commitments, notably a version of operationalism that officially restricts the attention of psychiatrists to the relations between observables and a commitment to an untenable form of reductionism.

The issues also include the place of psychiatry in the sciences of the mind and in the social world. The two sets of issues are interrelated, because much of psychiatry touches on central preoccupations in how we see ourselves, and its picture of humanity has great influence. And the social, ethical, and political issues depend for their resolution on the more obviously scientific explanatory and classificatory ones. I focus on the epistemic and metaphysical issues in psychiatry, but I need to address two concerns here. They take their inspiration from a view about the proper relation of science and the wider culture, and if they are correct then the two-stage picture needs to be abandoned. The concerns turn on the question of whether science can be as sharply separated from the rest of the culture as the two-stage picture seems to think. Before I get to the scientific issues that are my main interest, then, I want to look at how we should think about psychiatry as a science and its place in the two-part picture. In the rest of this chapter I do two things. In section 3.7 I defend the revised conceptual framework against the objection that my laxness about foundations has left in its wake an inquiry that is so vaguely specified, and countenances such heterogeneity in its subject matter, as to not be a scientific program at all. It may have scientific components, but the search for breakdowns in the mind/brain is not one science; it is none, or several. This objection says that the messiness introduced into the picture means that psychiatry has no standing as a science, because neither the folk conception of mental illness nor its current scientific conception refer to any natural category. The objection advocates the abolition of psychiatry. It makes heavy play of a requirement that the subject matter of a science is a natural kind, and that is what I reject. In doing so, I set up my second topic. In section 3.8 I discuss the relationship between the scientific and nonscientific purposes to which concepts of mental disorder answer. I argue that there is no single concept of mental disorder that can serve all our projects, so if mine does not, then it is no objection. But I do say something about the various projects, their relationship to each other, and their relationship to science. In other words, the rest of this chapter set outs what the two-stage picture looks like in the light of my conceptual revisions. The rest of the book is about the extent to which the revised two-stage picture is possible.

3.7 The Abolition of Psychiatry?

3.7.1 What Is Left of Psychiatry?

I have said that our current folk conception of psychiatry is deforming the science, and that because the mind is a bunch of capacities rather than a natural object, we are left with a diverse set of ways in which the mind can, colloquially put, break down. Now, "breakdown" does not look like a name for a scientific category, and we cannot know in advance of detailed empirical work exactly how diverse a set of phenomena will end up gathered together under the rubric. So why suppose there could be a science of something so diffuse? Have I, in fact, argued not for the reform of psychiatry, but for its abolition? Even supposing that I am right to say that, we should go ahead and ignore the conceptual barriers in practice, why suppose that the ensuing practice will be one science rather than several, and why call the ensuing practice(s) psychiatry rather than neuroscience? And some people will wonder whether, if there is no clear concept of mental disorder, we have any right to carry on talking about it.

If psychiatry is just concerned with brain diseases, and if there is no theoretical basis for carving the psychiatric subset of brain diseases out of the overall class, then we might as well abolish psychiatry and just do neuroscience. Psychiatry, on this view, has no future as an independent discipline. As I noted earlier, psychopathologies (like GPI) have historically disappeared from the province of psychiatry once we have learned their biological etiology. The proponents of biological psychiatry are envisaging the discovery of biological etiologies on a massive scale. So some thinkers are starting to wonder if psychiatry has any future as an independent discipline or will simply be annexed by neuroscience (Hobson and Leonard 2001). The rise of neuroscience threatens to finish psychiatry as an independent subject.

Although my view is a little more complicated than this, I do claim that psychiatry is just clinical cognitive neuroscience, not a theoretically separate field. It stands to the biology of the human brain as cardiology stands to the biology of the human circulatory system. Let me make three qualifications straight away. First, my claim has not been fully supported by the preceding discussion, as I am well aware. The discussion has, I hope, shown that the current distinction between clinical neuroscience and psychiatry has little to show for itself. It has not, though, established that no such distinction can be made. I take up that issue in chapter 4, where I discuss the medical model in psychiatry.

Second, in saying that psychiatry is clinical neuroscience, I am not at the moment committing myself to any particular theses about what neuroscience is or how explanation in neuroscience works. I do not say that all explanation of psychopathologies will involve specifying processes at a cognitive level of explanation, or at a molecular level of explanation, or in any other way. The nature of explanation in psychiatry is taken up in chapters 4–8.

Third, I do not assume that the result of the general synthesis of existing psychiatry and cognitive neuroscience will produce one big elegant theory. It is more likely to produce piecemeal theorizing employing generalizations or causal models of limited scope at different levels, but within an overall perspective provided by the synthesis I recommend. Our best theoretical accounts of normal mental functioning are elaborated at many levels. Some conditions may be explicable in terms of rich models that are articulated at every level with interconnections between them. Other disorders may be explained at only one level, via a partial articulation of the whole possible model. Ultimately, what is wanted is a story in which diseases are conceived of as templates—I call them exemplars—that we can fill in at a number of levels of explanation, with the relations among levels, and the different causal processes involved, established wherever possible.

In order to develop such models we must expand the concept of mental disorder across a domain of psychopathology much wider than that of common sense, or even that of contemporary psychiatry. We should let the question of the scope of the mental simply fall into place as a scientific question—not perhaps a purely empirical one, but a scientific one nonetheless. We want our thinking about mental illness to be integrated with the wider sciences. Undoubtedly there will be phenomena that we will continue to think of as paradigmatic mental illnesses. Schizophrenia and manic depression are obvious ones, and the science will investigate numerous conditions that bear both folk-psychological and causal-explanatory resemblances to those central cases. So think in terms not of a clear concept, but the ostensive definition of mental illness as things that are relevantly similar to *them*, where *them* ostends schizophrenia and some major affective conditions, and what counts as similar, and via what methods, is settled by the scientific search for models of psychological capacities and the ways they fail. The revisionist bullet that lumps perceptual deficits in with depression and schizophrenia should indeed be bitten, not as a piece of conceptual analysis but as a methodological heuristic, since psychological models will include models of the visual system and its role in thought and action.

The details depend on the nature of explanation in psychiatry. We should let the science of psychopathology work with whatever concepts and generalizations it needs. To do this, we should try to develop as full a scientific understanding of the way healthy minds work as possible, and to try to understand departures from normal working. The ensuring taxonomy of conditions will map poorly onto folk thought. As a result, the second abolitionist thesis can seem attractive.

The second abolitionist thesis starts from the rethinking of commonsense conceptions of mental illness that I have urged. The second thesis looks at the extreme heterogeneity that stands revealed when existing conceptual barriers are shown to be porous, and goes on to wonder whether that extreme heterogeneity makes a science of mental illness impossible. My conception of psychiatry's business is of a framework built around a loosely related set of breakdowns and deficits in our psychology that may not share any unifying features beyond the fact that they are breakdowns and deficits. The worry now is that there cannot be a science of this heterogeneous category. In rejecting the orthodox program's agenda for science, it may seem that I have foreclosed the possibility of a scientific inquiry.

I go onto make the notion of breakdown more precise in later chapters, but not in a way that assuages the worry, since I insist on the heterogeneity of mental breakdowns. The category is a rag-bag, even if the members of the category can be understood precisely. So the worry remains: can there be a science of breakdowns?

I think there can, even allowing for the exceptional heterogeneity it will need to acknowledge, since the science speaks to important aspects of our thought about ourselves. We need concepts of mental illness that have distinctively normative dimensions. But we need the science available as input into vernacular contexts in which explanation, prediction, and other purposes of inquiry remain germane, alongside normative concerns. So the science of psychiatry (I will continue to use the term) will retain its subject matter only in part because of narrowly scientific concerns, and in part because of its wider importance as a source of input into the various projects that have a place in the second stage of the two-stage picture.

3.7.2 The Unity of Psychiatry
Psychiatry is concerned with the diverse set of ways in which the mind/brain can break down. This diversity is unlikely to comprise a natural kind. If the mental illnesses are that diverse, can there be a science of them? Some scholars may say there cannot, since a science needs to have natural kinds as its subject matter.

Although I agree that "the ways in which a mind/brain can break down" is definitely not a description of a natural kind, I do not say we must conclude that psychiatry cannot be a science in consequence. Failure to have a natural kind as a subject matter is not a unique property of psychiatry but one shared by other sciences. Archeology, for example, is also concerned with the interdisciplinary investigation of many different kinds of things, including tools, human remains, monuments, and the physical evidence of climate change. These obviously do not make up a natural kind, but this is irrelevant to assessments of archeology's scientific status. Archeology has this diverse subject matter because it serves a particular set of interests we have. And psychiatry too is a science that possesses whatever unity it has because it answers to a set of human concerns. Its subject matter is whatever we need to give the science its point, which is the theoretical understanding of disruptions to normal psychology.

My quick analogy with archeology does not settle matters. One still might wonder what exactly gives psychiatry a scientific status, or whether the interests that it serves would be served better if psychiatry fragmented into several successor disciplines. You might wonder, too, how I can announce that psychiatry can be held together by human concerns after I have spent so much of this chapter denouncing our habit of trying to make psychiatry conform to commonsense constraints. It might seem that I introduced a clanking naturalism that rejected folk conceptual input, and that now I'm trying to weasel out of it.

But my skepticism about the hegemony of folk psychology does not imply that folk concepts have no place in a scientific enterprise, or even that they are limited to specifying an initial explanandum via ostension. Naturalists can take a harder and a softer line about folk theories. The hard line says that our folk view of the world should be discarded. It generalizes the idea underlying the charge that although we may not think whales are fish, we are just wrong and that ends the debate. The softer line admits a variety of folk constraints on our scientific projects that might preserve some role for folk thought. The softer line is more appealing when it comes to mental illness.

Hard-line naturalism opts for what Mark Johnston (1993) calls "fiendishly pure theoretical purposes." Philosophers who take the hard line tend to hold that since lay thought is not scientific, we should ignore it when trying to get an accurate picture of the world. For example, cultures all over the world distinguish similar divisions in nature and recognize local species (Atran 1990). Yet concepts in "folk biology" do not map onto scientific categories. People who are uninformed about systematics (i.e., nearly all of us) do not have a concept of "crocodile" that implies that crocodiles are closely related to birds, which they are. But cultures all over the world do recognize a sharp distinction between trees, bushes, and grasses. Botany does not. Atran credits Linnaeus with the key realization that folk taxonomies "often violate natural families and hence block a coherent sequencing of genera within and between families" (1990, 188). Folk biology tells how we intuitively organize nature, not how nature really is organized. And I have said something similar about not just folk psychology but about existing scientific conceptions of mental illness.

Paul Griffiths (1997) makes an argument along these lines for emotion.[11] Griffiths argues that the folk concept "emotion" does not pick out a natural kind but rather three different phenomena—modular "affect programs," cognitively sophisticated central emotional processes, and socially constructed behavioral repertoires. Griffiths concludes that there can be no science of emotion if by that one means a science aimed at explaining a part of our mental life that falls under the folk concept "emotion." The same arguments ought to apply to mental illness, on my picture. It seems that Griffith would give the same advice about improving our science of mental illness as the Pittsburgh city fathers reportedly received from Frank Lloyd Wright when they asked him how to improve their town: "abandon it."

Nothing said so far implies that a given disorder is not a natural kind, just that "mental disorders" lumps together natural kinds of very different types: complaints about the concept of mental disorder undermine the superordinate concept but may leave particular concepts intact. Therefore, one could argue that much psychiatric science can go on as before but insist that such a variegated collection of phenomena does not form a scientifically respectable kind.

In some areas, the naturalist hard line makes perfect sense, especially when (as is often the case) we have almost no pre-theoretic views about a topic. What is our folk theory of fermentation, for example? But in some areas, especially where our views about ourselves are affected, it may make sense to take a softer line. The softer line says that although our commonsense view might be extended or corrected by science in dramatic and surprising ways, we could nonetheless preserve a role for our thinking about ourselves even in science. Perhaps some non-epistemic projects might override scientific considerations in some circumstances. Despite the variety of ways in which our psychology can go wrong, we might seek to retain a concept of

11. Murphy and Stich (1999) connect Griffiths's argument with psychopathology more fully.

mental disorder or, perhaps, a variety of concepts of mental disorder in different contexts, acknowledging that there are non-epistemic as well as epistemic reasons for keeping the category. The picture is thus of two sorts of inquiry—one is a scientific attempt to understand the nature of our brains, psychological endowment, and social functioning, and the other is reflection on the ends we want that inquiry to serve. But the categories that the science investigates are defined by nonscientific concerns.

The softer line leaves the science to go where it wants with respect to a class of phenomena, but it provides a nonscientific rationale for treating those phenomena as a class. Our non-epistemic interest in having a given category groups the phenomena into that category. Our interests group together a set of entities for nonscientific reasons, but the entities themselves can be scientifically investigated. We can implement the softer line if we turn the scientists loose to discover what can go wrong with a mind/brain and then let the rest of us ask what to think and do about it. In the case of the gourmet lesion, we would then be able to point to a destructive process in the brains of patients and then go on to ask, about this dysfunction, whether it is bad or not. The answer is that maybe we do not think it is so bad, bearing in mind that not everybody is in the priviledged economic situation that makes being a gourmet possible. Often when we ask whether someone is *really* sick or *really* mentally ill, we are asking about the situation in which they find themselves. Is it bad to have such a life, or to have those beliefs? It is surely an attractive option to distinguish these questions more clearly from questions about whether the causes of their situations are psychological dysfunctions.

This way of looking at things assumes, of course, that we can make the distinction between psychological dysfunctions and "problems in living" or other undesirable features of life. I have not investigated whether, or how far, we can do that. At the moment I am concerned with defending the principle of a scientific enterprise that takes the softer naturalist line. To make the defense requires clarity about the enterprise I have in mind. The first step in the enterprise is to make the distinction between psychological breakdowns, explicable by science, and undesirable aspects of our nature and situation.

If we can make the two-stage picture work, we can arrive at an objective answer about whether someone is or is not dysfunctional, and then deliberate about how to regard the dysfunction. On this view, "mental disorder" is a concept like "pest," "weed," or "vermin." Weeds and vermin are not natural kinds, but they are made up of natural kinds that can be explained empirically. Furthermore, whether something counts as a weed or a vermin depends on human interests in a way that allows the class to grow over time, or vary across projects. Foxes are vermin if you're a farmer, but other people think of them differently. As we develop new habits of life, species come to be pests because they interfere with the new habits. Some insects prey on vines: we care about vines because their fruit is essential to civilized life, so those insects are pests. Other insects are not pests because they prey on species we do not care about. But, if we discovered a medicinal use for a plant that humanity never had previously husbanded or exploited, the insects that inhibited our use of the new plant would become pests. Concepts that are sensitive to human interests in this way are open-ended—things may fall into them (or drop out of them) as human interests change over time. Folk thinking does not determine in advance whether a species is a pest, nor does it make scientific investigation of a species of pest into a normative endeavor.

I have rejected, then, two views: both the view of the orthodox program that science should obey our folk concept of mental disorder in a way that fixes inquiry in advance, and the naturalistic hard line that just investigates scientific concepts and regards those investigations as providing results to which folk thought should defer. Instead, we should try to combine the scientific study of the mind and its habitual ways of going wrong with a variety of nonscientific projects that determine different, contextually relevant, concepts of mental disorder. The law, for instance, is a highly regimented and specialized context in which scientific findings are relevant but must be balanced against other considerations. The law also shows that it is a mistake to identify epistemic projects with scientific projects, as Griffiths seems to (1997, 200–201). The law tries to find things out, and surely epistemic interests are ubiquitous in everyday life. Different projects require different, proprietary concepts of mental disorder that further specific projects. Psychiatric concepts help to determine who is mentally ill, and that is different from the legal concept, which answers the question of whether a person is responsible for his or her actions. Indeed, we might distinguish the psychiatric concept from a wider therapeutic concept, which determines not whether a person is mentally ill, but whether they need some sort of care. Our feelings about human well-being do help determine the class of mental disorders, which cannot be a purely scientific class. However, given the heterogeneity of the class of mental breakdowns, can we extract a univocal concept of mental disorder that nonscientific projects can use? I think we cannot, because the nonscientific projects are themselves various. In the next section I argue that this heterogeneity means that there cannot be just one concept of mental illness.

3.8 Normative Concepts of Mental Disorder

3.8.1 Introduction

I have argued that our nonscientific requirements establish the scope of the science of psychopathology. In this section I revisit the two-stage picture, in which psychiatry discovers what can go wrong with the mind/brain and hands over the results of that inquiry as input to various nonscientific projects. It is at the second stage that our values and interests come into play. This is an objectivist picture, but without the objectivist's customary privileging of common sense and single-minded focus on dysfunction as the cause of mental illness. Because of the poor fit between science and common sense that my treatment suggests, I do not think that one concept can find a home in the various projects that would like to use a concept of mental disorder. A couple of examples come to mind immediately. Mental health professionals need to know which among the departures from our theoretical understanding of the mind are conditions we might want to treat, and to distinguish them from other problems afflicting people that may be helped by psychotherapy but which do not constitute mental illnesses. There are even different possible projects we could recognize within the larger therapeutic project. Military psychiatry, for instance, like military medicine in general, has always been torn between the therapeutic imperative of restoring casualties to health and the military imperative of just patching them up enough to return them to duty (Shephard 2001). The other instance is the law. Lawyers and juries need to know the circumstances in which departures from theoretical expectations about human function might excuse or mitigate criminal conduct.

There is no reason to suppose that a listing of human breakdowns on their own would furnish the proper categories for our therapeutic or legal projects. Although we should welcome the explanatory resources and greater understanding that the science brings with it, we should not simply abdicate our philosophical and social responsibility to work out for ourselves how we want to be. We should guard against assuming that scientific findings have a special status that trumps all other forms of argument.

3.8.2 What We Need

My ideal picture is that science makes findings available to a variety of nonscientific endeavors that can in turn think carefully about the wider purposes they serve and evolve concepts to fit. In doing this, one would hope that we could acknowledge that science has a role in psychotherapy, say, without regarding psychotherapy as a form of science. Psychotherapy is a characteristic institution of the modern era with an enormous cultural influence (Woolfolk 1998; Woolfolk and Murphy 2004). Although some therapeutic endeavors, such as the rehabilitation of stroke victims, do look like straightforward applied science, psychotherapy as it inhabits our culture is not a scientific enterprise. Decisions about whom to treat seem to have an irreducibly normative aspect and to be entangled with considerations about the shape of life and the nature of human flourishing. We should be careful, too, about becoming constrained by the belief that there is a scientific blueprint for normal people to which we must all conform.

But one would hope that the understanding of mental disorder and human nature that psychotherapists rely on would not clash with the understanding of humanity derived from the sciences. Usually it is disastrous when psychologists ignore science in order to push claims about humanity that have little or no empirical support (Dawes 1994), as when debates over child abuse were disfigured by pernicious pseudoscientific claims about memory or child psychology (Loftus and Ketcham 1996; Ofshe and Watters 1996; Yapko 1994). So science can inform therapy—we can evaluate treatments and draw on research into conditions in order to practice therapy. The claim that therapy is not a science should be sharply distinguished from the claim, which all parties may happily concede, that therapy may draw on empirical study and have measurable, scientifically evaluable effects. Those properties of therapy do not make it a science. Baseball scouts may develop statistical dossiers on players and coaches can study human physiology. Baseball has measurable physiological and psychological effects on its players, and economic effects on players, owners, and communities.[12] But there is no Nobel prize in pitching. So therapy may have scientific affiliations, and a base of scientific knowledge on which to draw. Yet that does not make it a science rather than a *techne*, nor does it establish the primacy of scientific considerations over other ones.

Jurisprudence has a different sort of interest in the deliverances of science. It is not interested in who needs to be treated but in those situations in which mental disorder excuses an individual or renders that individual incapable of self-government. Plato discussed legal insanity in the *Laws*, and Common Law references to mental illness as an exculpation date back to at least the thirteenth century (APA 1983; Moore 1980; Robinson 1998). The circumstances under which an individual may be taken, legally, to be of unsound

---

12. When I first wrote this, I conceded that scientific journals probably were not interested in the quantifiable aspects of ballgames. Then along came Abbasi and Khan (2004).

mind do not match up with the conditions that a mind might exhibit that depart from theoretical expectations: "not every nervous tic or neurosis constitutes an excusing condition. Nor does a general condition excuse in a particular case without some specific connection being shown." (Neu 1980, 84).

Determining whether individuals should be excused from criminal responsibility, permitted an affirmative defense of diminished capacity, involuntarily committed to mental institutions, denied insurance benefits for mental health services, or lose the right to make their own decisions because of incompetence all call for scientific knowledge within a legal context. But even though they are legal contexts in which concepts of mental disorder crop up, they are dissimilar endeavors. "Sanity," affirmed the American Psychiatric Association (APA 1983, 683) "is, of course, a legal issue, not a medical one." Distinctive juridical concepts of mental disorder have been generated. Some share features with the DSM account, but important differences exist as well. Indeed, Woolfolk (2001) notes that one legal definition stipulates that a disorder with a "biological" cause cannot be a mental disorder.

The APA's position on psychiatric testimony was in fact very close to the two-stage idea. The APA contended that psychiatric evidence should speak only to a defendant's "diagnosis, mental state and motivation" (1983, 686) but not attempt to determine legal insanity, which is a job for lawyers. DSM-IV-TR (xxxiii) notes that its diagnostic wisdom can provide information and check speculations about an individual's mental state, but not decide whether someone is legally sane or insane. The courts have concurred (Winick 1995). The expert opinions of psychiatrists feed into a complex process in which juries must think about responsibility in a legal context. This is in effect an institutionalized version of the picture that I have been saying is attractive: we want scientific input into normative structures. Psychiatry can supply the science, but not the normative principles.

In general, though, the current situation in psychiatry embodies with considerable precision the worst of two worlds. It has clear pretensions to fiendish theoretical purity, as if our psychiatry were being shaped by unadulterated scientific fact. At the same time, the actual process shaping our thought is a blood-curdling mixture of social pressure, inarticulate normative concerns, and conceptual confusion. We should acknowledge the dimension that broadly ethical and political concerns bring to our thought about psychopathology, and take that ethical dimension as seriously as we take the science. Our developing thought about psychopathology should be shaped by the goals of medicine, therapy, and the law—and the wider practical and moral interests those goals answer to. The scientific picture cannot dictate how we should think about dysfunctions in the contexts of psychiatry and the law; they are applied disciplines with their own concerns. Conversely, we do not want our thinking about mental illness to be immune to scientific findings. We do want the resources of science available to help us understand ourselves.

You will have noticed, however, that the two-stage picture, worked out on these foundations, seems to have pretty comprehensively bought into the fact-value distinction. It assumes that facts about human psychological organization and disruption are positive facts. Positive facts do not presuppose the truth of any claims about value. This absence of value claims is what the literature on the foundations of psychiatry usually means by "objective." The positive facts that psychiatry discovers via mechanistic explanation then are supposed to feed into value judgments. I have suggested that the value judgments we make at the second stage will be conditioned by the demands of distinct projects, so that there will be several different concepts of mental disorder, not one. At this second stage facts and values intermingle, but there is supposed to be a value-neutral body of facts at the first stage. Is this feasible, and is it desirable? I address the feasibility of the search for positive psychiatric facts in the rest of the book, especially in chapter 5, where I tie it to hard empirical questions about human cognitive architecture. But is it desirable to separate facts and values in this way? This is a large topic, but let me quickly suggest an answer.

The two-stage picture in psychiatry portrays a world in which a science driven by exclusively epistemic values like significance and simplicity derives its results and then hands them over to the wider world for moral and political appraisal and technological application. Many philosophers of science recently have denied the attractions of this separation of scientific and social activity (Dupré 2002; Kitcher 2001; Longino 2001). One of their concerns is that the power science possesses threatens other parts of the culture. Therefore, it needs to be tamed by introducing political, and other frankly evaluative perspectives, into the allegedly value-neutral scientific debates, as we decide what science we want to have.

I prefer the opposite view, in which the eight-hundred-pound scientific gorilla is confined to its cage, not taken out and domesticated. The attraction of the two-stage view for someone worried about technocracy is that the cognitive and biological sciences are kept at a distance from the various nonscientific projects that we otherwise wish to engage in, and that the opinion of psychiatrists in the normative issues arising in these areas carries no more weight that the opinion of any other citizen. If we mix up the science and the other concerns, the scientists may not be diminished in authority, but indeed handed more. I cannot go into this in detail, but I want to acknowledge the concerns. On my view, we do better to try to enforce the fact/value gap in this area, not blur it. We need to recognize a variety of projects in which concepts of mental disorder may feature, and we need to keep science relatively independent of those projects. For the benefit of both science and our other concerns. In the next chapter, I take up the science again, by looking at the medical model.

3.9 Conclusions and Previews

At the beginning of this chapter I noted three issues that cause trouble for the conservative objectivist version of orthodox conceptual analysis in psychiatry: common sense should not direct psychiatry, the proper psychiatric conception of the mental will develop as inquiry develops but is not to be found in psychiatry as it stands, and mental disorder is not always malfunction, let alone evolutionary malfunction. Despite these criticisms of the objectivist version of the orthodox program of conceptual analysis, I obviously find much in it to commend. I agree, above all, that diagnosing mental disorders should entail the provision of appropriate causal explanations (without, as yet, providing the details). A qualified objectivism survives as my preferred framework.

I suggest too that there are both theoretical and historical reasons for doubting the scientific utility of current psychiatry. But, as was true of the orthodox program, we should not conclude that there are no interesting insights in the DSM approach. (And I reiterate that not every theory in psychiatry is wrong or confused.) In particular, we should not hold it against the DSM that it does not obey the dictates of common sense. That is not to say that the DSM's taxonomy is correct, nor that it has the correctly revisionist attitude toward common sense. The DSM does seem to promote itself as giving the extension of a concept of mental disorder that is recognizably congruent with common sense. But it should not: unless we discard the notion that a scientific taxonomy of psychological malfunctioning picks out commonsense "mental disorders" it is impossible to avoid the conclusion that at least some coffee drinkers and insomniacs are mentally ill, and that the study of the visual system is not part of the science of the mind. The first claim does not fit with common sense, and the latter does not fit with scientific practice.

So I have taken over core components of the objectivist program as a methodological, rather than conceptual, program. The program is wedded to mechanistic explanation—the explanation of normal and abnormal mental life in terms of the mechanisms that underlie it. More generally, the two-stage picture envisages mechanistic explanations of psychological phenomena replacing folk psychological explanations. We arrive at a comprehensive set of positive facts about how the mind works, and then ask which of its products and breakdowns matter for our various projects. The limits of the two-stage picture coincide with the limits of mechanization. It is likely, as I show in chapter 5, that the limits of objective ascription of psychological dysfunction depend on hard empirical questions about the nature of our "cognitive architecture." Objective assessments of outputs may be possible only when we have a mechanistic characterization of a cognitive capacity and a clear account of its function. Assessments of rationality, for example, are likely to remain subjective. We do not have a generally accepted theory of rationality, and we do not have even the glimmer of a mechanistic theory of rational belief formation in terms of subpersonal information-processing (or cellular, or any other) mechanisms.

But even if the mechanistic picture cannot be carried out fully, it can be carried out in part. And those areas of human nature that resist a full mechanistic explanation still can be explained using various theories and models in the natural and social sciences. But those theories will be hard to disentangle from folk models of normal functioning, and hence partly normative wrangles about human functioning are likely to be a perpetual feature of the human sciences. In the next chapter I start to develop the picture of explanation in detail, by looking at the medical model.

The Medical Model and the Foundations of Psychiatric
Explanation

Insanity has become a strictly medical study, and its treatment a branch of medical
practice.
—Henry Maudsley, *Body and Mind* (1870)

As for this phrase the law of nature, I confess I read it a hundred times before I understand it once.
—John Donne, *Biathanatos* (1608)

4.1 Introduction

I have advocated the synthesis of psychiatry and cognitive neuroscience and
the use of that synthesis as a body of positive fact that can be exploited in
therapeutic, legal, and other projects. This synthetic neuropsychiatry, as I
noted in the book's introduction, is a version of the medical model, which I
look at in more detail in section 4.2. This chapter takes the revisionist objectivism about mental illness that I defended in chapter 3 and starts to tie it
to the details of the medical model, where it is most at home. Although the
medical model is often viewed as a reductionist program with close ties to
molecular biology, I argue that it is not tied to any particular view about
explanation. It is the view that symptoms should be traced to underlying
causal processes, or the idea that diseases have bodily sites. This perspective
is compatible with great explanatory pluralism. I discuss two specific ways
of cashing out the basic intuition that psychiatry is a branch of medicine:
a reductionistic, molecular interpretation of psychiatry in section 4.3, and
a multilevel approach that aligns it with cognitive neuroscience in section
4.4. I defend the second approach, which stresses the interrelation of
several different levels of explanation. Along the way, I introduce the rival
"biopsychosocial" (a horrible word, but the customary one) approach to
mental illness. On the most defensible understanding of the medical model,

I insist, the medical model and the biopsychosocial model are basically in agreement.

If psychiatry is clinical cognitive neuroscience, explanation in psychiatry must conform to the explanatory practices of the cognitive neurosciences. These practices are not best thought of as subsuming phenomena under laws, but rather as the search for regularities of limited scope at different levels of explanation. These regularities typically reflect the actions of various sorts of mechanisms that regulate the relationships between different components of models of abnormal biological and psychological processes. These models of abnormal processes are derived from models of normal processes.

In insisting on the utility of generalizations at different levels of explanation I follow recent philosophers who have written about the biomedical sciences by rejecting a global reductionism that tries to eliminate higher levels of explanation in favor of molecular biology alone. But although I reject a global reductionistic project in favor of a multilevel one, I still want to make room for reductionism. Psychiatry's recent emphasis on molecular biology must mean something. Any discussion of psychiatry must come to terms with the dominance of molecular/cellular methods in contemporary neuroscience (but also must make sense of the increasing search for the cognitive mechanisms underlying social behavior). So section 4.5 develops a view of reduction in psychiatry, based on the way reduction is understood in the contemporary neurosciences. It has two (not closely related) parts. First is reduction via decomposition and localization (Bechtel and Richardson 1992; Zawidski and Bechtel 2004) of cognitive capacities into brain areas with a fairly high-level cognitive characterization. This process shows that we must qualify theses about the strong autonomy of the cognitive level, since attending to brain function may reveal unsuspected complexities in the structure of cognitive operations. That is, although generalizations can be made at different levels, they sometimes cross levels, and cognitive generalizations cannot be formulated independently of each other in quite the way that traditional cognitive science approaches (Lycan 1990; Marr 1982; Pylyshyn 1984) seemed to think. Second, I discuss more narrowly molecular reductionist strategies, which usually employ model organisms. The picture here does not fit well-established philosophical discussions of reduction, as Ankeny (2002), in particular, has pointed out.

Many theories of reduction expect higher-level generalizations to be restated in terms of lower-level ones. In many cases, too, the expectation is that if the higher-level generalizations cannot be restated in lower-level terms, then the higher-level theory will be replaced by a lower-level one. But this view of reduction does not fit the clinical neurosciences. So in discarding

Kandel's reductionist picture, I am rejecting the traditional conception of reduction. But I argue for a different picture that is compatible with the retention of explanations at higher levels.

In psychiatry, reduction makes matters more complicated, rather than simpler. Instead of supplanting higher-level generalizations with lower-level ones, it adds new, lower levels of explanation to an existing structure. This both extends our causal understanding of why the upper-level generalizations work and increases our knowledge of points in the system where it might be tweaked to change the outcomes. We cannot know what goes on in the brain until we figure out what happens at the molecular level, and the more causal relations we understand, the more opportunities we have for therapeutic interventions in a system. However, in addition to this (welcome) gain in complexity from a clinical viewpoint that seeks to learn more about therapeutic strategies, low-level generalizations must be treated as idealizations. They are developed by manipulating variables of interest in models (usually model organisms) designed to minimize extraneous causal relations. When we apply them to actual populations, the extra information we need to add to the model often requires a broad generalization to be turned into several successor generalizations of lesser scope.

My discussion of reduction is aimed mostly at establishing an important role for molecular biology in psychiatry, but I am skeptical about one perspective on the molecular. Many theorists regard molecular, and other genetic, explanations as more fundamental than higher-level explanations. But in section 4.6 I show that the idea of fundamental explanation is very hard to make precise. If we understand fundamental explanation as the identification of the sole or chief cause of a condition, fundamental explanations rarely are obtainable, even when other things are equal. In psychiatry, the paradigmatic fundamental explanations are genetic because single-gene disorders typically are conditions in which the gene has its customary effects almost regardless of the state of other parts of the system. But most mental illnesses do not fit this picture and therefore lack a fundamental explanation. They result from interrelated causal processes at different levels of explanation and at different timescales.

Explanation is more or less fundamental, depending on the extent to which it is robust and multidimensional. A robust explanation is one that continues to apply when one moves from an idealization to the real world, and is thus largely unaffected by variation in context (cf. Woodward 2000). This fits the case of single-major locus disorders very well, since in those cases the effect of the errant gene usually occurs in almost any circumstances and hence the explanation holds not just in the idealized circumstances assumed by a model, but also relative to a wide range of actual conditions. Other genetic causes depend on too narrow a range of other factors to count as fundamental in this sense. A multidimensional explanation explains many (at the limit, all) of the symptoms of a disorder, albeit usually as a distal explanation in which the genetic effects are mediated by proximal structures; this too fits the model of single-gene disorders well, since the presence of symptoms depends on the gene, although the causal chain can be quite elaborate. Robustness and multidimensionality make for an explanation that is psychiatrically fundamental without qualification. There's a weaker reading I also employ sometimes: different explanatory interests rather than metaphysical primitiveness can make one causal-explanatory factor more fundamental than another. An explanation can be fundamental with respect to one question (say, in epidemiology) but not with respect to another (say, the precise form a disorder takes in Smith rather than in Jones).

By the end of the chapter, we will have a basic outline of the structure of explanation in psychiatry. But before moving on, I pick up the last of the four assumptions that we drew from the harmful-dysfunction picture in chapter 3. The assumptions were: (1) mental illness is a failure of normal function; (2) normal function is a product of natural selection for component mechanisms in the brain; (3) brain function can be decomposed into these component mechanisms; and (4) the ensuing cognitive architecture is shared by all humans. I have already argued against the first two of these assumptions, which fall out of the harmful-dysfunction concept of mental illness but do not have a very wide sway in psychiatry. I noted that many psychiatrists and psychologists distrust the claim that mental illness is necessarily a matter of dysfunction. I also denied that the biomedical understanding of function is an evolutionary one.

In this chapter, as part of my treatment of reduction, I consider the second two assumptions. Section 4.7 defends the decomposition assumption in terms taken from contemporary neuroscience. As noted in the last chapter, if the decomposition of the body into organ systems could be mimicked by the decomposition of the mind into functional components, the medical model would be easier to apply in psychiatry. The big obstacle to mental decomposition emerges in the final section of this chapter, where I look at the fourth assumption, of a shared human nature. I reject strong claims that there is no point talking about a shared human nature, but I draw two cautionary morals: first, the undoubted fact of psychological variation shows the need for cultural factors to play a role in psychiatric explanation even at the level of generalizations across people, not just at the clinical level where one assesses individuals in terms of their biographies. I discuss that in chapter 7. The second point is more substantial: variations in human behavior often reflect variations in central systems. Central systems are the sites of rational, executive control. If rationality cannot be understood mechanistically, then the project of a purely positive psychiatry cannot be consummated. This is a sufficiently large roadblock to the whole project that I discuss it at length in chapter 5, where I look at the prospects for a scientific psychiatry that synthesizes traditional psychiatry and cognitive neuroscience and generalizes the decomposition assumption. The success of that project depends on hard questions about human cognitive architecture.

That rather anticlimactic point can wait, though, until some more groundwork has been laid. So now I turn to the medical model. Nearly all psychiatrists these days agree that psychiatry is a branch of medicine. Quite what that means, though, is uncertain. The core of the medical model is the use of a disease concept drawn from general medicine. That disease concept is compatible with a great variety of theoretical outlooks. More important for my purposes, it is compatible with many theoretical outlooks existing in the same explanatory structure.

4.2   The Medical Model

4.2.1   Introduction

Guze, one of the medical model's chief architects, defines it as "using in psychiatry the intellectual traditions, basic concepts, and clinical as well as research strategies that have evolved in general medicine" (Guze 1992, 129). When they discuss the medical model, some theorists, especially skeptics (e.g., Scheff 1999), say the approach goes beyond the idea that mental illnesses have biological causes to the further claim that drugs can cure any mental illness. This assessment has some merit as a sociological depiction of current practice. But the dominance of psychopharmacology within the medical model does not follow from its basic assumption, to wit, that mental illness, like any other kind of illness, "represents the manifestations of disturbed function within a part of the body" (Guze 1992, 44).

No one denies the importance of drugs in medical treatment, but they are not all-important. A doctor treating a physical ailment, after all, might well prescribe not just medication but exercise, a change in diet, or some other nonpharmacological treatment. There is no necessary tie between the medical model and an exclusively pharmacological approach to treatment. This is true on any account of the biology of mental illness. The factor that best explains a condition may not be the factor that is most easily manipulated in therapy. As a general point, explaining an outcome depends on facts about the world, but finding the best remedy for that outcome depends on additional facts, including human capacities for redressing the situation and our opportunities to exercise those capacities. We might, for instance, explain a firm's declining market share by pointing out that its latest products are rubbish, while admitting that the best way for the firm to recover its lost share might be a clever and aggressive marketing campaign. Improving its products might be expensive, complicated, or time-consuming and offer no benefits that would not accrue from shrewder advertising. What causes the problem and what is best targeted to rectify it may not be the same. Even if a condition is best modeled via molecular representations of its causes, nothing follows about how best to treat it. But the causal knowledge is still an important constraint on the remedy. For the advertising to succeed, the firm might need to know what characteristic of its product causes the public not to buy it in the first place.

Scheff alludes to another red herring when he says that a further distinguishing mark of the medical model is a commitment to systematic classification. The medical model did indeed reform the DSM, but a stress on sorting patients into diagnostic categories is not distinctive of the medical model. A different theory of mental illness could embrace categorical taxonomy just as readily. Spaulding, Sullivan, and Poland, for instance, argue that the shift from psychoanalysis to biological and pharmacological psychiatry "was accomplished without changes in the key premises of the underlying medical model" (2003, 8). Given the comparative neglect of diagnosis in psychoanalysis, compared to its central place in recent psychiatry, this seems a stronger claim than is warranted. But it illustrates that what is distinctive about the medical model is not a biological theory of psychiatry, but a view of disorders as diseases. That view can be shared by different outlooks. The full-blooded medical model sees mental disorders as medical categories, instances of systemic dysfunctions in bodily systems.[1] As we have seen, some psychological processes such as learning introduce complications to that picture, but the basic idea remains: we understand the taxa of psychiatry as failures of some part of a psychological system to make the

1. Indeed, during the gestation of the DSM-III, the term "mental disease" was used but then withdrawn when the American Psychological Association threatened to sue the American Psychiatric Association. There is thus a sense, though nobody likes to mention it, in which the medical model represents the vindication of Thomas Szasz, whose skepticism about mental illness always has been predicated on the idea that to be a genuine illness a psychiatric condition must be a physical disease or lesion—a brain problem (Szasz 1987).

contribution to overall function that a correct account of that system would predict.

It is correct, though, that the medical model has come to be associated with molecular neurobiology. Guze, for instance, insists that the brain is "at the hub of psychiatric thinking" (1992, 59). The difference between Guze's medical model and the "biopsychosocial" account favored by some others (e.g., Engel 1977, 1981, 1992; Wilson et al. 1996) is that while both approaches take into account a variety of social, psychological, and neurological properties and causes, the biopsychosocial model is not hierarchical. It specifies no metaphysical or epistemic priority to any one element, whereas Guze's medical model is explicitly brain-centered. Guze acknowledges social forces and nonbiological causes may produce mental illness, but the medical model "specifies a hierarchy among its elements for conceptualizing and practicing psychiatry" (Guze 1992, 59) with the brain at the top. Looking at these two approaches in more detail shows what a difference the hierarchical picture makes. Guze's talk of hierarchy only makes sense as a claim that some brain processes furnish fundamental explanations. An elaboration of this position, as discussed in the following section, is Eric Kandel's (1998) view in which the fundamental process is gene expression. But I instead assert the superiority of Nancy Andreasen's (1997) multilevel view.

4.2.2 The Medical Model and the Biopsychosocial Approach

Guze (1992, 22) distinguishes three epistemic strategies in medicine: (1) the epidemiological, which aims "to account for the incidence, prevalence and varied manifestations of illnesses in terms of environmental and demographic factors"; (2) the biological, which concerns "the clinical manifestations of disease in terms of the structure and function of the body at many different levels"; and (3) the clinical, where one examines a particular individual and translates the results of the first two strategies into specific hypotheses about the patient. Guze contends that we can apply all three strategies equally well in psychiatry and that they let us use all the methods that any student of psychopathology might wish to employ—every possible causal factor fits into the overall structure.

Proponents of the biopsychosocial approach (which stems from Engel 1977, 1981, 1992) also stress the interaction between biology, psychological factors (such as learning history), and the social factors disclosed by epidemiological studies: "a single model of behavior is insufficient to permit full understanding of [mental illness], unless it gives weight to all three systems" (Wilson et al. 1996, 63). So what is the argument? Wilson et al. fault the medical model for neglecting nonbiological factors, treating all psychopathology as due to neurological and genetic abnormalities, and regarding therapy as exhausted by medication. Guze regards such claims as caricatures of his view and insists that the medical model gives due weight to nonbiological phenomena.

If the criticisms that biopsychosocial theorists make of the medical model miscarry, what of the criticisms that the medical model makes of its rival? Guze (1992) does not discuss the biopsychosocial model directly, but he does discuss the idea of sociocultural causes of psychopathology. He identifies sociocultural theses about causation with the antipsychiatry movement. Antirealist sociological theories of mental illness often do deny that any genuine disorders exist. But biopsychosocial theorists are realists about mental illness. The idea of sociocultural causation is consistent with a biological view of psychiatry, as long as it cashes out appeals to culture in terms of a material mechanism that affects behavior. The contemporary understanding of the brain is of a social and cognitive organ (Adolphs 2003; Quartz and Sejnowski 2002). That looks tailor-made for synthesizing the virtues of both the medical model and the biopsychosocial approach by providing neurological mechanisms that explain how the social and psychological can change behavior or disrupt normal processes. In chapter 7 I gesture at an account along these lines by a simple appeal to the mental representation of the social.

Guze's second objection to sociocultural causal hypotheses is that the medical model, unlike sociocultural approaches, follows the rest of medicine in recognizing the body and bodily processes as fundamental. Variations in brain function are "the heart of biological psychiatry" (1992, 59). Guze's metaphors of hub and heart are unclear. Proponents of the biopsychosocial model do not doubt that the brain is always involved in psychopathology—to doubt that would be to give up on materialism. It is also apparent that general medicine recognizes that some social environments, lifestyles, and personal histories may dispose one to illness. Guze's point must be that in all these cases medical doctors direct their attention to the organ systems that are the sites of pathology. This stress on proximate mechanisms as the fundamental causal pathway is a better candidate for the bone of contention between the two camps. But it depends on the way in which the proximate causes are understood.

Both the medical model and the biopsychosocial approach look for cognitive and sociological causes as well as biological ones. Both are ecumenical materialisms with a categorical view of mental disorders. So what, then, is the dispute about? The next section looks at some methodological proposals in detail. The difference between schools of thought in psychiatry depends on whether one kind of explanation is fundamental.

4.3 Molecular Psychiatry

4.3.1 Introduction
It seems that Guze's presentation of the medical model is consistent with the biopsychosocial approach, but this may be a function of the abstractness of Guze's presentation. Understanding how the medical model translates into particular research agendas requires more detailed proposals. When we do this, we see schisms among the believers. Section 4.3.2 details two recent attempts to outline a framework for psychiatry that reflects such differences. The first is Eric Kandel's attempt to push psychiatry closer to molecular biology, the second is Andreasen's attempt to articulate a multilevel cognitive neuropsychiatry. I prefer Andreasen's approach.

4.3.2 Kandel: Biological Psychiatry as Molecular Biology
Kandel's framework (1998, 460) is based on five principles. His first principle, that mental phenomena derive from operations of the brain, is just materialism, which is common ground and gives his view no advantage. For example, abnormal brain activity can be changed by behavioral therapy as well as by drugs. One study found that OCD patients differed from controls in having both increased activity in their right caudate and intercorrelated activity in the orbitofrontal cortex, caudate, and thalamus. Normalization of caudate activity and loss of correlation between the three areas was achieved both by SSRIs (selective serotonin reuptake inhibitors, the drug family that includes Prozac) and by exposure-based behavioral therapy (Baxter et al. 1992; Schwartz et al. 1996). Some people find this astounding, but for materialists about the mind, the idea that brain function can be changed by successful therapy should be entailed by their beliefs. After all, everyone knows that talking to people can change their thinking, and materialists believe that a change in thoughts involves a change in biology.

Kandel's second principle is that genes are "important determinants of the pattern of interconnections between neurons in the brain" (1998, 460). He goes on to conclude that genes thereby "exert a significant control over behavior" and that mental illness has a genetic component (1998, 460). Again, although many scholars find talk of genetic control over behavior misleading, nobody denies the claim that a genetic component exists in at least most mental illnesses, and I discuss what that commits us to in a moment.

Kandel's (1998, 460) third principle is that learning alters gene expression, so social or developmental factors also make important contributions to mental illness. Fourth, alterations in gene expression change neuronal connections and hence initiate and maintain behavioral abnormalities. Fifth, psychotherapy, when it is effective, changes gene expression so as to alter the structure of the brain.

Although Kandel countenances social and psychological causes of alterations in gene expression, he says that "all of 'nurture' is ultimately expressed as 'nature'" and that "genes determine the phenotype" (1998, 460). But if genes determine the phenotype, why in explaining the phenotype do we need to worry about causal mechanisms at levels other than the molecular? As Phillips puts it (2000, 686), Kandel's new framework is the injunction that "psychiatry must get with the molecular program" because fundamental explanation refers to the molecular level. Understanding the mind/brain at higher levels will perform only the heuristic function of characterizing the process that molecular biology will then explain, so higher-level theories will have no explanatory autonomy. Next, I defend this reading and show why it sets an implausible agenda for psychiatric explanation, but I also defend a kind of reductionism that better fits the current picture in psychiatry.

Kandel's talk of genetic control notwithstanding, his view is emphatically not the familiar conception of genetic determinism, which I understand as the claim that one's mature psychology is settled in advance by whatever genes one has at birth. Kandel's is the different, and undeniably correct, view that changes in behavior and neurophysiology from moment to moment involve alterations in gene expression in the brain. Gene expression—the switching on or off of genes in particular cells—responds to environmental influence. So, for Kandel, the effect on the phenotype of nongenetic factors is mediated by gene expression, which itself is changed by those nongenetic factors.

Let me drive this home. We have known for a long time that all our cells contain the same genes and that cells differ from each other in terms of the subset of those genes that produce proteins in a given cell. More recently, we have discovered how this differential production depends upon experience. As I said, all materialists know that psychology changes and is changed by brain activity. Kandel is drawing our attention to the material basis of this process—the production of proteins by some genes in our brain cells and the cessation of protein production by others. This is not genetic determinism in any traditional sense, given that gene expression is just as much an effect of environmental causes as it is a cause of behavior in turn. Nonetheless, Kandel's stress on gene expression does raise issues about the relation between theories of molecular and nonmolecular processes. What worries better-informed critics of views like Kandel's is not genetic determinism but what he implies in his own account, viz, that all genuine explanations of psychopathology will occur at only one level of analysis, and that nonmolecular factors will be relegated to a subsidiary, heuristic role in explanation: "biological psychiatry" looks for molecular explanations of mental illness.[2] In this intellectual structure, other levels of explanation are neglected or given only a subsidiary role as characterizations of that which is to be explained in molecular terms.

Kandel's view is that since all other variables (environmental, developmental, cognitive, etc.) have their effects on symptomatology via gene expression, we need only look at gene expression in explanation. In effect, we can explain mental illness by going no further down the causal chain than the last link—molecular variables are the only ones of interest since any other variables have a relationship to the symptoms of mental illness in virtue of their effect on gene expression. This stress on proximal causes, however, only answers some of the questions that we might want to ask about mental illness.

From the fact that all psychological phenomena involve gene-expression it does not follow that higher-level phenomena do not exist or do not matter, nor that generalizations across those phenomena are not autonomous, since they prescind from fine-grained differences in the molecular processes within individuals. Brain structure and development across individuals have enough plasticity to make higher-level generalizations, which smooth out those differences, indispensable.[3] Similarly, variation occurs across individuals at the

2. This is particularly the case in schizophrenia research, which has been dominated by molecular (and, increasingly, genetic) approaches at least since Snyder et al.'s much-cited manifesto caught the mood in 1974. Verhoeven and Tuinier (1999) complain that biological psychiatry has been handicapped by an excessive interest in molecular mechanisms to the exclusion of other aspects of neuroscience. I should note that in a follow-up to his original paper, Kandel (1999) discusses cognitive neuroscience as a natural ally of psychoanalysis (by which he seems to mean any form of psychotherapy). This reference to cognitive neuroscience might seem to tell against my interpretation. But when Kandel discusses details, he does so exclusively at a molecular level, and when Cowan and Kandel (2001) advocate the reformation of psychiatry via neuroscience, molecular genetics is the only neuroscience they discuss. Kandel gives his considered take on neuroscience in general in his Nobel Prize lecture (Kandel 2000).

3. For a clear and well-informed account of the difficulties involved in building purely genetic explanations even of simple cognitive capacities in well-understood model organisms, see Schaffner (1998).

level of functional organization within the frontal lobes, which means that we also may need to employ generalizations that cite cognitive capacities but abstract away from the details of the realizations of those capacities in different people.

The very questions we ask affect the generalizations that we need to cite in explaining why a person or a population meets some diagnosis. Chapter 7 discusses eating disorders and suggests that the explanation we need to cite in a given case depends on the relevant question: social factors may explain why eating disorder levels vary across populations or take particular forms in particular cultures, but they not tell us why, out of all the girls in a given family, only one develops an eating disorder. To explain her case, her membership in a class of people who share a particular brain chemistry or childhood trauma may be more relevant than her membership in a specific culture.

Reflecting on cases like this, Miller argues that they pose a problem for molecular approaches that is not "solved or avoided by placing trauma or poverty or learning history earlier in the causal chain than biology. If trauma fosters psychopathology via some mediating biological process, then the psychopathology is not fundamentally due to something biological" (1996, 620). Miller seems to think the nonbiological trigger is fundamental in these cases. Now, an obvious response to Miller is that people differ in vulnerability to environmental agents for biological reasons. Not everyone responds to trauma by becoming bulimic, and those who do may share a biological property. But that response does not establish, against Miller, that neurobiology is fundamental. It suggests that nothing is fundamental. The trauma may require a biological factor to have its pathological effect, and vice-versa. A full explanation must mention both biology and trauma, even though particular questions can be answered by citing one or the other. We should consider neither fundamental.

Meehl (1977) contrasts two cases that make the point that a fundamental explanation is sensitive to the context (and, in this case, norms) of inquiry: PKU and depression. Phenylketonuria (PKU) occurs when children with two copies of a recessive gene are unable to metabolize the amino acid phenylalinine. A normal diet includes phenylalinine levels above the clinically significant threshold, so if the PKU genotype is not diagnosed early, a buildup of unmetabolized phenylalinine occurs, leading to profound mental retardation. (Children diagnosed early enough receive a special diet which mitigates the effects of the gene.) Meehl points out that it seems natural to call PKU a genetic disease even though a diet with more than a certain amount of phenylalinine is a necessary condition for its development. But the necessary amount is nearly always present so, in effect, only the genetic predisposition seems salient. Meehl contrasts this case with another, that of an elderly man who is genetically somewhat at risk for depression, and who develops depression after the death of a spouse. In this case, the environmental factor that interacts with the genes, unlike a normal diet, usually is not present. In the depression case, the genes have their effect only under unusual environmental conditions, so it does not seem right to call this a genetically caused episode of depression. But with PKU we can talk about genetically caused mental retardation, because the gene expresses itself in any normal situation. Equally, however, it is wrong to cite only the environmental cause in the depression case because the widower would not have developed clinical depression had his genes been different.

I will elaborate these ideas in an account of fundamental explanation that permits genetic explanations to be fundamental, but only in a small set of cases. First, though, I want to give a sense of the other explanatory options by shifting to views of psychiatry that grant more autonomy to other levels of explanation. We must acknowledge other levels of explanation because a complete molecular explanation would require, as Schaffner (1994) points out, a molecular specification of environmental causes too. And as Schaffner says, specifying and measuring environmental variables in purely molecular terms "would be a very long-term project" (1994, 287). No one is likely to advocate a molecular description of environmental causes—Schaffner is reminding us of the limits of molecular reduction in psychiatry. Molecular biology cannot aim to replace generalizations about environmental factors with generalizations about the molecular basis of those environmental factors. That is an important point if the reductionistic program in psychiatry is interested in reduction in the classic philosopher's sense of replacing a higher-level theory by restating its generalizations in terms of statements in a lower-level theory (Nagel 1961). If a complete molecular explanation requires this, then Schaffner is right to point out its intractability. Furthermore, we have no reason to want a molecular reduction in this case, since in tracking environmental causes we are dealing with a different kind of causal process than we do when we track molecular processes inside the organism.

Kandel (1998) seems committed to a replacement view of reduction on which generalizations about, say, a cognitive system will be replaced by generalizations in terms of gene expression plus, presumably, other information about the cells that make up the system. This is the picture that is familiar to many philosophers of psychology, and it makes sense when levels of explanation are, putting it simply, different descriptions of the same causal process. But biological and environmental factors are different causal processes, in a way in which the computational and biological properties of the visual system are not. Biology and computation are not doing different things inside the brain: they are ways of representing the same process that are useful for different purposes. But gene expression and long-term unemployment are different processes, even though being unemployed affects the brain.

Kandel's (1998) stress on gene expression, then, threatens to obscure crucial differences between causal processes that explain different aspects of a condition. This assumes that Kandel is referring to molecular phenomena by "gene expression." On the other hand, if what Kandel means by "gene expression" is "behavior, neuroanatomy or whatever else seems relevant," then his picture is one in which our mental life is instantiated in the brain but is caused by psychological, social, and cultural factors and is explicable at several levels of explanation. And that picture is no different from the approach of the biopsychosocial model. It seems, then, that biological approaches to psychiatry either collapse into the biopsychosocial model in an attempt to make room for nonbiological explanations (on a narrow construal of "biology"), or, if they try to remain biologically pure, they develop an implausible fixation on molecular and genetic explanations.

To assert this is not to reject materialism or deny that genes are involved in mental illness. It merely recognizes that our mental life can be described at many levels of explanation and that we cannot tell in advance the level at which the best explanation is to be found. Some questions about disorders will be answered in terms of genes or ion channels. But there is every reason to believe that other questions are better answered in terms of social pressures, computational processes, learning histories, or evolutionary mismatches. These need not be competing explanations. They all need to be brought together. But they are explanations that appeal to generalizations that are not easily reduced to molecular ones. This point is clearly recognized in Nancy Andreasen's (1997) intellectual framework for psychiatry.

4.4 Cognitive Neuropsychiatry

I have suggested that a stress on molecular explanations of mental illness amounts to the view that molecular explanations are more fundamental than others, and I take that up again in a moment. I also have wondered whether the model implies that molecular biology should replace higher-level theories. This section addresses the latter question, and it suggests that explanations drawn from different levels will continue to have a role to play. After looking at multilevel theories, I discuss reductionism in more detail in the next section, and then return to the question of fundamental explanation with the resources for a fuller treatment.

I chose Kandel as my paradigmatic reductionist. As the multilevel theorist, I look at Nancy Andreasen (1997). Andreasen is struck by the convergence of complementary models of schizophrenia and depression from a variety of fields, including neurology, the cognitive sciences, and clinical psychiatry. In both cases, she argues, we can understand the conditions as cognitive pathologies instantiated in particular brain circuitry. This does not rule out, however, the study of mind and brain "as if they were separate entities" by "multiple and separate disciplines" using a "different language and methodology to study the same quiddity" (1997, 1586). But of course, as Andreasen's own discussion shows nicely, the issue is not just a matter of separate vocabularies and methods. Distinct disciplines attend to different properties of the mind/brain, and a fully adequate study of psychopathology must make room for all of them, plus others. Andreasen's view of schizophrenia resembles that of other researchers who try to integrate neurocognitive and biological approaches. Michael Green, for instance, argues that schizophrenia is "essentially a disorder of neural connections," but draws on genetic, neurological, developmental, and cognitive approaches to explain it: "neural problems lead to neurocognitive deficits. The neurocognitive deficits lead, in turn, to misinterpretation and confusion, and eventually to functional impairment" (Green 2001, 27).

Andreasen conceives of her synthesis as a form of biological psychiatry, but in contrast to Kandel's molecular reductionism, she sees psychiatry as a cognitive discipline. Indeed, in her view it is one of the cognitive sciences: "the discipline within cognitive neuroscience that integrates information from all these related disciplines in order to develop models that explain the cognitive dysfunctions of psychiatric patients based on knowledge of normal mind/brain function" (1997, 1586). I think this is the best short description of psychiatry that we have.

It may be difficult to see the difference between the biopsychosocial approach and Andreasen's pluralistic "scientific psychopathology," which incorporates not just neuroscience but a variety of other disciplines. Does it put the brain at the hub of psychiatry, as Guze insisted? Yes, because all the disciplines Andreasen mentions study the brain in one way or another. In Kandel's hands, Guze's slogan might be better phrased as: molecular genetics is the hub of psychiatry. But that, I have argued, is unduly reductionistic, and if we adopt Andreasen's more relaxed materialism the difference between the biopsychosocial and medical models disappears, and we are left with the attempt to relate clinical findings to a psychopathology that decomposes a neurocognitive model of a disorder into its component computational parts and then localizes those components in particular brain circuits.

The medical model fits Andreasen's view well, because it shares with general medicine a conception of disorder as disease that is based on the assumption that disorder represents breakdown in normal biological processes. So Guze's assumptions transpose easily into a cognitive key. The cognitive level has no clear analogue in nonpsychiatric medicine. But the logic of the medical model does not exclude the cognitive. Consistent with the view that disease stems from pathological processes in bodily parts, mental disorder can be understood as the product of disruptions to normal cognitive, affective, and other psychological processes, not just abnormalities in gene expression. Cognitive explanations allow us to understand, just as other forms of explanation do, how the brain can break down, and it is that understanding that the medical model is premised on, not particular ways of attaining the understanding.

Some conditions may indeed be better explained in terms of brain chemistry, but that does not imply that psychiatry should not incorporate a cognitive dimension. To reiterate, the claim is that the theoretical account of normal mental functioning is elaborated at many levels. Ultimately, what we want is a story in which disease models are conceived of as templates to fill in at a number of levels of explanation, with the causal relations among levels established wherever possible. Some conditions may be explicable in terms of rich models that are articulated at every level with interconnections drawn in, too. Other disorders may be explained at only a few levels, as a partial articulation of the whole possible model.

I illustrate this idea by applying it to schizophrenia in the next chapter. But first I want to discuss the question of explanations at different levels in more detail. As Schaffner (1993, 1994) in particular has stressed, biomedical theories typically cross levels, with effects from one level being related to causes from a different level, and so on. This comes about in part because, typically, levels of explanation in biomedicine are more intimately related than claimed in some traditional theories of interlevel relations in philosophy of psychology. That intimacy reflects the practices of explanation in the life science. In section 4.5 I discuss two ways in which reductionism works in the life sciences. I begin with the idea of decomposing cognitive capacities and locating them in physical structures. Doing so addresses the decomposition assumption, which is the third of the four assumptions of the harmful-dysfunction model. That leaves only the universality assumption, which I discuss at the end of this chapter.

I also suggest in section 4.5 a way of seeing molecular research strategies in biomedical science that preserves a central role for molecular biology but does not grant it the imperialist ambitions that I detected in Kandel. Molecular strategies, especially those based around model organisms, have two features: they add a lower level rather than replacing higher levels, and they are designed to reveal causal relations that can be targeted by therapeutic, usually pharmacological, interventions.

4.5 Two Kinds of Reduction

4.5.1 The Decomposition Assumption

This section sketches some reductionist strategies in those parts of the life sciences that have psychiatric significance. I deal with decomposition and localization first (Bechtel and Richardson 1993; cf. Wimsatt 1976) and provide some indication of our current practice when it comes to discovering how brain structures realize psychological capacities. That communicates the sense in which I endorse the decomposition assumption and further justifies my account (in the previous chapter) of the constraints under which functional analysis labors in psychiatry: it must relate components to the overall system.

Explanation in mature psychiatry will require a model of the normal realization of a psychological capacity in the brain and deform that model to show how abnormalities at the level of sign or symptom depend on the failure of normal relations to apply between components of the model. Hence, in this section I begin conveying a sense of how these models usually are developed. In the next chapter I look at some obstacles, and in chapters 6–8 I discuss some concrete proposals for explaining different conditions using this framework. Chapters 9 and 10 apply the framework to classification. At the outset, I should rebut one common objection to this approach. While I claim that psychiatric explanation relies on an overarching theory of normal functioning, this is not threatened by that fact that developing that theory of normal function will itself require inputs from psychopathology. Our understanding of the normal mind will be greatly helped by understanding how it breaks down, but good explanations of how it breaks down will have to be based on an understanding of how it functions normally. The two projects interact.

We find this interaction between theories of normal and abnormal function in cognitive neuroscience at the moment. Cognitive neuroscientists assume that an explanation of a deficit is made relative to an account of normal functioning: "normal information-processing systems are the domain over which any disorder of psychological function must be defined" (Marshall and Halligan 1996, 9; see also Stone and Young 1997). The study of patients with specific deficits has been widely used to build theories, since understanding how cognitive systems go wrong is an important guide to their normal function. The relation has also gone the other way: theories of normal function have been developed with the goal of applying them to pathological states. The two projects should develop together and mutually inform one another. The interaction of theories of normal functioning and theories of pathology is exactly what we should expect in psychiatry as a whole on the approach I recommend. Developing theories of the mind will produce models of normal functioning to explain the deficits that patients display and make specific predictions about the sorts of deficits that will be uncovered. Pathologies thus serve as both a source of evidence for theories of normal functioning and as explananda: "the patterns of impaired and preserved performance in individual cases (who may or may not have been assigned [a] taxonomic label) can be assessed for compatibility with normal models of psychological functioning. Those latter accounts can in turn be revised when data from pathology cannot be accommodated by the current version of the model" (Marshall and Halligan 1996, 7).

Modern neuropsychologists have a standard way of elaborating the models that explain normal and impaired performance, a method that they share with some nineteenth-century theorists who saw cognitive neuroscience as continuous with psychiatry (Shallice 1988). The typical picture involves "box-and-arrow" diagrams ("boxologies") that represent putative cognitive components and the information-processing relations between them. These are models of causal relations—graphical causal models. Glymour (2001) has discussed the mathematical problems involved in discovering which among collections of such graphs actually apply to people, regardless of whether the evidence comes from single case studies or groups of patients. The methods make a number of assumptions that, when qualified, (Glymour 2001, 136–137) may permit clinical data to discriminate among putative cognitive architectures. Glymour's interest is the causal claims that are warranted on the basis of hypothetical architectures and behavioral response data. His examples do not attempt to relate functional components of the putative architectures to anatomical structures: that is, they decompose but do not localize. Glymour does not argue that it is a virtue of a theory of the mind that it decomposes but does not localize.

However, other theorists do seem to assume that mere decomposition is a virtue, often because their interest is in the intrinsic properties of mental processes, which they take to be computational and hence knowable without any physical investigation. On this view, if you get the decomposition right you understand the nature of mentality, regardless of the particular details of its embodiment in a species (Fodor 1983; Pylyshyn 1984). In cognitive science, the view that psychological states could be characterized in terms of an abstract specification of computational relations came to characterize a relentlessly top-down picture. The top-down picture decomposes the mind into computational processes based on abstract characterizations of how information needs to be processed to perform a task, and then looks for the brain structures that correspond to the various stages in the performance of the task (Marr 1982). This view has great influence in psychology; evolutionary psychology, for example, defines the explananda of psychology in terms of abstract specifications of information-processing tasks (Cosmides and Tooby 1997).

If asked how explanation in cognitive science proceeds, a philosopher of psychology probably would give an answer couched in terms of levels of explanation. This picture comes from the work of David Marr. Marr advocated viewing psychological theories at three levels of explanation (Marr 1982, 24–25). The highest level specifies, in functional terms, the computational task accomplished by the psychological system with which a theory is concerned. The topmost level, Marr said, would show the computational problem that is solved by the particular information-processing algorithms the system uses;[4] these are described at the second (middle) level. The information constituting the system's input and output are represented at this level, and here too we specify the algorithm that transforms the input into the output. The lowest level is that of implementation, which asks how computational processes are realized in the brain. Marr called this the level of detailed computer architecture. The lowest level describes the physical imple-

---

4. There is some dispute over the interpretation of Marr's top-most level. My own preference is for Frankie Egan's reading (1995), but that does not affect the rest of the discussion. The view that the essential properties of mental phenomena as studied by cognitive science are computational was for years entangled with functionalism in the philosophy of mind, to their mutual cost. Although I am defending the view that explanation in neuroscience has shown the top-down picture in cognitive science to be a bad method of studying the human mind, I am not suggesting that no form of functionalism can be saved.

mentation of the algorithms in the brain. These three levels are approximations, in the sense that they specify kinds of description for a specific purpose
rather than unique descriptions. So levels can exist within levels. Pylyshyn
(1984) for example, decomposes the algorithmic level into two levels: that
of the algorithm and that of the functional architecture, which are best
viewed as a set of rules for carrying out a given computation and the implementation of those rules in a particular computational system. Hence, the
three levels Marr distinguished are best seen as specifications of three kinds
of explanation that attend to different properties of a computational cognitive system. So we have functionally specified levels of computation, algorithm, and implementation.

Marr's followers study the visual system by first specifying abstractly the
tasks that the system needs to perform. Marr (1982) imagined that we could
understand vision as a set of tasks independently of any information about
the biology used to perform them. But the mainstream in cognitive neuroscience never has accepted this top-down picture, because investigation of
the actual visual system reveals a decomposition that would not have been
suspected on grounds of abstract task specification or even adaptive efficiency. Rather, the decomposition of cognitive activity and the identification
of cognitive parts—physical structures in the brain that carry out information-processing jobs—mutually inform each other (Bechtel and Mundale
1999; Zawidski and Bechtel 2004).

Mishkin et al. (1983) cited lesion studies of monkey brains as evidence
that visual information is processed along two main routes (my discussion
of the philosophical morals of this work draws heavily on Zawidski and
Bechtel 2004). The first route defines the "What" system: information in this
system proceeds ventrally from the prestriate cortex down to areas TEO and
TE. Lesions in the system cause failures to recognize previously familiar
objects, and the very large visual receptive fields of the neurons in the ventral
pathway suggested to Mishkin and colleagues that not only did the neurons
identify the physical qualities of a visual object, they did so in a way that
enabled reidentification of those objects to take place irrespective of the
objects' location in the visual field. And to do that, they argued, the system
must forget information about the object's position in the visual field.

The second pathway uses the information about location. The second
pathway is the "Where" system, which runs dorsally up into the posterior
parietal cortex. Lesions in that area of the cortex prevented monkeys from
responding appropriately to an object when the response was sensitive to
the object's position in space, suggesting that the second stream contained
information about the relevant location of objects, not their visible qualities.

When imaging techniques and other experimental methods were added to the mix, the final picture was more complicated than was originally supposed. But neither the complications nor the original separation of the What and Where streams would have been predicted via attempts to reach an abstract task decomposition of the mind. Decomposition must be constrained by information about the whereabouts and relationships of cognitive parts for it to succeed. This approach does not rule out the significance of theory, nor render box-and-arrow diagrams redundant. Frith says that the "cognitive approach in psychology is essentially theory-driven. Theories are first presented, preferably in the form of 'box and arrow' diagrams, then detailed hypotheses are derived and tested experimentally" (1992, xi) And historically, intentional or behavioral theories have marked the starting point, and neurological investigation has attempted to understand the realization of a mental taxonomy taken off the psychological shelf, not one built up from brain investigation (Hatfield 1999).

The Marrian approach views the intentional level as autonomous. The intentional level harbors abstract characterizations of mental phenomena that could support generalizations across species and, indeed, across artifacts. The generalizations are intended to apply to anything displaying the elements of mentality. That theory does seem to be too general. It must be interwoven with neuroscience and amended in the light of how the brain works. The computational realization of the capacity might vary as much as the physical realization was supposed to vary in traditional functionalism; there is no reason to suppose that the division of object recognition into What and Where systems is a universal or necessary feature of visual systems.

Decomposition cannot be the autonomous analysis of the abstract tasks that minds carry out. But the decomposition assumption is vindicated as part of the twin processes of decomposition and localization. It reflects the belief that mental capacities can be distinguished functionally, but the final decomposition of our mental life into components should be guided by the interrelation of cognitive hypotheses and physical facts. We should reject top-down approaches to decomposition, as assumed by classical computational theories that thought decomposition could be done by reflection on the demands of cognition. Compared to the Marrian picture, this is a reductionist view that puts much tighter constraints on the relations between levels—our understanding of realization feeds back into and constrains our understanding of the abstract demands of cognition. In effect, we should expect, at higher-levels, to see generalization across idealized representations of cognitive parts. Rather than relating abstract psychological capacities to each other, higher-level generalizations relate capacities understood as functional descriptions of fairly coarse-grained brain areas. This will be the case even though variation exists in the way capacities are realized in the brains of different people. The normal system will be an idealized depiction of human mind/brains as made up of cognitive parts.

I make this prediction with some confidence because it echoes the situation that already exists with respect to model organisms. Lord Kelvin (for whom the Kelvin temperature scale is named) thought we could not understand anything until you could express it in numbers. Likewise, the modern biologist thinks we do not understand biological systems unless can intervene so as to perturb them in predictable ways. To really understand a system, we need to see what structures are involved at the molecular level and how they can be manipulated. The end result of localization is molecular localization. I now turn to consider reductionist strategies at the molecular level, ask what they are for, and assess how far they can give gene expression a general fundamental role, as opposed to a fundamental one in answering some questions about some conditions.

4.5.2 Molecular Research Strategies

I agree with theorists who assert that at higher levels of cognitive explanation we should expect to see a more complicated relation of mutual constraint between the intentional and the neuroanatomical than is foreseen by the traditional picture in philosophy of psychology. The behavioral and cognitive neuroanatomy that is developed at the higher levels ultimately will need to be integrated in a structure that extends down to molecular explanations. Something about molecular explanation makes reductionism especially important in the life sciences, namely its greater potential for developing controlled experiments that permit explanation and intervention. But however desirable genetic reductionism is, it cannot play the unifying *explanatory* role Kandel expected of it.

Molecular reductionism in the life sciences should be seen in the context of its dependence on animal models. As Rachel Ankeny (2000, 2002) has shown, the upshot of exploring a model organism is the development of a "descriptive model," which is an idealized representation of a particular species, like *C. elegans*, or subset of that species, like a *C. elegans* hermaphrodite. From this model, scientists derive a representation of the aspect of the species that they are interested in, such as the wiring diagram of the *C. elegans* nervous system. What we gain is a representation of a normal system—even though no single actual organism may have exactly that set of genes—that can be used as the basis for explorations in organisms that depart from the canonical representation in various ways. Differences in behavior or development can be compared against the idealized "wild type" and, in theory, traced back to the differences in wiring between types, and the genetic differences those wiring differences depend on. The actual picture, even in a very simple organism such as the small worm *C. elegans*, is of great complexity in the causal pathways between genes and phenotype (Schaffner 1998), which suggests that Kandel's language of genes as determinants of the phenotype requires extensive revision.

Animal models in psychiatric research provide analogies with human systems that can be targeted by systematic intervention. Detailed knowledge of the nervous system of another species, researchers hope, holds out promise of a better knowledge of analogous processes in humans, and better opportunities for testing interventions that one might make in humans.

For example, researchers working with records of the 1957 Helsinki flu outbreak have found an elevated incidence of schizophrenia among children of mothers who were infected during the second trimester. No such increase has been found among children exposed during their first and third trimesters *in utero*. Some, but not all, other studies support the conclusion about the second trimester, and similar results have been obtained for measles, rubella and other mental illnesses, including autism (Green 2001, 31–32; Patterson 2002). Patterson and colleagues (Patterson 2002; Shi et al. 2003) explored the possible pathways between infection and behavioral change by infecting mice with the flu and measuring their offspring on a variety of tests. The results showed that mice born to infected mothers were less likely than controls to interact socially and explore novel environments. The mice had another property of great interest to schizophrenia researchers: they exhibited a lowered PPI that could be corrected by antipsychotic medication. Let me explain.

The startle reflex includes several measurable responses, such as eyeblinks. The strength of eyeblinks can be measured by electrodes placed around the eyelids. A prior exposure to a quieter, unstartling noise (the "prepulse inhibition," or PPI) lessens the subsequent startle response. But not in schizophrenics, who typically are deficient in PPI; that is, their startle response is not lowered to a normal degree by the prior stimulus (Green 2001, 77). Patterson and colleagues found the same effect in the offspring of infected mice and also found that it could be corrected by clozapine, to which the mice were abnormally sensitive. PPI deficits also were found in the offspring of mothers who had been injected with synthetic RNA to induce an antiviral immune response. This immediately suggests a line of research on the interaction of maternal infection (as an environmental risk factor) and genetic predisposition to schizophrenia, as well as the possibility of tracing the mechanisms in the immune system that mediate the relationship, and developing technologies aimed at targeting that relationship in therapy.

Why, on the basis of research like this, do I say that molecular theories have a special place? After all, this is only one avenue of research and obviously this is not going to explain everything about schizophrenia. Nor does it speak to higher-level explananda, such as the delusions and thought breakdowns involved in first-rank schizophrenic symptoms. The reason for thinking molecular models are special has largely to do with the amount of precision they supply. Of course, it is not possible to subject human beings to deliberate infection and experimental exploitation in this way (although given the lobbying clout of the drug industry, you never know what might change). But the animal models are simpler than the human biology that they stand in for, and hence they allow for much greater control. It is possible, by contrasting wild types and experimental types, to keep everything except the effect of one gene constant and trace the effects of that change, or to manipulate elements of the system while holding others constant. Descriptive models of aspects of human biology also are being developed, of course, with the same ends in mind. At the moment, the greater complexity of human biology makes it difficult to advance much beyond data gathering, but the hope is that the human data can be tested against manipulations of animal models that share relevant properties, or close analogues of them. Patterson's lab, for example, currently studies the brain activity of mice who receive different sensory stimuli, to see if definite patterns of activity can be correlated with different inputs. The hope is then to monitor neuronal activity in the brain by using "immediate early" genes (genes that turn on expression when the state of the cell is altered)—this is a familiar technology. The object is to see whether giving the mice a drug that induces hallucinations in humans induces activity in the visual or auditory cortex of the mice in the absence of sensory input. If this is the case, we ask if a mouse model of mental illness also displays activity in these areas in the absence of sensory input. Perceptual activity without specific input is used as the operationalization of hallucination. Most interesting questions about human cognition remain when we move from endogenous perceptual activity in mice to hallucinations in humans. But the experiment is not intended as a complete explanation of hallucination, merely a way of asking precise questions about its biological basis, and manipulating the perceptual system so as to make its causal pathways more perspicuous.

Compare the relative ease and precision with which one element of a descriptive model of the mouse nervous system can be manipulated to the extreme difficulty involved in changing some of a person's beliefs without any psychological ramifications, and you can see why reductionist molecular strategies offer a manipulative power that higher-level theorizing lacks. We simply do not understand cognitive capacities at computational or other high levels in ways that are analogous to descriptive molecular models in terms of power and precision.

But the existence of these genetic reductions does not mean that higher-level generalizations can be replaced by them, as some philosophical models of reduction expect. Most environmental causes of mental illness, for example, do not admit of molecular understanding in the way that *in utero* influenza infection does. And most aspects of human cognition are nowhere near reducible to the molecular. Furthermore, as I suggested earlier, the variation across individuals at different levels makes reduction difficult even in principle. Rather than matching up generalizations at different levels, we are constructing idealized models at different levels, answering to the demands of theory construction at each level. There may be no way to smoothly reduce the model at one level to the model at another level, because at each level a series of idealizations move the models away from reality in different directions: the elements of a descriptive molecular model of the nervous system and the elements of a box-and-arrow diagram at the cognitive level may not line up well enough to admit of a reduction, rather than a rough indication of how the cognitive parts that correspond to the boxes and arrows work at several levels of biological analysis.

We should continue to see, then, the multilevel causal models that Schaffner (1993, 97–99) identified as the rule in biomedical science; causes described in genetic vocabulary will be related to effects described in terms of behavioral tests, for example, and generalizations will cross levels, relating elements of a model at one level to elements of a model at another. What we achieve by molecular means is not the elimination of higher-level generalizations, but the discovery of new low levels in a way that interacts with the existing structure but leaves it basically in place.

So gene expression is special, but it is special because it gives us an especially precise understanding of some biological processes that can be experimentally manipulated. That does not make genetic explanations fundamental—as I said, the easiest factor to manipulate may not be the causally fundamental factor, and in any case most mental illness result from interacting factors operating at many levels of explanation. The exceptions may be cases where genetic fundamentalism (genetic determinism) is appropriate, but these are rare. In the next section, I introduce those cases and explain in more detail what I mean by fundamental explanation.

4.6 Fundamental Explanations

4.6.1 Introduction

The epistemic benefits of genetic reduction, especially in animal models, are considerable, but they do not replace higher-level theories or make gene expression an explanatorily fundamental factor. This section explains what I mean by fundamental explanations, suggests the requirements for an explanation to be fundamental, and asks if any psychiatric disorders meet the conditions.

Fundamental explanation is a slippery concept. For instance, Sarkar (1998, 46) connects what he calls epistemological fundamentalism to the drive for replacing a higher-level theory or description with a lower-level one. He defines fundamentalism in terms of three conditions. Unfortunately, one of his conditions is that the rules of the reducing realm are more fundamental than those of the reduced realm, which still leaves us wondering what makes something scientifically fundamental. Sarkar's introduction of the term refers to deepening insight and correlating disparate understandings "and so forth." This is not very precise, but that is unsurprising because the idea of a fundamental explanation is hard to make precise. But it is clear that some important role for reduction, as somehow more basic, is intended.

The notion of explanation as unification (Friedman 1974) often accords an important role to reduction on the grounds that a collection of facts or separate theories could, after a successful reduction, be derived from the reducing theory. The reduction gives us just one hypothesis in place of several. But if a fundamental explanation is a unifying one, it is not necessarily reductive, just an explanation that replaces a large number of hypotheses with one successor that shows why they hold. Darwin's theory of natural selection unified a large number of hypotheses without being reductive in the sense of explaining higher-level phenomena in terms of lower-level ones. As Hatfield (1999) points out, biological hypotheses in neuroscience have always been guided by psychological hypotheses about cognitive relations: "the study of the brain's functional organization is guided by descriptions of psychological function" (2000, 268). So, which is more fundamental: the reducing theory that explains of what the cognitive mechanisms are physically made, or the higher-level theory that guides the enterprise? This is not a question that admits of a unique correct answer. The reductive explanation may answer some questions that the higher-level explanation could not, but the higher-level explanation may still retain its autonomy in other cases.

To make the notion of a fundamental explanation more precise, I claim the crucial mark of a fundamental explanation is the ability to explain a lot even in the face of messy, real-world detail. The genetics of Huntington's Disease may serve as a case for which we do have a fundamental explanation, as contrasted with the genetics of major depression, for which we do not. The difference is not just a matter of the availability of a molecular redescription of higher-level phenomena. It is matter of the robustness of a generalization. And very often molecular generalizations do not have the necessary robustness. In fact, molecular generalizations typically hold only relative to environmental factors that obtain outside the organism. The genetic basis of Huntington's Disease captures an important sense of fundamental explanation, but very few cases like it exist in psychiatry.

4.6.2 The Clinical Genetics of Huntington's Disease as a Fundamental Explanation

The study of the genetic basis of mental disorder once relied on twin studies and other methods of behavioral genetics, and it typically made use of the dubiously valuable concept of heritability. Heritability is usually defined as the variation of a given phenotype in a given population that's due to genetic factors. This is not a measure of the causal contribution that genes make: if people vary in the number of limbs they have it is because of environmental, not genetic factors. So limb number has a heritability of zero, but that does not mean that genes are not causally involved in building arms and legs. The difference between heritability and causation has been pointed out many times (see especially Lewontin 1974), as have the drawbacks of twin studies, including twin studies in psychiatry (Kaplan 2000; Robert 2000), and these days even behavioral geneticists admit that we cannot get anywhere without discovering in detail how genes actually work (Plomin and McGuffin 2003).

My case of a fundamental genetic explanation is one involving a causal process, not a heritability estimate. The entire causal story is not yet known, but we do know that Huntington's Disease is associated with a gene, IT15 or "huntingtin," on the short arm of chromosome 4, characterized by an abnormally long CAG trinucleotide repeat (see Ashizawa et al. 1994 and Gottesman 2002, 294–295). A very strong correlation exists between repeat length and age of onset of the disease—people with 39 and 40 repeats have an average age of onset of 66 and 59, respectively, whereas for people with 49 repeats the average age of onset is 28 and, for 50 repeats, 27. The gene is not fully penetrant, meaning that one can have it and not show the symptoms. But this occurs only in rare cases and with low values of the repeat. All Huntington's patients so far assessed have between 36 and 121 repeats, with 90% of them having 39–50. Some people at the lower end, with 36–39 repeats, are asymptomatic.

In the case of such "single-major-locus" (SML) conditions, it makes sense to think of the genetics as the fundamental explanation. That does not mean that a deterministic relation exists between huntingtin and the symptoms of Huntington's Disease. Huntingtin is not, as I said, fully penetrant. Moreover, if a deterministic relation is one in the presence of the gene necessarily leads to the condition, then even for high values of the repeat the relation will not hold. One could, for instance, die of something else before the disease develops. However, the relations between alleles and SML conditions are determinate enough for us to call the genetic explanations fundamental in the sense that they identify genetic causes that, under normal conditions, suffice for the diseases. I explain in a moment exactly what the qualification "normal conditions" means. First I want to contrast the huntingtin story with another story about a genetic contribution to mental disorder. Other genetic contributions are much more complicated, and lack the neat causal association we see in Huntington's, in which a change in value in the genetic variable produces a corresponding increase or decrease in a phenotypic value.

Note too, that the explanation is multidimensional, in the sense that the diverse symptoms of Huntington's all appear to depend on proximal processes that are caused by the gene. The repeat does not just explain something about Huntington's; it seems to cause the neural damage on which the set of symptoms depend. Huntingtin, however, lacks one quality that you might expect an entity cited by a fundamental explanation to have. It is not general. Conversely, a cause can be general, in the sense of being involved in many symptoms and conditions, without ever being explanatorily fundamental.

Compare the robust and multidimensional causal properties of huntingtin with a different cause of mental illness. Stress is involved in episodes of many different conditions, and its presence or intensity affects the form that symptoms take. But stress levels alone do not predict who will come down with, for example, depression, since reaction to stress depends on many other factors. In that sense, stress is not a robust explainer—a model that assigns great causal significance to stress will need to make a large number of additional assumptions. And stress may not be multidimensional. It may explain, say, why insomnia features as a symptom in Fred's latest depressive episode, but it may not explain many of his other symptoms.

However, stress as a cause of mental illness is general in a way that huntingtin is not. Huntingtin is implicated only in one disorder, and stress is implicated in many (including many physical disorders). So stress is a general cause, but it is neither robust nor multidimensional. It is not robust because it is sensitive to small perturbations in context, and it is not multidimensional because in each case it may explain only a small number of the total symptoms.

So many causes of mental illness are ill-suited to be fundamental explanations. Many conditions appear to resist fundamental explanations. In contrast to the neat story about gene-phenotype relations in Huntington's, for instance, we can look at the long and uninspiring search for a genetic basis for depression. Several researchers have reported finding a statistically imposing linkage between depression and some stretch of DNA, but these reports have not been replicated (Kaplan 2000, ch. 7). Even if we could find a gene, though, the relationship between huntingtin and symptomatology found in Huntington's seems not to apply to depression. The probability that an individual will suffer a major depressive episode is sensitive to a large number of risk factors, with many variables intervening between genetic factors and depressive symptoms, including psychological factors including temperament and various accidents of one's biography. Kendler et al. (1993) attempted to summarize these relationships to better predict depression in women, and concluded that 9 predictor variables (3 independent, and 6 in three levels of intermediate dependency) represent the interactive causal factors. The ensuing hideous complexity is represented in figure 4.1. Even this is a simplified model, though, since Kendler et al. left out some variables, could not test the predictive value of gender (as the sample was wholly female) and, most important, chose to assume that the relations between causal factors in depression is entirely additive, which is surely false.

The complexity involved in the depression case and the relative simplicity of the huntingtin-Huntington's case (h-H) makes it difficult to make sense of the notion of fundamental explanations in the former, but defensible to employ it in the latter. This does not, let me emphasize, mean that molecular approaches to depression are of no value, nor that molecular generalizations cannot be integrated with other generalizations. But the complexity does imply that the idea of a fundamental explanation in psychiatry, which I suggested is Kandel's goal, is generally misguided. The exceptions occur in those cases where the relation between symptoms and cause, as in Huntington's, nearly always holds. Let me now say more about what "nearly always" means.

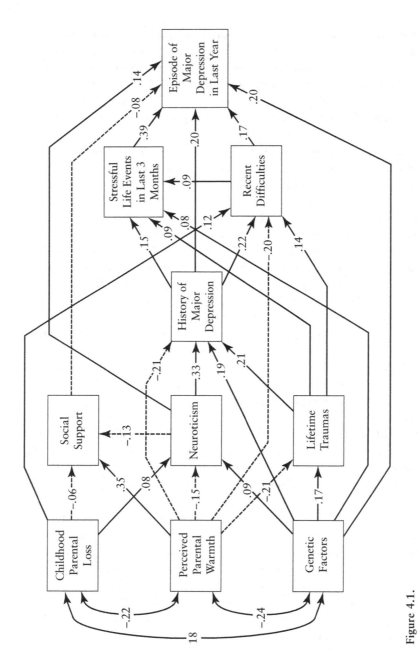

Figure 4.1.
Statistical relationships among risk factors for depression in women (Kendler et al. 1993)

4.6.3 Psychiatric Generalizations

Philosophers usually assume that, outside physics, all lawful scientific generalizations are true only if other things are equal.[5] An example is the law of supply and demand, which says that alterations in the price of a good result in changes in the quantity of the good demanded and the quantity supplied. For a second example, consider the geological law that a meandering river erodes its outside bank. This second example was used by Fodor (1987, 4) to criticize Davidson's (1980) claim that there are no psychophysical laws, but Davidson calls it an illustration of his position, because it shows that the laws of geology are not strict, unlike the laws of physics (Davidson 1993, 10). The "laws" of economics and geology are non-strict in Davidson's sense in that they hold only for the most part, subject to other factors. If the price of a good drops but its potential consumers suffer a catastrophic loss of income at the same time, then the law of supply and demand may not predict what will happen, because something has intervened to prevent the usual effects of the cause (the increase in demand) from occurring. Similarly, considered as a strict law, the claim that a meandering river erodes its outside bank is in fact false. It is false because many factors can prevent the cause from having its usual effect: suppose that "the weather changes and the river freezes; or the world comes to an end; or somebody builds a dam; or somebody builds a concrete wall on the outside bank; or the rains stop and the river dries up . . ." (Fodor 1987, 5).

The moral is that in the special sciences—all the sciences except physics— laws are true only if other things are equal. If this is how psychiatric laws work, we should expect that a given cause, such as a genetic endowment or a neurodevelopmental defect, does have a typical effect, but one that occurs only in the absence of intervening causes. David Lykken (1995), for example, argues that the biological basis of psychopathy is a genome that makes some individuals hard to socialize, and that this genome typically leads to the development of psychopathy in early adulthood. But the claim that the "hard-to-socialize" genome causes psychopathy holds only for the most part, "because really talented parents or, more likely, a truly fortuitous combination of parents, neighborhood, peer group, and subsequent mentors, can socialize even these hard cases" (Lykken 1995, 12).

---

5. Whether *ceteribus paribus* laws are really laws at all (Schiffer 1991; Woodward 2002), and how they are to be understood if they are laws, are vexed questions. I think the points I make here are unaffected, except for terminology, by the outcome of these fights.

Lykken has not posited a strictly lawful or deterministic connection between the inborn biological endowment and the adult behavior, because whether the genome exerts its typical effect depends on other factors. But Lykken's understanding of the connection between the "hard-to-socialize" genome and the adult behavior does exhibit the characteristics that, as we have seen, non-strict laws in the special sciences exhibit: the law says that, other things being equal, the "hard-to-socialize" genome causes psychopathy. Lykken's view is that other things are often equal, but not enough to make psychopathy simply genetic, as opposed to an example of Meehl's "diathesis-stress" story, on which a genetic predisposition interacts with environmental factors to produce a syndrome. It is important that variation in symptoms is correlated with different values for environmental variables as much as it is with genetic ones. The extent and severity of a psychopathic phenotype varies depending on both the genetic load and the socialization process.

Many explanations in psychiatry appeal to causal laws that are not strict laws, and which show this dependence on environmental contingencies in the presence of a genetic predisposition. Andreasen et al. (1999), for example, reject the disaggregation of schizophrenia into subtypes in favor of an approach that cites a fundamental neurodevelopmental abnormality, which they call "cognitive dysmetria" or a "disruption in the fluid coordination of mental activity" (1999, 911). Dysmetria can produce a wide variety of symptoms. Andreasen et al. acknowledge the variety of deficits that schizophrenics may exhibit. Their view is that the underlying cognitive disruption can produce one or more symptoms from the total set of schizophrenic symptoms depending on what other factors are present. There is therefore no strict lawlike connection between dysmetria, or its genetic and developmental precursors, and any given symptom, such as thought disorder. But, since the presence of dysmetria is, on this account, necessary for thought disorder, we learn something important from thought disorder, namely that cognitive dysmetria is present. We learn about the presence of the cause even though we do not learn this from a strictly lawlike relation. This prominent role for necessary, but not strictly lawlike, causal relations is widespread in the special sciences, including the biological sciences that proponents of the medical model try to emulate. If psychiatrists can find statements relating physical causes to psychological effects that have the character of laws in the special sciences, then some basic assumptions of scientific psychiatry are vindicated.

I have discussed cases in which a particular etiology causes some set of symptoms, other things being equal. This connects with the idea of a fundamental explanation as follows: a more fundamental explanation is one citing a factor that reliably produces an outcome despite different values for the other relevant variables. A less fundamental explanation cites a factor that makes a difference to the outcome only under a very restricted range of circumstances. At the extreme, a generalization might be true in so few possible cases as to be effectively false in the real world. What economists call the "resource curse" or the "Dutch disease" might be an example. Suppose we assume that natural resources make a country more prosperous, other things being equal. But time and again the discovery of a considerable and valuable natural resource turns out to be a drag on economic development. This can happen because the government steals the money from the citizens instead of reinvesting it, or because other avenues of economic development are neglected, or because other countries buy the resource and drive up the value of the home nation's currency to export-hampering levels (which happened to the guilder after the Dutch found a huge offshore gas field), or for other reasons. Other factors so seldom are equal in the real world that the claim that natural resources are good for an economy can be true relative to some range of factors in a model, but almost always false in fact.

Reconsider in this light the h-H and depression cases. We want to say, I think, that the h-H case is quite a lot like the geological case of a river eroding its outer bank. If the natural system is left alone to develop normally (without outside intervention), the expected outcome is well assured. If the huntingtin gene were 100% penetrant, the analogy would be even better. Meehl (1977) envisaged a Mendelizing and 100% penetrant Huntington's gene as his parade case of a specific etiology in the strongest sense: the disease occurs with probability one in the presence of the cause, and probability zero otherwise.

Meehl's expectations about the Huntington's gene turned out to be incorrect[6] But if the system is left alone, we can predict with great certainty based on just the one variable (repeat length) that the outcome will be Huntington's, and the system nearly always is left alone. All the great variety in the culture, psychology, and other physical attributes of people with the huntingtin gene seems to make almost no difference to the h-H relation. The normal unfolding of the system is compatible with many variations in every aspect except the genetic. This means that the generalization that huntingtin causes Huntington's is terrifically robust (Woodward 2000). When we move

6. Although, of course, most of the time they might as well be correct, especially from a clinical standpoint. The relationship of generalizations to individual cases is taken up in chapter 10.

from the idealized h-H case to the clinical level, we put in more and more detail, but as the additional factors are added the h-H relation, especially above 40 repeats, continues not just to hold, but to exhibit the relation between repeat length and age of onset. Adding more real-world detail does not change that relation. We can imagine intervening to block the genetic action directly (e.g., by retroviral injection into brain cells [Kitcher 1997, 115–116]), but at present we know of no environmental factors that might mitigate the development or reverse the effects of Huntington's, so we are left with the tyranny of the gene. Even though the brain of a Huntington's sufferer, like any other brain, is the outcome of a huge number of causal interactions, it seems that, to a strong approximation, none of those interactions can change the brain in a way that alters the h-H relation. Only if the system collapses (as in premature death from some other cause) will the h-H relation be prevented: basically, nothing affects your chances of getting Huntington's once you have a long enough repeat.

This imperviousness to real-world variation makes the Huntington's long repeat a fundamental explanation. Obviously the explanation picks out a process that goes on in a context—but there are hardly any ways of specifying the context that make a difference to the explanation that cites the h-H relation. We can invent a few exceptions: you can not put a string of CAGs in a petrie dish and give the dish Huntington's Chorea. But the exceptions are usually beside the point, because we are interested in the h-H relations in humans. And from human to human the physical context is usually the same, because it is a matter of facts about human biology, which, thanks to evolutionary and developmental processes, come out more or less the same every time. And the differences do not effect h-H.

Most molecular explanations will not work like this, which is what makes Kandel's search for fundamental explanations at the molecular level generally quixotic. Even if we can reduce some mechanism to its molecular base (as Kandel did in his own work on memory in *Aplysia*, the smooth-bodied sea hare) the relations between the molecular reduction base and the symptoms of a disorder simply will not be robust in the way the h-H relation is robust, because when we put in real-world details, the typical outcome very often fails to unfold, since it can be inhibited, masked, or even reversed.

The h-H relation obtains between two relata, each of which is a reduction base of some higher-level process. The h-H relation relates a molecular characterization of a gene that predicts a phenotype (a Gene-P, *sensu* Moss 2003) and the brain processes occurring in the unfortunates who have the phenotype. This reductive character is important, since it supplies detail to help us see what in fact realizes the relation that we previously knew about between cognitive decay and a putatively Mendelizing gene. But it is not the reductive character of the h-H relation that makes it fundamental. It is fundamental because once we understand it we know something that applies across a great range of physical setups and produces the same outcome in each.

Causes other than genetic ones could be fundamental explanations in this sense of greater imperviousness to real-world detail. We could have an equally robust relation between a cognitive impairment and a stroke or some other trauma, for example, if variations elsewhere in the sufferer seemed not to change the cognitive effects of the stroke. That explanation too could be both multidimensional and robust, if it explained all the symptoms that develop after the lesion develops, and does so in a way that is almost impervious to variation in the rest of the system.

So in my sense, fundamental explanations in psychiatry are those that (1) cite maximally robust causal relations, including relations that cross levels of explanation, and (2) explain all the symptoms of a condition. The property of being a fundamental explanation does not attach to an explanation at a given level, but to an explanation of a particular form. We should expect to see few fundamental explanations in psychiatry, and many models that explain disorders in terms of interacting, nonfundamental, explanations. In some cases, we might ask a question about some property of the disorder, such as an epidemiological difference, that is best explained by only one causal factor. So in that sense there could be numerous explanations that are fundamental relative to one explanatory interest. So, contrastive explanations, which say why something rather than something else occurred, might quite often cite a variable that tracks a process that is fundamental in that context. But fundamental explanations *simpliciter* are likely to be very rare in psychiatry.

This is all by way of establishing the point of molecular projects in psychiatry, understanding Kandel's perspective, and appreciating the explanations likely to dominate a reformed cognitive neuropsychiatry. The upshot is that psychiatric explanations will range over multiple causal processes, and each of these may cross levels, rather than track phenomena at only one level. Like Andreasen, I anticipate a psychiatry that draws on a conception of normal functioning at several levels of explanation. But it is hard to deal with because of its very inclusiveness. Price et al. argue that the education of both neuroscientists and psychiatrists should "emphasize basic neuroscience, genetics, neuroanatomy, neuropathology, neuroimaging, neuropsychology and cognitive neuroscience, behavioral phenomenology, neuropsychopharmacology and psychological interventions" and that the disciplines need to "include perspectives from social scientists, ethicists, philosophers, religious representatives, patient advisory groups and the legal community" (2000, 14). On the one hand, this sounds right, and suitably comprehensive; on the other hand, it sounds akin to recommending an education in just about everything except tax law and folk dancing.

How can we make sense of mental illnesses at the confluence of so many different natural and social processes? In chapter 6 I sketch out how explanation works in the synthetic psychiatry I envisage. In chapter 5, though, I qualify the enterprise before I begin it, mentioning some obstacles that confront both the explanatory structure and the assumptions of the two-stage picture in which I try to place it.

Before that discussion begins, however, I want to continue elaborating on the last of four assumptions I took from the harmful-dysfunction analysis of mental illness found in the work of Baron-Cohen. So far, I have qualified the first three assumptions with the contentions that (1) mental disorder should not be assumed to be necessarily the product of malfunction in a cognitive part rather than the product of drives or learning; (2) that the functions of cognitive parts are not best conceptualized as evolutionary products because their discovery respects the demands of theory construction in cognitive neuroscience; and (3) that theory construction in cognitive neuroscience proceeds not by top-down methods but by decomposition and localization. I now turn to the last assumption, that of universality. Are our cognitive parts the same across all humans? The foregoing treatment of reduction comes in handy in answering that question. My answer is: yes, sort of.

4.7 The Universality Assumption

4.7.1 Introduction
One of the four assumptions that make up the framework under investigation—the framework that represents the empirical commitments made by the harmful dysfunction picture—is that humans share a common universal psychology. Paul Griffiths has dubbed this "the doctrine of the monomorphic mind" and, along with other theorists, he has urged its implausibility (Griffiths 1997, sec. 5.5; cf. Hull 1989). Now, everyone agrees about the variability in human psychology, as measured by any tests we have as well as by common sense. Some ways of making this concession have only a minimal effect on the pretensions of psychiatry to be a general science of the abnormal mind, and some ways ruin those pretensions. Let me outline each possibility and then state its significance. The first source of variability in behavior is the different performance, in different people, of common cog-

nitive parts. This is not a big issue, since such differential output is universally acknowledged. But it does raise questions of how normal performance is to be measured, which I address later. Second, if the different performances of different people reflect differences in underlying cognitive parts, with some people having one set and others having a different set, the situation is more complicated but still manageable. At the limit of difficulty we find the idea that human nature is so diverse that no theory of normal functioning is possible. One could combine these three possibilities and hold that each applies to some areas of human psychology.

4.7.2 Performance Differences

People vary in their measurable properties. Some of these properties, such as height and IQ, are naturally continuous; others, such as whether a person votes or has a particular blood group, are not. Normal functioning occurs within a range, and the range is relative not to all people, but to all people within a class. People differ in visual acuity, for example, and the same individual differs across a lifetime. Many objects in nature change over time as a result of normal development: protodwarf stars become red dwarves that become brown dwarves. This change over time is a natural process that depends on hydrogen fusion followed by the transformation of hydrogen isotopes into helium. At each stage, we can specify a range of normal properties that stars in that stage exhibit. We may hope that the effects of aging on psychology can be dealt with similarly.

A cataract, for example, is a clouding of the crystalline lens in the eye. A wide range of injuries and illnesses, including diabetes, can cause cataracts, but the commonest cause is the normal aging process, as old cells die and become trapped in the lens. Most people over 65 have cataracts, so we can say that the onset of cataracts is part of the normal development of the visual system. However, that process of normal development also permits us to make judgments about abnormality if we find unusually large numbers of dead cells in the eyes of a young person with diabetes.

The acknowledgment of variability, including age-dependent variability, does, though, raise the issue of how we approach health and normality. If we say that an eighty-eight-year-old has cataracts, and thus is normal, are we calling that person healthy? George Vaillant argues that we are not (Vaillant 2003). Mental health, on Vaillant's view, is concerned with flourishing: it is not a matter of functioning above a baseline below which people are mentally ill. It distinguishes people who are well above the baseline, where one reaches a higher standard altogether. People are not just sick or normal, but sick, normal, or healthy.

The idea that problems in living are distinguishable from disorders reflects the same idea. A problem in living is an obstacle to flourishing that affects someone who is in the normal range.[7] A disorder prevents someone from being within the normal range. Whether this is generally plausible is not something that I have yet established, but it is a basic assumption of much psychiatry, and has much to recommend it. We are prepared to take steps, such as surgical cataract removal, to help the elderly, without thinking of normal aging as a disorder. But when aging is abnormally accelerated, we regard it differently: children with Hutchinson-Gilford Progeria Syndrome are normal at birth and have normal intelligence, but they go through all the stages of human aging at an astonishing rate and are nearly always dead by seventeen. By that time, they are in advanced senescence. Nobody knows what causes this very rare condition, although it seems to relate to defective DNA repair. The point is that even though we are ignorant of its etiology we think of Hutchinson-Gilford as obviously different from normal aging, and obviously caused by some underlying pathology.

So although variability in measurable psychological performance is ubiquitous, it does not threaten the medical model in principle. Our existing medical practices acknowledge it. However, psychological variability might in practice take a form that does threaten the extension of medical practice to psychiatric explanation. The threat arises because observable differences may not always reflect underlying discontinuities. Current practice is predicated on the belief that categorical boundaries can be drawn among people who lie along a dimension of functioning based on some more fundamental discontinuity in underlying systems. Even biological medicine does not always work this cleanly: a biopsy score or MRI image may represent some naturally continuous quantity, like blood flow, suggesting that dysfunction in biological systems is not, at the level of gory detail, a matter of being, as it were, switched on or switched off. Determinations of dysfunction often reflect decisions about cutoff points made with respect to measurements of the underlying systems.

However, in biological medicine we often have a causal explanation of why systematic relations should exist between measurements of symptoms and measurements of underlying processes. But in psychology, even when we have a reliable result we are often unclear why it obtains (Meehl 1957). A prediction in psychology can be correct while still being in this sense, as

---

7. Obstacles to flourishing can be removed by medical treatment just as deficits and diseases can, and this raises the issue of "enhancement technologies" in which medicine is used to benefit people who are already healthy (Elliott 2003).

Bishop and Trout (2004) say, "ungrounded." Whereas we can see on theoretical grounds why a biopsy score is a good predictor of the health of a prostate gland, it is difficult to see why an interest in mechanics magazines should be associated with paranoia (Dawes 1994). The hope is that causal theories in psychiatry can work to ground predictive instruments in the way that they do in general medicine.

The project of relating symptom-level measurements to underlying ones will hit trouble if there are no underlying abnormal processes to provide a causal anchoring. The result will be dimensional, rather than categorical, diagnoses. Unlike a categorical system of diagnosis and classification, a dimensional system sees conditions not as qualitatively distinct from normal function, but as extreme variants of normal functioning.

The question of whether mental illnesses are on the whole dimensional or categorical is a vexed one, which I return to in chapter 10. For now I conclude that human variation does not show the general inapplicability of models of systemic malfunction in cognitive parts, any more than it shows that biological medicine is mistaken in adhering (on the whole) to the same model. But, might the existence of variation in psychology show that variations exist in humans at the level of cognitive parts? Could we have different mental organs?

4.7.3 Variation in Cognitive Parts

A second variant of the claim that psychology differs across individuals goes beyond the recognition of different performance to argue that people have differences in cognitive architecture. The claim is not best understood as a claim about differences in brain morphology. Everybody has a brain, and although some people are born without some components of gross brain morphology, such as an amygdala, this is very rare and the subjects tend to suffer as a result. (People who are born normal but suffer severe brain damage in youth are a different case. Often their brains can rewire to compensate, in ways that are poorly understood.)

Metabolic differences account for greater activity in brain areas, and that this can differ across individuals. Also, the brain is plastic enough to develop differently across individuals in response to experience, as in the famous study that found, relative to controls, significantly increased gray matter volume in the hippocampi of London taxi drivers (Maguire et al. 2000). The hippocampus of a taxi driver is enlarged because it stores large-scale spatial information to use in navigation, among other things. These sorts of findings are consistent with the idea that cognitive variation is variation in performance only, with such variation reflecting differential development in response to experience, just as athletes and dancers may have muscles that are highly developed relative to sedentary people.

The force of the worry about variation in components, rather than just performance, is the claim that there is no unique shared set of psychological capacities across humans. Hence, decomposition of our psychologies into their component parts will return different results for different people. The claim is not that these differences will sort people into sets that track anything like traditional racial groups, or even biological sexes, since the individual variation that the view takes as evidence exists within such groups. The intuitive appeal of the view is clear. There is considerable variation in observed psychology, and considerable genetic variation, within human populations, as well as cultural differences that obtain more often across populations. Given that genes are involved in building brains, and that brains cause behavior, there is something strange about the idea that great genetic variation exists at the bottom and great behavioral variation exists at the top, and great environmental variation all around, but perfect psychological uniformity in the middle.

This is obviously a suggestive argument, but less clear is what it entitles us to conclude. Proponents of cognitive variety often use analogies: David Hull (1989) for example, points out that people belong to different blood groups. But the obvious response from proponents of the monomorphic mind is, yes, but, everyone has blood of some sort, and that is the important point. Nobody has sap or phlegm in their veins, so at the appropriate level of generality, we are, in fact, all the same. The favored analogy of the monomorphizers is that just as our physical organs are shared across individuals, so are our mental organs.

But that comeback is only as persuasive as the claim that there are mental organs, and it remains at the level of analogy. Clearly, this is an argument that can only be settled if we move beyond analogy, by the empirical discovery of what our psychological capacities are. If the mind/brain is decomposable into a number of distinct modularized capacities, which are not all instantiated in everyone, then the organ analogy would survive in a weakened form. Instead of one blueprint that assumed that same cognitive structures exist universally, we could develop a set of related blueprints: some with one system, some with another, at the different sites where cognitive systems diverge.

But as well as functional components, there is likely to be a central system that is responsible for nondemonstrative inference that uses the outputs of specialized systems (Fodor 1983). Even though we all share it, any uniformity might be dwarfed by the immense variety of factors that bias the system's operations in one way rather than another. They could include the particular cultural information that the systems imbibe in development, the different alleles that the developing systems inherit, and a host of other factors. In this case, without some clear grasp of underlying computational structures, we can effectively correlate measurable variation only with causes that are much more distant than proximal mental structures—the whole panoply of developmental resources and accidents that forms the individual. And we lack, as the next chapter shows, a clear grasp of the computational basis of nondemonstrative inference. This has consequences for explanation, since we lack the computational, as opposed to the folk-psychological, theory of how the systems work, and thus the multilevel picture is missing a level: we can relate neuroanatomy to intentional explanation directly, but not via the computational realization of the intentional. This also has implications for the normative status of psychiatry, since in this case the normative assumptions of everyday psychology (or, in some cases, social science) are just taken over by the neurosciences. That dependence of the science on prior normative judgments robs the first stage in the two-stage picture of its claim to be merely positive fact.

Variation in central systems does not mean, however, that we cannot form generalizations across humans that hold for the most part, nor that we cannot investigate properties of our psychology that cause the generalizations to fail for specific reasons in some subsets of cases. At the extreme, however, there could be variation so great that all our generalizations become mostly false in practice. That prospect is the upshot of the third claim about variation that I address.

4.7.4 No Such Thing as Human Nature?

I have looked at two claims about human variation that do not threaten the explanatory project of psychiatry. Human behavior varies along numerous measurable dimensions and humans vary with respect to their psychological mechanisms. Both kinds of variation are consistent with the possibility of a general science of the mind/brain. The third position denies that any general science of human nature is possible. Our intrinsic biological and cognitive properties matter little, on this view, compared to the great diversity of relational properties that humans possess: we are related to our cultures, both present and past, and various accidents of our personal histories. But, not only is the variation in outcomes as a result of developmental resources different, the relationship between intrinsic and extrinsic factors is interactive, not additive. We should not think of cultural factors as extras that are responsible for imposing differential expression onto a shared biological nature. To do that "implies that there is something important that remains constant after this change has occurred, namely biological human nature" (Dupré 2001, 95).

This picture entitles us to expect at most descriptive human behavioral ecology, as particular behavioral patterns in particular cultures are theorized, but it rules out interesting generalizations across the species. This position seems hard to square with the manifest success of folk psychology in describing people from all parts of the world and the manifest similarities (not identity, but similarities) in biological and cognitive systems that do, in fact, affect behavior. It is quite correct that we will not be able to explain everything about people without using culturally specific, and indeed biographically specific, information. But that does not mean that no interesting generalizations can be formed that cross cultures. It is too soon to give up on the hope for a science of human nature, but we must acknowledge that the science will comprise a family of more or less detailed models, with the less empirically rich ones probably having wider application. Making the model work in a particular culture, or on different people in the same culture, means taking a general, large-scale model and adding information derived from more specific, local models. At the limit, we bring together Guze's epidemiological, biological, and clinical strategies, to explain not a diagnosis relative to a population but a particular individual. The relevant parts of that person's biology make up a model of maximum specificity, since it might only apply to that one patient (along, perhaps, with a few patients, perhaps relatives, who share enough properties of interest.) We may expect a general science made up of a family of more or less specific models of people. We might only manage such a patchy science, but it still could include generalizations of very wide scope.

Given where and how the science of human nature is conducted, though, it is natural to worry about the "wild type." The models of different aspects of our psychology that will inform theory-building will be based more closely on people from the rich Western democracies than from other places. Within those societies, but even more urgently outside them, we must not let the basic models either come to be seen as a blueprint to which everyone must conform, nor as a set of methodological expectations that bias the collection of data in other settings. That warning concludes my tour of the medical model as I want to defend it.

4.8 Conclusion

The core of the medical model is the understanding of disease that we see in general medicine, and not a commitment to a particular type or level of explanation. The medical model shares with biomedicine more generally the idea that observable pathologies are causally dependent on discontinuities among underlying systems, usually understood as failures of some part of the system to make the contribution to overall function that a correct descriptive model of that system in its normal state would predict. The fact that psychiatry concerns the organ of cognition introduces some complications: it requires us to make room for explanations that usually do not occur in studying, say, gastroenterology, notably the intentional and computational. But physical systems are describable at several levels of explanation anyway, so the provision of some extra ones should not be a novelty.

These multilevel models are not reducible to molecular models, but their precision and manipulability, especially with respect to related systems in model organisms, earns them a special place in theory construction and testing. Also, if one thinks of fundamental explanations as especially robust and multidimensional, molecular explanations are the best candidates for fundamental explanation in psychiatry, but only for a few conditions. Lastly, human variation means that theories in psychiatry will be based around a family of models of more or less generality and more or less empirical detail, and explanation in a particular case will involve bringing these models together, with the more general meeting the more local.

This chapter, then, is the beginning of the articulation in terms of actual science of the revisionist objectivism I defended in chapters 2 and 3. I supply some more detail in chapter 6, but first I want to look at some important obstacles both to this picture of explanation and to its suitability as the basis of the two-stage picture. They concern the explanation of irrationality.

5 | The Limits of Mechanistic Explanation in Psychiatry

And then a plank in Reason broke, and I fell down, and down.
—Emily Dickinson, *I Felt a Funeral, In My Brain*

5.1 Introduction

The picture developed so far advocates taking the explanatory principles of clinical computational neuroscience, with their patchy, interlevel character, and applying them across the board in psychiatry. I have talked about this conception of psychiatry as cognitive neuroscience within the two-stage project, which tries to separate matters of positive scientific fact from matters of evaluation. The success of the two-stage project depends on the ability of other cognitive neurosciences to understand the mind. This chapter argues that we may not achieve a scientific understanding of the cognitive structures underlying our rational capacities. Although the logic of the medical model remains untouched by this chapter, its foundation will look different. The justificatory role of the two-stage picture cannot be maintained for many psychiatric investigations. We cannot point to a failure of mechanisms as establishing an objective diagnosis unless we have a theory of what the mechanisms normally do. When the mechanisms are the "central" or "executive" systems that underwrite rational capacities, we cannot attain a theory of normal performance just by looking at the systems. We rely on a normative theory of rationality. For the cases in this chapter, judgments of malfunction and judgments of norm-violation coincide. This is reminiscent of the antipsychiatry skepticism I have already mentioned, but there are some important differences. First, the norms are not moral norms, but epistemic or rational ones. Second, it is not a general complaint about psychiatry: its scope depends on how far we can answer some important questions about the organization of the mind, and so in that sense it is an open question whether the two-stage program is feasible. Third, even if psychiatry has largely normative foundations, it does not follow that the relevant norms are ones that are imposed by some sections of society on others, nor that we cannot reach agreement on them. For some conditions widespread agreement is likely, whereas for others disagreement will persist.

The search for the neurocomputational basis of human rationality is what I call the *mechanization project*. It is different from the project of naturalizing reason, which tries to understand the human capacity to constitute and follow norms in a way that respects the fact that we are cognitively limited animals, but is agnostic about mechanisms. The mechanization project asks about the cognitive and neurological basis of our rational capacities. In section 5.3 I go over Fodor's arguments that such a theory is not possible, and ask where that leaves us. We have more chance of success in the naturalization project of understanding reason as an aspect of our animal nature than in the mechanization project of establishing the neurocomputational basis of those animal capacities. In some cases the possibility that rationality will be understood naturalistically, but not mechanistically, raises few worries, but in other cases it raises several serious issues. For instance, agreement that delusions are irrational is very widespread, and theories that try to defend the rationality of delusions are not plausible. But the normative component in the study of delusion is unlikely to cause much scientific trouble; even advocates of the rationality of some delusional beliefs admit that something is wrong with delusional individuals and advocate scientific attempts to discover and reverse the problem. But there are other cases where disagreements about what is instrumentally rational do make a difference to both explanatory theories and conclusions about the status of the condition. One case is addiction, which I discuss in section 5.5. I doubt that even empirical findings could resolve the disagreements about addiction, which are intrinsically normative. Last, in section 5.6, I turn to psychopathy, where both rationality and moral decency are at stake. In this case we have an intermediate situation where possible agreement depending on how some empirical findings, and some philosophical theories, turn out.

5.2 Rationality and Mechanistic Explanation

5.2.1 Rational, Arational, and Irrational
The science of human behavior aims to demarcate rational from irrational behavior and explain both in terms of underlying mechanisms. To consummate the medical model in psychiatry we need both a theory of rationality and a theory of the mechanisms underpinning rational thought and behavior, so that departures from both can be assessed, explained, and treated.

Chapter 4 sketches a version of the reductive program that, if it could be carried out, would provide a theory of the mechanism. But where does the theory of rationality originate?

Now, not everything that a rational person does is itself a rational act: even the best of us are sometimes irrational, and we also do much that is *arational*. Human beings can lift a spoon, infer an object's physical structure from its motion, and correctly pronounce words. None of these are rational capacities in the way that figuring out what to choose or believe based on available opportunities or evidence is rational.

A second type of arational behavior is behavior that is specifically governed by norms. Often, the norms distinguish among ways of behaving that are all rational. For instance, people in different countries, and in different classes in the same country, dine at different times. It is not irrational to eat your evening meal at 8 p.m. rather than 6 p.m. Nor is it irrational to eat it with chopsticks or fingers rather than knives and forks. (I discuss norms and other sources of cultural variation in chapter 7.) When I say that the first component of the positive project aims at demarcating rational behavior, I include behavior that follows norms, but I do not count differences in norms as differences between rational and irrational behaviors: they are different ways of doing the same task. So I continue to think of the first component of psychology, broadly construed, as an enterprise aimed at explaining human behavior in rational terms. A science of psychopathology must explain failures in arational capacities; I am just putting them aside for now.

The overall psychological project, then, conjoins a (naturalistic) rationalization project and a mechanization project, an understanding of what rational people do and an understanding of the computational and anatomical basis of their rational capacities. The theory can fail to apply to a person in two ways: first, the theory could be false. Second, the theory could be true but the person could fail to be rational. Since human judgment is not infallible, people sometimes fall short of what reason requires, and since human judgment is not epiphenomenal, these departures from rationality influence behavior. Sometimes this happens because a normally rational person makes a mistake. But many mental illnesses involve disordered reasoning. The assessment and treatment of these disorders requires a (usually inexplicit) model of what good reasoning is. If rationality is a normative notion, the model must be in part normative. And since it is normative, it undermines the pretensions of the medical model to purely positive science. In this section I discuss some of the conceptual issues involved in the project of naturalizing reason, and ask if a mechanistic understanding of its neurocomputational basis solves any of these problems.

The naturalization project can draw on long-standing experimental and theoretical traditions in the social sciences, but the results that the social sciences generate are compatible with very different stories about underlying mechanisms, and indeed they are often deliberately noncommittal about mechanisms. The social sciences typically employ concepts of instrumental rationality. Instrumental rationality is concerned, above all, with the most efficient ways of realizing goals in the light of beliefs about the world. This includes picking the right action relative to beliefs and desires. It also includes many constraints on beliefs and preferences, such as the demand that preferences be transitive, and some consistency requirement for beliefs. What a naturalistic theory must do is to explain how good reasoning is possible, where standards of goodness are relative to animals like us, but only to some of us, as Thomas Schelling notes when he draws attention to the "large population disqualified by infancy, senility or incompetence from being represented in our theory of the consumer" (1984, 59). What counts as rational behavior depends on and is sustained by our beliefs about what reasonable people do. Hence, the mentally ill are called mentally ill, sometimes, because of their failure to exhibit the performance required of instrumentally rational subjects. Philosophers and social scientists differ over these requirements.

A long philosophical tradition regards humans as inherently rational, though this view has never denied that lapses from rationality occur. From this perspective, chronic irrationality is a mark of a failure to function properly as humans should, and hence a sure sign of mental illness (Megone 1998). In the last few decades, though, psychological research has identified a wide range of systematic violations of familiar canons of rationality, which has led to the claim that average humans lack the capacity to engage in properly rational thinking and instead exploit a variety of cognitive shortcuts that lead us to into logical errors, mistakes about probability, and stereotype-driven thinking, all of them accompanied by a worrisome degree of overconfidence (Gilovich 1991; Kahneman, Tversky, and Slovic 1982; Nisbett and Ross 1980; Piatelli-Palmarini 1994).[1]

One response to these findings is that it employs standards of rationality that are too demanding for humans. The human visual system is never criticized for its failure to be as good as that of a bird of prey. So we should not criticize human beings for failing to, for example, think logically, or to conform in their reasoning to the canons of probability theory. A naturalistic theory of reason must make allowances for human fallibility, but it cannot

1. For some skepticism about aspects of this tradition, and the claim that humans may be closer to economic rationality than it suggests, see Smith 1991.

just describe how people reason, because performance does vary greatly across people, and if we count them all as normal then we lose any right to make distinctions among them, and everything counts as rational.

For instance, rather than insisting that a rational agent have no contradictory beliefs we might admit that rational humans do have some contradictory beliefs. In that case we still could impose a minimal requirement such as: if one becomes aware of contradictions among one's beliefs, one ought to resolve them if it is not too costly to devote resources to their resolution (Cherniak 1986; Harman 1995).

The extent to which normal people depart from standards of good instrumental reasoning, and what those standards are, is disputed. But good evidence shows what normal people are like when they reason. By "normal people" I mean those who are antecedently agreed not to be pathological. Clearly, then, appealing to this evidence will not settle disputes about who is normal, since only the behavior of normal people counts as data. In some cases the assumptions underlying our picture of instrumental rationality are likely to be shared widely enough to make this normative foundation surmountable as a practical matter when it comes to explanation and assessment of mental illness. We can agree that schizophrenia is a mental illness even if our agreement rests on shared norms; the normative does not have to be contested. But it many cases disagreement over a condition will reflect deeper disagreement over both instrumental and substantive norms, leaving us with debates about rationality even with minimal assumptions in place.

Instrumental rationality draws on many epistemic capacities, such as updating beliefs and plans in the light of new evidence and otherwise adapting to feedback from the environment. This is where the theory of instrumental rationality makes contact with the psychological theory of central systems, since general belief fixation and goal-directed thought are what the central systems are all about, as opposed to specific mental activities such as language comprehension. Furthermore, we have compelling reason to believe that the brain areas responsible for central systems are the frontal lobes (Goldberg 2001). That, of course, is a considerable achievement. But the absence of a computational account in the middle means that we are lacking a level of explanation. This chapter is about what that lack means. Before we move on to the connections with cognitive science, we need to look at a much broader conception of rationality that is continuous with ethics and plays an important role in our thinking about mental illness.

5.2.2 Beyond Instrumental Rationality
Instrumental rationality is silent about the worth of goals. It just assesses the means to their realization. But we might think that some goals are irrational.

Rational choice theory, which is part of my paradigm of instrumental rationality, often is faulted on the grounds that it denies that preferences can be given rational evaluation. Amartya Sen (2003, 39–40) has an especially wince-inducing instance of this criticism. Imagine someone who is cutting his toes off with a blunt knife. The self-mutilator assures us that severing his toes is what he wants to do. If we stick with a purely instrumental concept of rationality, all the advice we can offer him is that he should use a sharper knife. And perhaps even a perfectly rational goal could be faulted on broader grounds, for example if it crowds out other goals. Having only one end in life, insisted Rawls, "strikes us as irrational, or more likely as mad" (1971, 554).

It is no wonder, then, that Sen concludes that instrumental efficiency is not all there is to rationality, and that we must insist that preferences or goals can be subject to reasoned scrutiny. This does not mean that we must think of ends as inevitably given via deliberation. Our goals often seem to come to us unbidden. But even if generally we do not come up with ends via deliberation, they are still things we can deliberate about, once we have them. This richer conception of rationality is continuous with the general assessment of human life as good or bad, flourishing or pitiable. Assuming instrumental rationality, as is done by folk psychology and the mathematized folk psychology of the social sciences, may enable us to examine human behavior systematically and relate behavioral phenomena to neuroanatomy. If that is possible, we will have a theory of human behavior, including social behavior, which will be of immense help in understanding mental illness. However, there are debates about rationality touching many mental illnesses that the theory of instrumental rationality will not resolve, since in these cases a richer conception of rationality often appears to be at stake. This richer conception is a matter of how we evaluate someone's life, rather than merely instrumental efficiency. To understand human behavior fully, including its abnormalities, we need more than just the instrumental conception of reason.

Instrumental rationality, unlike theories of human flourishing, can be studied in a way that gives us some hope of integrating it into the explanatory structure of the medical model, by relating cognitive phenomena to neuroanatomical mechanisms. There is some hope for agreement about instrumental reasoning, if only because of its importance to folk psychology. But there is very little hope for agreement, let alone agreement in scientifically tractable terms, about the good life.

That has immediate consequences for the idea of a positive psychiatry. Two consequences in particular are interesting. The first consequence of the normative nature of rationality for psychiatry is that we cannot *always* divorce questions about function and malfunction from questions about the status of malfunction. Part of the motivation of the picture of psychiatry developed earlier was the desire for a realm of positive fact within which science operates, and a separate but equal realm of moral and social evaluation within which conceptions of human flourishing have a home. If questions about the characterization and explanation of mental illness are strongly normative in this sense of being tied to questions about human flourishing, the hope for a realm of positive fact has vanished.

This difficulty is evaded by the tendency to search for theories that explain the behavior of people whom the theorist already has decided are mentally ill, often uncontroversially. Schizophrenics, for example, are paradigmatically mentally ill, and so the attempt to uncover their defective central systems can seem like an unobjectionable application of the thinking behind the two-stage picture: we look for the mechanistic explanation of the observable pathologies. But why should the mechanistic explanation by itself show that someone has an unhealthy mind? In fact, people work within existing psychiatric categories, supplemented by some more or less explicit theory of rationality. They investigate people who fall within the categories or who depart from the theory, then find out what their brains look like. But merely finding a neurological difference between patients and controls does not establish that the patients are ill unless we have some previous reason for calling them patients.

If psychiatry were a mature and successful science this issue could perhaps be finessed, on the grounds that practically speaking things are going fine. But psychiatry is not especially successful, and in fact this issue speaks to its foundations. Looking at the neurocomputational underpinnings of addiction will help us to understand it. But investigating those neurocomputational underpinnings does not show, objectively, that addiction is a disease. To do that, it is necessary to show not just that addicts have different brains than non-addicts, but that the differences warrant our medicalizing addicts. The medical model as it stands tends to take existing diagnostic schemes for granted, which means that it can show us what is going on in the brains of members of a certain class. It cannot, without a broader conceptual arsenal, answer the question of whether some behavior is pathological or just dumb and squalid.

If we are looking for the underpinnings of behavior that we already have decided is evidence of an unhealthy mind, then the project is based on normative foundations supplied elsewhere. It is perhaps feasible, although the difficulties of establishing a comprehensive theory of central systems are daunting. But if the idea is that figuring out how central systems work will enable us to have an objective assessment of mental health, one that dispenses with vague subjective or conceptual assessments, then the project stands in need of a justification it has not been given. That is, it needs input from the theory of rationality.

The theory of rationality allows psychiatric explanation to be informed by an awareness of how goal-related actions work, and their sensitivity to, for example, the contents of beliefs and desires. Despite what some people seem to think (Turner 2003), the project of understanding reason is not in competition with scientific, causal, understanding. It is a form of causal understanding. The whole point of modern cognitive psychology is that the representations that are semantically evaluated and rationally related in folk psychology also are causally related. The coincidence of the causal and truth-functional is what a computational psychology is designed to capture, and making room for the computational allows us to address the psychiatric importance of the social within the overall scientific project. Meanings affect behavior only in so far as they are neurally represented.[2]

The next section takes up the relation of the theory of rationality to the theory of neurocomputational systems, in particular, "central systems" underlying belief fixation and revision.

5.2.3 Brain Science and Social Science

The story so far has suggested that only some interactions of neurally represented meanings count as rational, which seems correct, and that which computations count as rational is a matter for a normative theory, which threatens the foundations of psychiatry as envisioned by the two-stage

---

2. Bolton and Hill (1996, 74ff) noted this in their pioneering discussion of representationalism in psychiatry, but they fretted that it introduced a form of "intentional" causality that could not be understood as a normal case of causation in the sciences. This may reflect a more general worry about the lack of general laws in psychiatry (explicit in Turner 2003). As I have said, I do not think we should expect general laws rather than generalizations of limited scope, typically based on an idealized understanding of mechanisms. This is not a problem, since (1) general laws usually are unobtainable in the special sciences, and (2) explanation in the sciences usually is not a matter of subsumption under laws. Bolton and Hill also worried about understanding the particular point of view of a patient, and especially what is distinctively meaningful for that person. That is a separate issue, and it falls under the clinical perspective, rather than the explanatory one. I say more about the relation between clinical work and research in chapter 10.

picture. Much psychology offers accounts of what occurs when people depart from rationality in terms drawn from broadly folk psychological categories. These are *intentional* or *personal-level* explanations. An example is cognitive dissonance reduction (Festinger 1957): if I want something but learn that I have no chance of getting it I may cease to regard it as desirable. Elster calls this the sour grapes mechanism and offers this explanation of it: people "tend to match their beliefs and desires to their means so as to avoid the intolerable situation of wanting what they cannot have or finding inescapable circumstances undesirable" (1993, 54). Others will not see this as an explanation at all, but just as a description of behavior for which we should find an explanatory reductive mechanism. We often cite interacting beliefs and desires in this way to characterize irrationality, especially motivated forms of irrationality such as weakness of the will or self-deception. If this is an explanation, it is not a computational systems explanation—there are no grounds here for positing a "sour grapes" module. Similarly, as I claim in section 5.3, to explain delusion in terms of abnormalities of belief formation is to characterize a person's mind, not to cite an underlying mechanism that explains why that mind has that property.

Therefore, we also need sub-personal or computational explanations. The subpersonal level cites a mechanism in terms removed from, rather than continuous with, those of commonsense psychology. We look for a cognitive part: a brain area with a computational role. The hope is that we can generalize throughout psychology the strategy of decomposition and localization that I mention in chapter 4.

The difference between the naturalization and mechanization projects corresponds to these two levels, which also correspond to the philosophical distinction between the personal and the subpersonal levels. Rational choice theory, for example, is a personal level theory, since it takes the basic categories of folk psychology and translates them, with great and increasing mathematical sophistication, into a predictive and explanatory system that dominates the rationalization project in the contemporary social sciences. Increasingly, the social sciences and the evolutionary study of cognition offer promise of understanding our rational capacities as aspects of human nature, which is why the project of naturalizing reason is tractable. But naturalization projects in the social sciences come up with accounts that are emphatically not mechanistic. They are not theories of subpersonal mechanisms; they are neutral with respect to underlying systems in the brain. The discovery of those systems and their properties is the job of cognitive neuroscience. That is not to say that the social science models cannot guide brain science, nor that results from neuroscience cannot constrain personal level theories.

For the mechanization project to work we need to discover brain systems that stand to rational behavior as the dorsal and ventral streams stand to visual object recognition. And if it is to be perfectly objective functional analysis, it must be done in ways that do not import normative considerations from the personal level into theories of computational neuroanatomy. The hope is that computational description will give us the algorithms of reason just as they have discovered the algorithms underlying the construction of neurological models of physical objects.

This might be feasible given a prior agreement over what counts as rational (at least in some cases, if not in general). If we agree that schizophrenic thinking is pathological, then a theory of central systems will let us find the neurocomputational abnormalities that mark that problem. We might think, however, that if we *could* carry out a satisfactory decomposition of "central" or "executive" systems, we could use computational and neurological information to obviate the need for normative assumptions about rationality and derive a theory of abnormal reasoning by establishing the proper functioning of executive systems and recording departures from that proper functioning. Then we could ask whether those malfunctions are or are not pathological. This approach treats our rational capacities in the way that I treat our perceptual capacities in chapter 4, and it conforms to the two-stage picture's separation of positive fact and evaluative inquiry.

We might not be able to do this because models of normal performance for central systems, attained via decomposition and localization, may well not be available, as I suggest in section 5.3. Also, even if we do develop a neurocomputational account of central reasoning and belief fixation, it still will not tell us what is rational. As I note toward the end of chapter 4, any such theory will have to make room for individual variation. So a detailed mechanistic/computational account of central reasoning and belief fixation in Gordon will differ from the account of central systems that applies to Thomas. The hope is that since both Gordon and Thomas are normal, we can compare their frontal lobes with those of a psychopath, schizophrenic, or addict, and we will have a deeper understanding of just why they behave as they do. To make this move we need to know that Gordon and Thomas have central reasoning systems that are healthy. In order to know that we need to know how central systems are supposed to function. It is unclear how a detailed mechanistic theory of central systems, even if we obtain it, can resolve that question.

I can see only one way to rebut this second objection, and it will not always work. The rebuttal I have in mind takes the decomposition of central systems (assuming it can be had), adds the lesion method, and tries to align computational abnormalities with neurological ones. This method probably guides much research in cognitive neuroscience, though it seldom is avowed explicitly. It exploits the idea that physical departures from customary brain development and function must mark a computational breakdown, rather than just a computational difference across normal subjects. As I say in chapter 4, levels of explanation in the brain are not as distinct as traditional pictures of cognitive science have suggested, so if we think of neurocomputational theories, rather than just computational ones, the task may be a bit easier. A picture of normal information processing involves both understanding the relevant algorithms and understanding the nature of the tissue that performs the computations. If some computations are the result of physically abnormal brain structures, then the case for calling them irrational may seem stronger, since we now have a presumption that the organism is not supposed to be that way—a presumption that is independent of the norms of reason, derived instead from nature's own norms, as chapter 4 envisages.

As I mentioned earlier, this can seem straightforward in some areas of cognitive neuroscience. Few people deny that the aphasias or agnosias are pathologies, since in those cases we have a model of normal mechanisms that seem to apply widely, and in which individual variation is less dramatic. We also have in these cases evidence of lesions that supply a clear case of a cause intervening from outside the system. Defending the claim that something is wrong with someone's mind is easier when we can point to a lesion in the brain. When critics like Szasz contrast psychiatry with neuroscience or general medicine, they are contrasting psychological abnormalities resting on uncontroversial physical causes with other cases in which psychological abnormalities occur in a system that has not been damaged by extrinsic agents but has developed in some way that some (or even many) people think is abnormal, and who are they to say so?

The Szaszian disease model is oversimplified even for physical medicine, but it points to two important components of our thinking about disease: the decomposition of the system into components, and the presence of a causal agent. These correspond to the concepts of reduction and fundamental explanation that I discuss in chapter 4. If we return to the case of the visual system, we can see that not only is the reduction of the system into its neurocomputational components possible, but that in many cases we can identify a lesion which has many of the properties of a fundamental explanation. Somebody who is perfectly normal can be psychologically transformed by a lesion, which suggests that the lesion, like a gene in a single-major-locus condition, is doing most of the work. The 1848 case of Phineas Gage is a famous example (Macmillan 2000). Gage was a railwayman in Vermont whose whole personality changed after a 13-pound tamping iron was blown through his left cheek and out of the top of his head. Although he had been efficient and respectable, he became profane, aggressive, and incapable of following a plan of action. In the case of Gage, the extent of the physical damage makes it easy to sustain the thought that his changes in personality were pathological. Since normal brains are not supposed to have gaping holes in them, the difference between pre- and post-accident Gage looks objectively grounded.

The cognitive neuroscience of central (or "executive" systems) is full of similar cases, in which some psychological capacity or trait is wrecked or drastically changed by physical damage. The problem for classically psychiatric conditions is that we typically lack an obvious physical cause that can play the fundamental explanatory role. Most neurological abnormalities are more subtle and systemic than a simple hole in the brain. They are often abnormal levels of a neurotransmitter, or of a type of brain tissue, or some other biological variable, rather than a gross lesion. And the effects are often much less dramatic than the transformation that poor Gage underwent. Even if we can find a physical difference in brains, its mere presence does not resolve the conceptual issues. If alcoholics have abnormal brains, that alone does not imply that they are ill, since taxi drivers have abnormal brains too (Maguire et al. 2000). What is needed is an account of the brain abnormality and an account of the reasoning in alcoholics that jointly suggest a defective cognitive part.

If this strategy is to work then, it must be possible both to understand central systems as cognitive parts via decomposition and localization, and to find physical markers as compelling as lesions. Currently, we have no decomposition into cognitive parts, and although we can identify the difference that some lesions make to normal rational thinking, we are often without any convincing physical cause to cite. The search for physical markers, may, as many psychiatrists seem to think, become easier as we develop a greater appreciation of the brain's workings and are able to observe it in greater detail: perhaps some fundamental causes exist in the brain at currently unobservable levels.

My own guess is that these developments will have only a limited effect, and that we will continue to see psychiatric conditions as the result of a number of interacting causes, none of which are very strong. My attention here, since I deal with the molecular in chapter 4, is devoted to the cognitive. What chance do we have of a satisfactory mechanistic theory of cognition? To get it, we need a clear understanding of the components of central systems, understood as aspects of the fixation and revision of belief.

Computational understanding was essential to the theory of cognitive parts in the multilevel cognitive neuroscience of chapters 2–4: it lets us see what brain structures do. Although rationality might be naturalised, I doubt that it can be mechanized. If mechanization is feasible, it should be possible to explain the signs and symptoms of psychopathology as failures of normal algorithms, and to display the causal connection between the distinct signs and symptoms and the suboptimal cognitive parts sustaining them.

As recent work on delusion shows (section 5.4), some moves are afoot to do just that. This work tries to explain delusions via theories about central systems and abnormal or biased belief fixation. If the explanation arrives, as I mentioned earlier, we might not only have a computational level of explanation, but a principled level at which to understand thought as a computational phenomenon. Furthermore, it would accomplish an important job of reductive analysis in the life sciences, by making explicit as many causal relations as possible, thereby enabling us to maximize our opportunities for interventions and manipulations.

However, our best current science does not suggest that we can make the end run around normativity via computational theories of executive function, or even have good computational theories of rationality at all. Existing theories do not reflect any independent understanding of the computational basis of executive cognition. They just reify aspects of broadly philosophical, folk-psychological theories of rationality. These theories assume that thought is representational, on the grounds that representational theories surpass their competitors in explaining how thoughts can have both causal and semantic properties. But a general appeal to the representational nature of thought is not enough for a mechanistic explanation of cognition, which is premised on the provision of algorithms that realize cognitive capacities. The representational elements of psychiatric explanation often will be supplied by folk psychology.

In the cases of delusions, addictions and psychopathy—but not only in those cases—the normative disputes about whether the conditions are pathological or not probably will continue, as will norm-laden characterizations of the conditions as we build theories to explain them. The persistence of normative considerations reflects the fact that we lack an understanding of planning and belief fixation. A failure to explain central systems of belief fixation and revision is not inevitable, but I note the difficulties and conclude that many questions about mental illness are tied up with hard empirical issues in the study of rational inference and cognitive architecture.

5.3   Explaining Central Systems

5.3.1   Center and Periphery

Current theories of delusion in cognitive neuropsychology run together personal and subpersonal explanations to produce the illusion of a computational theory when all we have are folk-psychological hypotheses. The source of the trouble is the difficulty we have in explaining central systems.

Bermudez (1998, 460) distinguishes psychiatric disorders from neuropsychological ones on the grounds that neuropsychological conditions are disorders of peripheral modules and psychiatric ones are disorders of central systems. Of course, I think there should be no distinction between the disciplines, but Bermudez is also wrong about current practice, since neuropsychologists deal with delusions or problems in "executive function," which are clearly breakdowns in central processing. Conversely, psychiatrists deal with hallucinations and motor problems that implicate peripheral systems.

However, the distinction between modular and central processes does suggest a useful way of thinking about explanation and objectivity in the two-stage picture. Bermudez employs a concept of modularity that has become ubiquitous in the cognitive sciences since the publication in 1983 of Jerry Fodor's enormously influential book, *The Modularity of Mind*. Fodorian modules must have most or all of the following features to a significant degree:

(1) informationally encapsulated (i.e., having only a small proprietary database, not the complete set of beliefs available to the mind; the flow of information goes from the periphery to the center and not in the other direction)
(2) mandatory
(3) fast
(4) shallow
(5) neurally localized
(6) susceptible to characteristic breakdown
(7) domain specific

Other thinkers have relaxed Fodor's requirements. (Segal [1996] is a good taxonomy of conceptions of modularity). But theorists of modularity typically retain and seek to generalize the idea that positing psychological systems with a dedicated function and a proprietary database is the way to break the mind down into components that can be studied. Fodor, however, argued that this strategy could have only limited application.

The mind, Fodor argues, has three sorts of structures. As well as modular input and output systems it contains, first, transducers that transform sensory stimuli into the low-level representations that are the province of input systems and, second, central systems that do the thinking: these take the representations that form the output of modular input systems and employ them in domain-general ways as part of a network of concepts. Fodor insisted on domain generality in central systems because, in principle, any information could be relevant to any episode of belief revision. Central systems need to be almost infinitely creative and flexible (because thought is). Fodor famously regarded central systems as nonmodular and argued that "the limits of modularity are also likely to be the limits of what we are going to be able to understand about the mind" (1983, 126). In the last twenty years many thinkers have argued for the existence of modules that are not just peripheral in Fodor's sense (Fodor himself [1995] seems to accept that theory of mind may be modular), but the general problem of mechanizing the place in the mind where it all comes together and is evaluated has not gone away. Patients with frontal-lobe lesions suffer from a failure to integrate different bodies of know-how into purposeful activity, but our knowledge of where the integration happens has not yet told us how it happens.

Now, Fodor claims that we cannot understand central processes because we cannot study them computationally. There is an obvious sense in which this is too pessimistic. Several social and cognitive sciences study the work that central systems are presumed to do, including learning, decision making, categorization, inference, and so on. But the pessimism is sound in so far as we remain ignorant about whether these systems exist as discrete computers, let alone their inner workings. Without a computational theory we can naturalize reason, but we cannot mechanize it.

Since I make such a big deal of it, I should go over the difficulties Fodor saw in central systems in a little more detail before I return to the psychiatry. To begin with, I rebut two objections: first, Fodor's claims about belief acquisition and revision, which rest on an analogy with scientific reasoning, are not undermined by alternative (and more plausible) accounts of scientific reasoning. Second, although Fodor presented his pessimistic conclusions only in the context of classical, modular, and heavily nativist accounts of cognition, they also pose a problem for alternative accounts. The second rebuttal is not supposed to be a refutation, since I am not ruling out the possibility of a solution to the problem of central systems. I merely want to point out that it is everyone's problem. Fodor's pessimism, if it is well founded, makes it impossible to apply computational theories to disorders of central systems and leaves psychiatry unable to carry out the multilevel version of the medical model in many areas of mental illness.

5.3.2 Science and Thinking

Fodor assigns to belief fixation two properties that trouble computational theories. First, belief fixation is *Quineian*. It is sensitive to features of the total belief system such as conservatism or simplicity—that is, global features of the belief system as a whole. Fodor argues that properties like simplicity, context-dependence, and epistemic conservatism (or bandwagon-joining) cannot be reduced to intrinsic, syntactic properties of representations: they are relational properties. But computation in the brain, insists Fodor, is sensitive only to the syntax of mental representations (upon which the logical forms of thoughts supervene), and syntax is an intrinsic property of mental representations.

The second property of belief fixation that causes difficulties for computational accounts is *isotropy*: anything you believe could be relevant to anything else you believe. Ken Taylor (2000) has argued that isotropy reflects the broader plasticity of cognition, which lets a mind reorganize its representations by revising the inferential links between them. Taylor's idea of cognitive plasticity encompasses all the propositional attitudes, not just belief. Unlike Fodor's more austere view, which is restricted to the epistemic, Taylor's more general account allows for the importance of non-epistemic factors in the reorganization of cognition. But the upshot is still that a purely modular mind could not be plastic, because a mind containing nothing but modules cannot reorganize its component representations. Radical reorganization is ruled out because the range of thoughts a mind can bring to bear on a topic is restricted: the information contained in a module can be devoted only to that module's domain.

Fodor contends that theory construction and revision in science is Quineian and isotropic and that it also is a good model of belief fixation in general. So one way to block the pessimistic conclusion that belief fixation is computationally intractable is to dispute the analogy between theory formation in science and individual cognition. Peter Carruthers (Carruthers 2003) has recently done this (for a briefer version of the same complaint, see Pinker [2005]). Carruthers's objection to Fodor is unsuccessful, but Fodor's analogizing belief fixation to science is somewhat misleading, but Fodor's discussion of belief formation is significant for theories of the mind in addition to the neorationalist accounts. Perhaps despite appearances, all this is important for psychiatry.

Carruthers denies that science is a good model of everyday belief fixation on the grounds that science is social and institutionalized and normal thinking is not. But, although science is a social enterprise in which scientists form beliefs in concert with others, the same is true of commonplace belief fixation. It depends on conversation with others and the acquisition of knowledge from diverse sources. A geneticist may gain much of her knowledge from trusting journal articles and talking to other scientists, but we all form beliefs by reading things and talking to other people. Most beliefs are formed or revised socially. Carruthers is right to scorn the image of the lone scientist, but he seems to have forgotten that the lone gossip is just as improbable.

Carruthers also stresses that the social and cooperative nature of science enables different people to bring information to bear on the same problem. Carruthers claims that science is designed to make this possible by allowing individuals to bring specialized knowledge to bear on shared problems. But this objection is just as weak. Fodor's point is that anything you believe may be relevant to anything else you believe. To refute Fodor on this score by appealing to shared inquiries Carruthers must show not just that different scientists can bring different facts to bear on the same problem, but that an individual scientist cannot. If Mary the astronomer can understand why, to take Fodor's example, a result in botany might be relevant to a problem in astronomy, then Fodor's point goes through, because the significance of the one fact for the other is apparent to Mary's mind. It is quite irrelevant whether Mary herself or some bunch of botanists discovered the botanical fact. Mary only has to see its significance to astronomy. And this must happen, since if nobody sees the relevance of conjoined disparate facts then their mutual significance will go unrecognized, and science could make no progress.

So bringing disparate findings to bear on one problem may require social structures that generate the different findings, but it also requires individuals to spot the mutual significance of the facts. In other words, the social structure of science exploits the isotropy of individual belief. It does not correct for it; it relies on it.

Fodor's problem remains untouched by appeal to the social nature of science. He identified the problem of giving a mechanistic account of rationality considering that rationality is Quineian and isotropic. Some theorists seem to assume that the problem only afflicts views of the mind that share Fodor's taste for limited, peripheral modularity, nativism, and classical processing (which regards the mind as a traditional digital computer). In fact, the issue remains if we adopt a different view of cognitive architecture, if we become empiricists, and if we adopt a different view of computation. The difficulty arises for any view of cognition and therefore is likely to afflict all attempts to explain psychopathology by exploiting a mechanistic theory of normal psychology.

5.3.3 Other Architectures

The problems Fodor raised for central systems afflict even theories of the mind that do not believe in central systems and do not have a classical view of computational cognition. An example of a nonclassical view of cognition that does not escape the difficulties is Hurley's (2001) shift from (as she puts it) vertical to horizontal modularity. Hurley discusses approaches to cognition on which the mind contains "layer upon layer of content-specific networks" (2001, 8). Fodorian minds contain perceptual input systems that pass information on to a big central reasoner, the products of which are passed on in turn to output systems that initiate action. Hurleyan minds contain a collection of dynamic systems, each of which is a complete "input-output-input loop" (Hurley 2001, 8) with a particular job to do. Hurley suggests that the "what" and "where" systems (chapter 4), might be examples of such loops, as well as systems that recognize food and predators. The horizontal modules in Hurleyan minds each involve coupled, rather than distinct, input and output processes and a neural net tuned to specific features of the environment. Advocates of such approaches to the mind are at great pains to emphasize their differences with advocates of vertical modularity. But from the point of view of understanding rationality, the solution Hurley prefers actually is very close to that of some people who are firmly in the anticonnectionist modular camp. Both strategies argue that evolution has given our mind many different specialized systems and that the joint operation of all these specialists produces a general intellect. So Hurley says that evolution and development select for dynamic networks that incorporate aspects of the environment to do specific jobs and then also selects for the complex relations between the networks that produces rational responses. And the classically oriented Cosmides and Tooby (1997, 81) claim that evolution designed our minds to have myriad domain-specific modules attuned to features of the ancestral environment that are then "functionally integrated" to produce behavior. So despite their differences, these schools of thought agree that something must connect and integrate all the task-specific systems they attribute to the mind. Something functionally integrates the modules in Cosmides and Tooby's story or, in Hurley's story, ensures that "some layers get turned on and others get turned off, in a totality of ways that count as rational overall in the circumstances" (Hurley 2001, 10). In both cases the answer is not supposed to be a central system, but in neither case are we given any details beyond the assertion that the specialist systems are related in appropriate ways.

Theorists like Hurley want functional integration without a central control system. Other theorists do not believe in functional integration at all and hope that many local heuristics, cued automatically, can do the job. A heuristic is a quick-and-dirty rule of thumb that can be used to circumvent exhaustive searching throughout the mind for a correct answer. A heuristic, like a Fodor module, is triggered by a specific and restricted input. But since heuristics only operate on particular information, the right information must somehow come their way for them to be triggered—for the rule of thumb to work, one must recognize the situation as one in which it applies. As Fodor points out, for that to happen the relevant cues must be discriminated in "the output of your sensorium," which means that "the distinctions your sensorium makes include *whatever distinctions your mind can make at all*" (2000, 74). Fodor calls that the "Empiricist Principle" that there is "nothing in the mind that is not first in the senses" (Fodor 2000, 74). That is, the input systems need to sort information and send it on to the right heuristics to make use of it. Whether it is the input systems that have to do this is not the point. Something has to ensure that the heuristics get the right information, and whatever that something is, it plays the role of a central system.

Rather than go on and on about the different proposals that have been made for evading Fodor's problem, let me just summarize matters briefly and then say how it all relates to psychiatry.[3] Fodor's problem about mechaniz-

---

3. But I should say something about connectionism, if only for enthusiasts. Connectionists model cognition by using networks of units that are allegedly analogous to neurons, and weights that measure the strength of connections between neurons. These weights can be changed by learning algorithms. Fodor dismisses connectionism (2000, 49) on the grounds that the identity of a unit in a network is defined by its fixed position relative to the rest of the network. This objection does not apply to more recent connectionist proposals, which use activation-based processing (O'Reilly and Munakata 2000, 380–381). Activation-based processing maintains active representations in memory and compares them along dimensions of similarity with features of the input. O'Reilly and Munakata (2000) advertise this as permitting cognitive flexibility: by simply switching among active representations, for example, new categorization rules can be applied. But, as they note, the control of active representations requires systems that go beyond the trial and error search that they themselves propose as a simulation of the dopamine system (O'Reilly and Munakata, 382). That is, they need a central system.

ing Quinean and isotropic rationality is not rebutted by an appeal to the social nature of science, nor is it solved by any current brave new worlds in cognitive science. Until we obtain a theory that goes beyond mere assertion that cognitive capacities can be functionally integrated and shows how they can be, we are stuck.

5.3.4 Does It Matter?

Two consequences of our haplessness in the face of central systems matter for psychiatry. The first is that we do not have the multilevel picture that, as I suggest in chapter 4, we aim for in neuropsychology. Some levels cannot be filled in, which makes the causal story gappy. Processes at the computational level remain unarticulated in the theory. This gappiness also restricts the number and nature of places in the system where we can intervene to change it. The second consequence is that our ability to explain human action remains shot through with normative considerations. Psychiatry cannot look at a model of the normal mind and read off the pathologies by seeing where other minds depart from normal.

I said that if we did have a good neurocomputational theory of central systems we might go some way to solving the problem of saying why some structures are healthy and some are not. The same is true in nosology. If we could explain psychiatric conditions in terms of malfunctioning psychological systems then we could classify them objectively in terms of underlying processes, just as medicine does with physical disease, rather than relying on behavioral and other surface signs and symptoms whose interpretation is permeated by normative assumptions. However, that relies on linking up intentional-level explanations and cognitive ones. We may not be able to do that, in which case the best we can hope for is a naturalistic theory of reason and a description of the biological processes that go on in the brains of reasoners, but not a computational story to link them.

I do not say that the mechanistic program of the cognitive neurosciences cannot be completed, merely that we cannot foresee its completion, and that the fate of both the two-stage picture and the explanatory pretensions of cognitive neuroscience are bound up with it. However, the absence of some mechanistic explanations of irrational behavior may be, in the context of psychopathology, unproblematic. Delusions cannot be characterized in terms of deviant computational structures, but we can hope to understand them via personal-level, naturalistic theses about rationality and to study their organic basis. There are other conditions where we will not agree about the personal level, though, and they will remain conceptual battlefields.

5.4 Explaining Delusions

5.4.1 Introduction

Attempts to explain delusions involve both assumptions about rationality and assumptions about the cognitive mechanisms of belief formation. Some theories attempt to make delusional subjects broadly instrumentally rational, though not rational on the thicker conception, but these theories fail. Other theories explain the irrationality of delusions in term of abnormal processes of belief formation. Without a satisfactory understanding of central systems, such theories also face grave problems as explanatory theories, even if, as a matter of how we evaluate delusional individuals, we can expect considerable agreement on subjects' irrationality.

Recall from chapter 4 the explanation of the "what" and "where" systems in terms of function and underlying brain anatomy. Although we do not yet have clear explanations of all the disorders that might afflict the visual system, we do have a good and fruitful model of normal function, and we know what systems to look at to explain breakdowns. There is a substantial difference between our ability to explain pathologies of visual systems and our understanding of failures of rationality, of which delusions are a prime case. With delusional states, and psychosis more generally, the challenge is to explain what is and what is not rationally impaired in terms of underlying mechanisms where the mechanisms are "central systems"; the systems responsible for our capacities to form and revise beliefs based on input from the systems that provide us with the information our beliefs require. The task is complicated by the fact that we usually cannot just say that in these cases the central systems have collapsed. We must explain a pattern of deficits and preservations among rational capacities, which requires a specification of failure in some respects and not others, and not just general collapse.

5.4.2 Are Delusions Acquired Rationally?

Although Maher (1988) has argued that delusions are often completely rational responses to anomalous experiences, conventional wisdom distinguishes bias and deficit models of the central problems afflicting central systems. Bias models hold that delusions arise from the interaction of abnormal experiences with a cognitive style that falls within normal human variation, not one that caused by an underlying deficit in belief-fixing machinery. Deficit models of delusions (Langdon and Coltheart 2000) argue that no reasoning style can be as immune to counterevidence as delusions seem to be and so there must be a *deficit* in reasoning implicated in their etiology, as well as abnormal experiences and biases in reasoning. A mixed model could incorporate elements of both bias and deficit models.

I want to establish that the familiar view of delusions as irrational states is in fact correct, then assess the difficulties involved in explaining them in terms of mechanisms. So I try in a moment to discredit bias models, which aim to make delusions look broadly rational. The best way to introduce bias models, though, is by pointing to a difficulty that faces deficit models and serves as a motivation (though not the only one) for bias models. So we begin with a difficulty for deficit models. The difficulty is that if a delusion marks a deficit in normal thought, it is not one in which all rational capacities evaporate. For example, people with Cotard's Delusion say that they are dead. One Cotard patient, WI, when taken to South Africa by his mother formed the belief that he was in hell (Young and Leafhead 1996). Gerrans (2000, 112) explains the Cotard Delusion in terms of "a seriously distorted reasoning process." That is, a deficit. But note that WI's reasoning was, in one sense, impeccable: he was dead and he had arrived somewhere warm, so he drew the obvious conclusion. Gerrans's point is a narrower one: he argues that Cotard subjects lack a particular cognitive capacity. They cannot recognize themselves as the bearers of their own experiences because of deficits in the normal signaling of the processes taking place within their bodies. But in fact even these processes are not wholly impaired. Cotard subjects often feel hunger, the beating of their hearts, and the pressure of full bladders (Young and Leafhead 1996). They just insist that these sensations cannot be evidence that they are alive, because they are not alive. Given this absence of rational impairment across the board, it is hard to argue for a total collapse of central systems of belief fixation in delusional patients, and it is hard to specify the deficit precisely. Perhaps, then, we should adopt a bias theory, which argues for a cognitive style that interacts with an anomaly to produce a delusion.

So much for the motivation for adopting a bias model. The bias-driven explanation of delusion is most fully worked out in the context of the Capgras Delusion. Individuals with Capgras believe that someone close to them—typically a spouse—has been replaced by an exact replica (Ellis et al. 1994). Parieto-temporal lesions occur in Capgras patients, and an influential idea compares Capgras patients with prosopagnosics, who may show occipito-temporal lesions (but see Feinberg and Roane [2000, 159], who claim that most Capgras patients do not show gross lesions). The assumption is that parieto-temporal lesions inhibit the usual emotions associated with a familiar face. Recent work suggests that part of the explanation for the Capgras delusion is that the face recognition system is an and-gate: it requires two sorts of input. The first sort is the input missing in prosopag-nosics, who are unable to recognize the faces of close relatives, or even their own face in a mirror. It has been suggested that the mechanism that pro-duces this input is either a template-matching system or a constraint-satisfaction network (De Renzi 2000; Farah 1990). However, it appears that there is also an affective response needed to underwrite face-recognition—a neural pathway that gives the face you see its emotional significance. Asked to compare to faces and judge if they are the same, prosopagnosics are much quicker to judge that familiar faces are the same, even if they cannot recog-nize them (Palmer 1999), which suggests that some information is getting through; it is proposed that what is preserved in prosopagnosia is missing in Capgras, and vice-versa. Full recognition only occurs if both these sorts of inputs feed into the face-recognition gate. Several authors have suggested that the system subserving this affective response is disrupted in Capgras patients. Ellis and colleagues (1997) showed that familiar and unfamiliar faces evoke equal autonomic responses in Capgras patients, thus suggesting that the affective response is indeed absent. They also showed that the deficit was specific to faces, since Capgras patients can recognize the voice of their loved one over the phone.

This cannot be the whole story, however. If a strange perceptual experi-ence were enough for a delusion, why do prosopagnosics not come up with some strange story about their surroundings? And if I thought that my wife were a robot, why would I not decide that something is wrong with me, instead of surmising, let alone believing, that she really is a robot? Given a strange experience, people may decide that the fault lies in themselves. Suf-ferers from the alien hand (chapter 3), affirm that their hand's independence from their own control is evidence that something is wrong with them. They do not form the delusional belief that someone is controlling their hand movements. Schizophrenics, though, suffer from delusions of control and do decide that they are playthings of alien forces. This second component, the placing of blame on an external agent rather than an inner fault, is part of the delusional profile, and so something extra must be going on in these cases, including Capgras. It appears that in Capgras, although the face recog-nition system is suffering from the converse pathology to prosopagnosia, it is interacting with reasoning and attribution systems that are also defective.

Stone and Young (1997) argue that the problem with Capgras' patients is that in addition to their agnosia they have a belief system that is too heavily weighted toward observational beliefs at the expense of background knowl-edge. If these speculations are correct, then there is in Capgras' Delusion a face-recognition system receiving only one of the two sorts of input that it needs and passing its own output on to a central system which is not broken but which is excessively positivistic.

Mixed models incorporate elements of both bias and deficit models, so they too face the problems with the view that delusions rest on reasoning biases. Davies and Coltheart (2000), for example, argue for a four-part account etiology of delusions that attributes both biases and deficits to reasoning systems. In their story, a delusion is caused by the retention at all costs of an attributional hypothesis to the effect that the source of anomalous experiences is external to the subject. The attributional hypothesis reflects a reasoning bias, and its retention in the face of other evidence reflects a deficit in belief revision (Davies and Coltheart 2000, 23). Davies and Coltheart (2000, 34–35) argue that all delusions can be understood as circumscribed beliefs that explain away the nature of an anomalous experience. Why circumscribed? Because subjects do not let the belief they formed to explain an anomaly ramify very far through the overall belief system, thereby inhibiting any dramatic revision to the total system.

Now, let me turn to the problems with seeing delusions as arising from reasoning biases. These problems, remember, afflict both bias models and the aspects of mixed-models that rely on reasoning biases to explain properties of delusions. Rather than repeat myself by discussing bias and mixed models separately, I treat all the objections to reasoning biases as a group.

Bias views usually assume the interaction of reasoning biases with anomalous experiences, so I begin with some arguments that tell specifically against this "two-factor" view. Then I will move on to narrower arguments that apply just to reasoning biases.

First, despite Davis and Coltheart's aspiration to provide a general theory, it cannot be the case that all delusions arise from an initial anomalous experience that interacts with a cognitive style. Bermudez (2001) raises this problem for persecutory delusions. A more troubling example might be De Clerambault's Syndrome (the subject of Ian McEwan's novel *Enduring Love*), in which the delusion that one is loved deeply by a distant and inaccessible figure. Subjects interpret all manner of ordinary experiences, like a twitching curtain, as evidence of another's secret passion, but it seems that the condition may arise without any initial perceptual anomaly at all, as in the original case of a Frenchwoman who traveled to London in the belief that George V was crazy about her (Enoch and Trethowan 1979). However, the absence of strange perceptual experiences in these cases does not show that biases are not generally involved in delusions, just that the combination of experience and bias is not generally involved. And in fact there is some evidence that subjects with delusions of persecution have a distinct attributional style, different from that of both patients with nonpersecutory delusions and normal subjects (Blackwood et al. 2001).

A second objection to the view idea that each delusion stems from an anomalous experience is that it predicts too many delusions. On this view, each anomalous experience should cause a delusion in people who have the relevant biases. Davies and Coltheart are saddled with this because they explain delusions as arising from interactions among anomalous experience and biases in reasoning. But plenty of visual disorders do not bring delusions in their wake. If the theory is correct we should expect to see prosopagnosics who explain their failure to recognize familiar faces in terms of a conspiracy affecting the outside world, or patients with cerebral achromatopsia who explain their lack of color vision in terms of a conspiracy to paint everything in shades of gray. Davies and Coltheart make no attempt to find such cases, and I know of none. In effect, Davies and Coltheart predict many more delusions than there appear to be, since plenty of people have odd experiences but not delusions.

Third, as Campbell (2001) points out, the bias view requires a very narrow range of permissible affective responses to a given stimulus. What is alleged to be missing in the case of Capgras is the characteristic affective response to a spouse or someone else subjectively important. But one's affective response to even the face of a familiar loved one might depend on features of one's current situation—when walking in on a spouse who is entertaining a lover, for example, face recognition would not produce positive affect on anyone's part.

These three objections tell against bias versions of the currently dominant two-factor view, which says that delusions arise from the interaction of abnormal experiences and reasoning styles. I now turn to views that argue against bias-based explanations of delusions that do not also appeal to anomalous experiences. My fourth objection, then, is that no attempt is made to identify the reasoning style or bias precisely, or to show that delusional individuals exhibit it in other areas of cognition. Fifth, Bermudez (2001) notes that the bias model does not rationalize the persistence of delusions in the face of constant counter evidence even if it works as a theory of their rational acquisition. This, of course, does not affect mixed models, which offer a separate, deficit, explanation for the retention of a delusion.

Bias models try to make delusional individuals broadly rational. However, we may note as a sixth objection that it is unclear why circumscription of delusions is supposed to be rational. It can be epistemically quite proper not to act on beliefs that are anomalous with respect to one's overall belief set: scientists do not throw out a theory on the strength of one anomalous result.

But it is not good epistemic practice to fail to act on the anomaly when the costs of doing so are restricted to revising a small set of beliefs. Most of what a Capgras patient believes is unaffected by the delusion, so we are not in fact dealing with something that has immense ramification throughout a total belief structure. Even if the cost of acting on the delusions is small, Capgras patients often make no attempt to act on their beliefs, by, for example, calling the police to find the spouse-counterfeiters.

To summarize, attempts to explain delusions in terms of more or less rational epistemic responses to abnormal experiences exaggerate the extent to which those abnormal experiences can be assimilated to the agnosias; overpredict the incidence of delusion; and fail to rationalize the persistence of delusion as an epistemic process. So rationalizing approaches to delusions do not stand up. The overall pattern is not one of instrumentally rational agents responding to anomalies. If attempts to make delusions seem instrumentally rational are flawed, what of deficit models that explain them in terms of irrational belief formation?

5.4.3 Deficit Models of Delusion

Like bias theories, deficit theories of delusion rely on a picture of belief revision and formation that is tantamount to a theory of rational inference. Deficit theories argue that delusions are not caused by normal cognitive styles, but represent failures of normal functioning in belief systems. As such they are, to say the least, underdeveloped, since we simply do not have a satisfactory theory of central cognitive systems, and we have no reason to simply posit them on the basis of folk psychological theories of thinking. This approach has some of the same problems as the bias model, as well as some more that are peculiar to it.

For example, the overprediction of delusions is a problem for two-factor deficit models as well as bias models: why should only some anomalous experiences affect reasoning systems? It looks as though the visual problem and the doxastic one are not as independent as we thought. If they were truly independent, there would be more delusions, as people with the reasoning deficit should be equally likely to undergo any anomalous experience.[4]

---

4. Jennifer Radden pointed out to me that this objection is not unanswerable as it stands. It could be that there is a property that only some experiences exhibit, and it is that property that affects the reasoning systems. Empirically, this seems unlikely, since the delusions do not appear to arise from perceptual experiences of only one sort. That is, although the doxastic and perceptual phenomena do not look to be fully independent, their mutual dependence does not seem to reflect the dependence of doxastic processes on some specific property of experiences.

However, my main interest is the failure of deficit models to actually specify deficits in terms that go beyond folk psychology and make contact with a theory of central systems. If Fodor is right, of course, explaining abnormal belief is a mug's game, since we cannot explain belief. And certainly when we turn to look at specific Proposals for what is going wrong with central systems, we seem to get no advance beyond folk psychology.

Langdon and Coltheart, for instance, argue that normal belief revision depends on two types of sensory information; some we attend to because of "heightened personal salience," and some we are automatically "oriented towards because it is discordant with our prior experience of how the world should be" (2000, 203). But this is just the view that people normally change their mind when they learn something important or surprising, which we already knew. The psychological clout comes when Langdon and Coltheart suggest that "two distinct mechanisms" exist to carry out separate monitoring tasks that correspond to these two psychological traits (2000, 203) But these putative mechanisms have no independent justification. They are (and this is ubiquitous in the literature) no more than names for some aspect of reasoning that a theorist has chosen to identify. The theorists are reifying commonsense capacities and assuming that systems exist to underwrite them. This error is ubiquitous in the thinking of psychologists influenced by evolution, who tend to think that if they can identify a psychological trait that people often exhibit, there must be a mechanism with the evolutionary function of enabling that trait (Murphy 2003). We saw the same assumption lead Wakefield to hypothesize the existence of a self-esteem system. If all he means is that humans think more or less well of themselves and this is connected with their other beliefs and desires, I have no objection. But Wakefield clearly means that an evolved computational system exists whose output is an estimation of self-worth, and we have no reason to believe that.

On the other hand, the failure of the face-recognition system in prosopagnosia (and in Capgras, if it really is present there) is explicable in terms of a computational system with a particular realization in the brain. What we want is that sort of understanding across the board, so that not just perception, but thinking, can be put into a tractable theory. But although we know something about the biological basis of the computational properties of the visual system, we still are groping about for the neurocomputational theory of rationality. In most areas of biology we know how to connect functions with organs: this knowledge is largely missing in cognitive neuroscience in just those areas that rely on our intuitive understanding of rational human nature, which might in any case be false.

Although a speculative decomposition is a worthwhile heuristic in many sciences, it needs to decompose the surface phenomena into simpler components (Cummins 1975, 66), and it is questionable whether we even see that in the deficit models of delusions, which just translate aspects of folk psychology into putative computational systems with no gain in simplicity or tractability. Furthermore, as a general theory, even deficit models may not fit the appropriate range of facts about delusions in general. Some delusional states do not seem to have the properties that beliefs should have, and they can exist alongside normal reasoning abilities. This suggests that explaining delusions in terms of systemic deficits in central mechanisms is unlikely. If bias models fail too, what is left?

5.4.4 The Nature of Delusion

The theorists mentioned so far assume that explaining delusions is a special case of explaining false beliefs. There are problems with this view. The stress on explaining delusions as odd beliefs often overlooks the total psychopathology of these patients, which may include widespread mental disorder suggesting other causes. Cotard's original patients did not all complain of being dead, and Cotard himself stressed that what they had in common was deep-seated, if not psychotic, hopelessness ("délire de negation"). The Cotard delusion seems to be "mood congruent," and thus a form of psychotic depression. It may be an affective disorder rather than one driven by face-recognition anomalies (Davies and Coltheart 2000, 30–31; Gerrans 1999). And Capgras's initial patient was deeply paranoid. It has been suggested that the delusions of misidentification, such as the Capgras Delusion, are the result of chronic dissociation or depersonalization, which can have several causes other than brain damage (Christodolou 1991; Feinberg and Roane 2000). And indeed Capgras-like symptoms do occur in schizophrenia, where depersonalization/derealization is common.

Separate, cognitive explanations of the etiology and persistence of delusion may well be required. Organic theories of delusion may correctly situate the presence of delusions in a wider syndrome, but they do not provide what we want explained in cognitive terms. Kapur (2003), for example, views delusions as the result of a dysregulated dopamine system. Kapur argues that dopamine mediates the salience of external states of affairs and that hyper-dopaminergic states invest unexceptional external stimuli with great subjective importance. The patient tries to understand why these things seem so significant, and the precise form of the delusion depends on accidents of the patient's biography. Kapur's theory does not attribute delusions to any particular cognitive style or reasoning bias, although he does describe delusions as the "cognitive effort" to come to terms with the effects of dopamine irregularity. The view is in that sense akin to the bias views, but notice the differences: it does not mention central systems but simply relates neurochemistry to phenomenology and some intuitive ideas about epistemic reactions to anomalies. Kapur's model is well suited to exploration in neural nets, which can simulate dopamine dysregulation via altering connections within or between layers. On the other hand, since the key variable is just dopamine activity, it will seem to many theorists that there is a missing level of explanation in Kapur's story: there is not, on these broadly organic accounts, any computational or intentional theory of delusion, although we still need to characterize the role that the organically produced delusions play in the patient's mental life.

Campbell (2001) argues that delusions are beliefs that structure subsets of beliefs in fundamental ways, which is reminiscent of Jaspers, and that they are caused by straightforwardly organic factors. This view denies that there is any rationalizing explanation of delusions to be given at all. I revisit views like this in chapter 6, but note two things: first, as Davies and Coltheart point out (2000, 8) this view is inimical to folk psychology, which cannot admit beliefs, as opposed to experiences, with a purely organic basis. This is not a problem unless we have some reason to believe that explanations of what causes delusions should be intelligible in psychological terms, and we have none. Second, if the cause of a delusion is just brutely organic, a delusion cannot be a rational state, since its cause is not evidence for it. But a view that says that delusions are organically caused still may reserve them some cognitive properties once they are part of the belief system. An organic theory of etiology does not rule out investigating why delusions matter to psychology. But neither bias models nor models of faulty information processing systems can provide that explanation.

In fact our thinking about delusions is guided by a commonsense view, and that view may be one that we can articulate within the confines of a naturalistic theory of reason, even if it cannot be mechanized via reduction to a neurocomputational theory of central systems. Start with the most basic issue: how do we know when a belief is delusional? One's spouse being replaced by a robot is a very strange belief, but it is not an impossible state of affairs: is it any more delusional than the belief that one has been possessed by a ghost? That belief is widespread in many societies, even if Western psychiatrists might want to treat it with drugs (Hale and Pinninti 1994). And what about the belief that a communion wafer can be transformed without perceptible change into the body and blood of a member of a triune deity?

A delusion is not just any false belief, or even a floridly false one. Nor is it a belief that is immune to revision, since many false beliefs are immune to revision even in the absence of clear counterevidence (think of the belief that physical objects are solid, which usually survives acquaintance with the atomic theory of matter). Successive classes of Caltech students have delighted in asserting that if delusions are false beliefs retained in the face of overwhelming counterevidence, then all religious people are deluded. If I had taught this material at Sectarian Protestant University, the students would have said that Darwinists are deluded.

In fact, whether or not something is a delusion is a matter of how it strikes us, and that depends on how well it comports with our understanding of what people are like, both in general terms and within our culture. It does not rest on an identifiable failure of some psychological mechanisms or a formal property of beliefs. Even if we can identify belief formation mechanisms, it is unlikely that any mechanism has *completely* broken down even when subjects are delusional, since even in the maddest psychotics we find some preservation of normal reasoning alongside the delusional reasoning. A deficit model needs to characterize what is wrong with delusional belief formation in a way that preserves the partial character of the breakdown. Since we have no idea how central systems work, this is not currently possible.

However, since we need to understand the cognitive role of delusional states, even theories of their organic etiologies will need to appeal to considerations of instrumental rationality to understand their effects as in Kapur's talk of cognitive effort. And we need to appeal to some general ideas about belief acquisition to explain why certain beliefs count as delusions in the first place. We will continue with the assumption that delusions are beliefs, which may be wrong (Stephens and Graham, in press). Many theorists have echoed Jaspers's (1959) view that delusions are a special sort of state, isolated from beliefs both inferentially and emotionally. Schizophrenic delusions can exist side-by-side with contradictory nondelusional beliefs and can have little impact on the patient's emotional life. Manic depressives act on their harebrained schemes, but delusions in schizophrenia often lack connections to action. Louis Sass argues on these grounds that schizophrenic cognition is characterized by " 'double or multiple bookkeeping' whereby the delusional is kept separate from the rest of experience" (1992, 275). As he says, schizophrenics often fail to act as one would expect them to if delusions are a species of firmly held belief; "the patient who insists her coffee is poisoned with sperm still drinks it without concern" (1992, 274). According to Sass, even hallucinations, which may seem to involve the perception of objects, may be revealed on questioning to actually be about images within the visual field, although they may be regarded as "more real" than veridical perceptions (cf., Bleuler 1911, 111–113; Jaspers 1959, 68–71, 105–106).

Delusions in schizophrenia have a peculiar quality of incorrigibility but often lack the guiding role in action and cognition that we would expect of incorrigible beliefs. This is not always the case; schizophrenics can reason practically on the basis of their delusions, but the relations between delusions, thoughts, and actions are very complicated. Jaspers (1959, 106) understood delusions in terms of two qualities: the delusional experience, which was something akin to experiencing a deep insight shared by no one else, and a change in the personality of the subject. What this change amounted to, Jaspers thought, was almost impossible to say, except that it invested the delusion with great importance, so that to admit its falsehood would have demolished the patient's world (1959, 105, 410). It was this feature that led him to argue that the common understanding of delusions as necessarily false beliefs was incorrect; delusions could even be *true*, he thought, provided they exhibited these qualities.

I think a naturalistic account of rationality can go some way to supply what we need to understand the troublesome features of delusions. But the little theory I am about to present does not fill in the missing level of computational explanation. Normal human cognitive development includes the acquisition of beliefs and other mental states, not just from the world, but from other people. If we come across individuals who have beliefs that are important to them but seem to be based on no causal contact with the outside world, nor on testimony, we are entitled to wonder about them, especially if no rational justification for the belief can be given. So even if you believe, as I do, that rational justification for religious belief will not be forthcoming, it is undeniable that normal maturing brains do pick up religion, along with many other false theories of the world.

That, in fact, is why religion is not delusional even if religious beliefs are false. Our evolved psychology, it appears, just commits us to the same epistemic mistakes, generation after generation. These mistakes are things like beliefs in the supernatural, naive essentialism, incompetence with probability, and optimism about the Red Sox.[5] They reflect facts about our normal psychology. So even though we have many beliefs that are epistemically ignoble, they reflect our normal human propensities. This may lead us into error, but it is not irrational, since trying to gather information without paying attention to the beliefs of others is impossibly costly. Even though

5. In light of the events of October 2004, I must now apologize for this sentence to all my in-laws in Massachusetts. But come on guys, could you blame me?

normal exercise of our rational capacities can produce hopeless error, it is our shared rational capacities that produce the error. *Inexplicable* error entitles us to a prima facie assumption that some beliefs (or other ways of attending to the world) are actually delusions. Delusions are not acquired from others, nor caused by reliable connections between the deluded person and the nonsocial world. Since delusions are acquired by neither of these means, it is hard to see how they can have a nonpathological basis.

Numbers matter with delusions. If only a tiny minority of humans were religious we might be more tempted to call them delusional, but the temptation would reflect the fact that in that case we would be less inclined to think that it was natural to human psychology to acquire religion. The fewer people believe in something, the less natural it seems. Most people have beliefs in the broad category that we call religious, although only minorities (often tiny minorities) have specific beliefs—like cargo cult beliefs, for example. But once we accept a generic human propensity to have religious beliefs, we can appeal to cultural or psychological contingencies to explain why a particular group has a particular belief. This leaves room for disagreement about borderline cases, but in general we are inclined to count all sorts of beliefs as nondelusional if they are attached to some system of religious belief.[6]

Generally, then, the more that we cannot supply an intentional, plausible story about the acquisition of a false belief that comports with our folk views about human nature and the sources of knowledge, the more likely we are to call someone deluded. My little theory is not a justification for any subpersonal mechanisms. It relies on the idea that we have evolved to acquire information through various fallible means, like perception or coordination with others. It says nothing about what mechanisms underwrite those capacities, nor how particular breakdowns lead to particular delusional contents. It does not explain delusions mechanistically via cognitive parts. There is no suggestion that what folk psychology recognizes as acceptable constraint on and sources of belief formation maps on to components of cognitive architecture.

What the little theory does is justify our prior pathologization of delusions by appeal to natural epistemic capacities, while admitting that those capacities, even when working as designed, lead us into error. It acknowledges that these normal systems can err, but then appeals, in the case of delusions, to our inability to see how, given our normal workings, someone could

6. Frith and Johnstone (2003) report that the standard way to distinguish a delusion from a religious belief is to contact an authority from the relevant religion and ask for verification that people in that religion do indeed believe that.

believe *that*. In other words, it relies on a conception of reason as natural and limited, but not on a mechanization of reason in terms of computational processes. A good theory of naturalistic reason could be of great help in assessing and explaining the mental ill even without a mechanistic understanding of reason. It could, for example, help to articulate a structure in which we understand rational norms as parts of our nature and know something about their neural representation, but in which we do not have a grip on an independent computational level in between.

But why have an intentional aspect to the story at all? What do we miss by not having computational accounts of reasoning? Ironically, we miss the distinction between organic and cognitive theories of delusions, since if we cannot tell a cognitive story of belief acquisition, we cannot distinguish a cognitive etiology from an organic one. All we can do is model belief acquisition as a function of, for example, the dopamine system's role in mediating salient stimuli. We cannot differentiate between the dopamine system's mediation of rational and nonrational causal connections without some theory of what rationality is. I argue that we can do the job of differentiating rational and irrational based on a naturalistic theory of reason, even if we have no computational theory of belief fixation. We are left with a theory of brain activity and a theory of rationality, but no grounds for distinguishing rational from irrational phenomena except those provided by the theory's criteria for what counts as rational.

Such grounds might be enough in many cases. A naturalistic theory of reason abstracted from the computational systems would not be a body of positive facts. But it could, if there were sufficient agreement over it, justify some conceptions of a mental illness over others, and organize research on the neurological basis of departures from what would be, in effect, a philosophical-cum-social scientific theory of reason. A naturalistic theory of reason might provide an account of rationality that would suffice to explain and ground our attributions of pathology, especially if some or all of the theory is formalized according to the familiar habits of the social sciences. In such cases, diagnoses of mental illness would indeed reflect the failure of some individuals to follow norms, as the antipsychiatry movement often has alleged.

But, in contrast to the antipsychiatry movement, we should note that this inescapable normativity is not a general truth about psychiatry. It applies to those diagnoses where we lack an objective grip on underlying processes. Furthermore, the relevant norms are not moral or political but rational. There is disagreement over what counts as rational, of course, but it might admit of more consensus, and easier testing, than claims about moral norms, and it is easier to anchor theories of rationality in defensible views of human nature. (So the theory must be a naturalistic theory even if it is not a mechanistic theory.) In some instances, we can expect sufficient consensus about naturalized rationality and epistemology to anchor a diagnosis. A diagnosis made with reference to scientifically and mathematically articulated elements of our folk picture of rationality would be objective, although its objectivity would not be tied to positive facts in the way that biological psychiatry expects. Rather, it would set a normative agenda that brain science could follow, by looking for the brain sites where certain kinds of processing seem to occur (with the processing understood not computationally but physiologically.)

However, complete consensus on a naturalistic theory of reason is unlikely, so widespread disagreement will remain about the pathological nature of some mental states the evaluation of which requires an assessment of the degree to which they are rational. Methodological and normative debates will consequently endure. In the next two sections, I discuss two other sorts of psychopathology that involve questions of rationality. In section 5.5 I look at addiction, and in section 5.6 psychopathy, discussing attempts to characterize them using general theses about reason, including formal theses, but not underlying mechanisms. In the absence of a mechanistic theory of rationality, our understanding of the disorders depends on normative claims.

5.5 Addiction and Irrationality

Addiction is linked to poverty, violence, and the shrinking of all one's horizons to serve a craving. Addictions seem like paradigmatic causes of irrational behavior, and becoming addicted often is viewed as the result of insufficient regard for the long-term consequences of one's actions. Decision theorists formalize this attitude to the long-term as a rate of discounting the future, or valuing goods less according to how remote they are. The level of discounting can vary as a function of remoteness, which means that several different discounting rules can compete in the explanation of addiction. Furthermore, on some views of the discounting rule under consideration, addiction may not be irrational at all. On these views, rational agents can knowingly become addicted to drugs, and, perhaps, remain rationally addicted to them.

The next section brings out the implications of these theories, they illustrate the issues raised by the rationalization project in social science as it applies to behavior. Having looked at these assumptions, we can ask how they relate to questions about rationality, examining the normative assumptions that the study of addiction must make.

5.5.1   What Is Addiction?

Although it is agreed that humans can become addicted to substances, there is dispute over whether we can become addicted to activities. Goldstein (2001) marks one end of the spectrum, with his insistence that one can become addicted only to seven categories of drugs, viz: nicotine, caffeine, cannabis, opiates, psychostimulants (e.g., cocaine and speed), alcohol and its relatives (the barbiturates and benzodiapezines), and hallucinogens (which include PCP and ecstasy, as well as LSD, mescalin, and others). All these drug families have similar and fairly well-understood effects in the brain: essentially, they extend the chemical signal that a neurotransmitter passes between neurons. Opiates, for example, interfere with receptors for the endorphin/enkephalin family of neurotransmitters, mimicking the presence of a member of the family and causing a huge number of the receptors to fire simultaneously. Goldstein's view is that addiction is a brain disease and that habitual behavior, however urgently one craves it, is not addiction (see also Leshner 1997).

At the other extreme, Becker and Murphy (1988, 675–676) assert that people can become addicted "not only to alcohol, cocaine and cigarettes but also to work, eating, music, television, their standard of living, other people, religion, and many other activities." Becker and Murphy's very broad conception of addiction threatens to erode any distinction between addictions and things we do more and more of because we like them. (For a broad psychiatric view of addiction as also including behavior, see Campbell [2003], who regards addiction as a volitional impairment that can apply to diverse behaviors.) An intermediate position, adopted by Elster (1999b) and Orford (1985), ties addiction conceptually to a suite of experiences on the part of the addict, of which the most important is craving, which Elster regards as the "most important explanatory concept in the study of addiction" (1999b, 62).[7] Another important concept in Elster's view is that of withdrawal, since withdrawal symptoms do not occur in all habitual behaviors and can thus distinguish some genuine addictions. Other members of the suite (Elster 1999b, 59) include tolerance, euphoria, dysphoria, the desire and inability

---

7. The DSM-IV-TR does not list craving among the symptoms of substance dependence, although the IDC-10 does include craving among the symptoms of its diagnosis of "dependency syndrome."

to quit, and "crowding out," which is the tendency for one's short-term plans, or even life, to revolve around feeding the addiction.

It is the connection between the phenomenology of addictions and the behaviors they entrain that is of interest at the moment, so I adopt Elster's intermediate view. Readers who prefer the exclusively brain-based model should read the ensuing discussion as directed at patterns of behavior and motivation that characterize addicts but do not define the addictions. I focus on behavior because I want to bring out the status of addictions as rational or irrational. I start by looking at a model that argues for the rationality of addiction and note some of the ways that this model can be threatened. I discuss, in particular, the suggestion that the irrationality of addiction can be formally demonstrated by decision theory, then note some problems with that view. Then I look at the idea that what makes addiction irrational is the failure of the addict's chief goals in life to withstand reasoned scrutiny, and argue that debates on that score capture what we mean by the irrationality of addiction but pose important problems for a view of addiction as a psychiatric condition, at least according to the two-stage model. Unlike delusions, addictions are unlikely to meet widely agreed standards of irrationality.

5.5.2 Addiction and Instrumental Reasoning

Becker and Murphy (1988) offer a model of rational addiction based on *tolerance* and *reinforcement*. Tolerance is the increase over time in the amount of a substance that is needed to obtain the same intensity of experience. (Whether there is a genuine analog of tolerance for behavioral addictions is controversial, even if the existence of behavioral addictions is granted.) Another consequence of prolonged use is that one's marginal utility from increased consumption rises: that is reinforcement. On this model, either abstention or addiction can be accounted for by altering the degree to which the future is taken into account by the agent: decision theorists formalize the weight given to long-term consequences as the rate at which one discounts the future.

Becker's addict is an individual whose consumption of a good leads to a greater consumption of it in the future. Becoming addicted is the consequence of a decision to self-medicate when under stress. The decision brings short-term payoffs that lead one to disregard future costs (which the model assumes the agent knows all about). However, as time passes the overall welfare of the addict is less than that of an abstemious individual. The problem is that addicts are better off in the short term if they consume, because current utility from use is greater than current utility from abstention, even though each decision to consume will lead addicts to be worse off in the long run. For the addict, welfare *now* is greater if one uses, even though welfare *from now on* is greater if one quits, and "the short-run loss in utility from stopping consumption gets bigger as addiction gets stronger" (Becker and Murphy 1988, 693).

The addict is in a mess. Each decision to use a substance gains in short-term utility over a decision to abstain. A series of such decisions leads to steadily decreasing overall utility. To get into this downward spiral, evidently, you have to largely disregard the future and consider only your short-term utility. To get out of the downward spiral you have to find a way to make long-term benefits outweigh the short-term costs of adjusting to abstention; Becker and Murphy regard this as no different than the motivational problem faced by a chronically disorganized person who wants to get organized but is reluctant to face the short-term pain of tidying the office.

Becker and Murphy's model has been faulted on a number of grounds, notably its failure to capture the phenomenology of addiction and its failure to predict patterns of behavior among addicts that are not accounted for simply in terms of the transaction costs of the activities (Elster 1999b, 55). I address here the formal conceptions of rationality that the model uses. Becker and Murphy's addicts are rational because they are fully informed about the negative consequences of addiction and they discount long-term costs very heavily. Thus, they have enough information to make a rational decision and their preference for the present over the future is a preference, and hence not subject to rational appraisal. These two stipulations—full information and the arationality of preferences—raise different issues.

The stipulation that addicts know exactly what they are getting into is very implausible, but there are other ways in which addicts can get into the downward spiral. There is an important distinction between the rationality of remaining addicted (in which the long-term costs of staying addicted are preferred to short-term costs of turning abstinent) and the rationality of getting addicted in the first place. People could be ignorant of their chance of becoming addicted or they could decide to gamble, weighing what they take to be their prior probability of addiction against the pleasures of (what they hope will be) casual use (Orphanides and Zervos 1995). Most users of a drug do not become addicts. Even for a strongly addictive drug like cocaine, only about one user in six gets addicted (Robinson and Berridge 2003). Sullum (2003) recounts evidence that smokers may have a better grasp of the health risks of smoking than nonsmokers do, and also produces anecdotes starring people who habitually use heroin with no ill effects, such as a woman who shoots up before she does the housework. Apparently, being high makes the chores less onerous. (So you might have been right if you wondered whether your colleagues had been smoking something before that last faculty meeting.)

People who take the initial decision to use might therefore be making a reasonable gamble. At least, people who take up a drug are not clearly irrational if they lack evidence that they are in a group of people whose nature makes them likely to become addicts. The irrationality of the initial decision to use also depends on the weight one gives to the future. Suppose there really is no accounting for taste, and that placing great weight on the short-term is just another taste. Then people are not irrational even if they start using and expect to become hopelessly addicted, since their long-term misfortune has no deliberative resonance for them when they make the initial decision.

So, on the rational-choice picture, whether addiction is rational depends on how agents discount the future. Hence, if we could distinguish rational and irrational discounting we would vindicate the intuition that addicts are irrational because they do not care enough about the future.

Becker's agents discount the future exponentially, but one discount rate that is widely believed to be irrational is hyperbolic discounting (figure 5.1). George Ainslie (2001) regards hyperbolic discounting as characteristic of addicts, explaining their ambivalence and their inability to stick to a

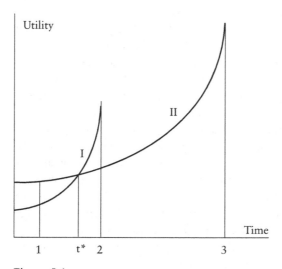

Figure 5.1.
Preference reversal because of hyperbolic discounting (Elster 1999, 171)

resolution in the face of temptation. Hyperbolic discounting is called irrational because it permits dynamic inconsistency. It works likes this: at t1 the
agent chooses between a small reward at t2 and a larger reward at t3.
Because of the discount rates of the two rewards, it turns out that before t*
the present value of the larger reward exceeds that of the smaller, but at t*
the curves cross, so that the value of the smaller reward now dominates. At
t2 the agent then chooses the smaller reward, since it is now valued more
highly.

The classic argument for the irrationality of hyperbolic discounting is that
it makes one a money pump. An observer can exploit the dynamic inconsistency exhibited by the agent's preference flip at t* "buying goods from
her before she is in the grip of a craving and selling them back to her at
inflated rates when the craving strikes" (Yaffe 2002, 198). There are problems with this argument. It ignores the whole point of discounting the future,
which is that future prospects are risky. Anything could happen between the
time the agent sells and the time the craving is due to strike. The addict could
quit before the craving returns, or arrange to be out of the country when
the craving strikes, or the price of the good might collapse, or it might be
stolen from the original buyer. Given all the uncertainty that exists at t1
about what will be happening at t2, it takes more to show that the initial
transaction is irrational.

Elster (1999b, 171) has another objection to the claim that hyperbolic discounting is irrational. Since dynamic preference reversal is so common, and
has been demonstrated in so many species (as Ainslie himself is at pains to
point out), it is unclear whether our naturalistic account of rationality should
call it irrational at all, for fear of imposing an excessive normative standard.
Indeed, Elster claims that if one adopts an instrumental conception of rationality it is not possible to differentiate discounting rates as more or less
rational. That brings up the second point I mentioned above. The degree to
which an agent discounts the future is basically a set of preferences between
future states and present states, and on an instrumental conception of rationality, preferences cannot be rationally evaluated. Elster (1999b, 146) sums
up the position:

That weight [of future gratifications relative to present ones] is not itself subject to
rational assessment. A time preference is just another preference. Some like chocolate ice cream, while others have a taste for vanilla: this is just a brute fact, and it
would be absurd to say that one preference is more rational than the other. Similarly,
it is just a brute fact that some like the present, whereas others have a taste for the
future. If a person discounts the future very heavily, consuming an addictive substance
may, for that person, be a form of rational behavior.

The argument may seem counterintuitive. I believe, however, that if we want to explain behavior on the bare assumption that people make the most out of what they have, the idea is exactly right. If some individuals have the bad luck to be born with genes, or be exposed to external influences, that make them discount the future heavily, behavior with long-term self-destructive consequences may, for them, be their best option. We cannot expect them to take steps to reduce their rate of time discounting, because to want to be motivated by long-term concerns ipso facto is to be motivated by long-term concerns.

Yaffe contends that Elster's argument in the first quoted paragraph is unsound. He glosses it (Yaffe 2002, 187) as follows:

P1. Someone's preference for the present is irrational only if that person could be motivated to change it.

P2. Someone who is motivated to have a preference for the future over the present already has that preference.

Conclusion. A preference for the future is never irrational; one could never both have such a preference and be motivated to correct it.

Yaffe thinks P1 is probably false and P2 certainly false. But in fact Elster does not endorse P1. He thinks that preferences are neither rational nor irrational. Yaffe thinks that P2 is false because subjects may have second-order preferences to the effect that their first-order preferences are different. One may prefer ice cream to wheatgrass juice, while at the same time wishing that one liked wheatgrass juice, profoundly disgusting though it is. In this case one may be motivated on grounds of health to have a preference that one actually lacks. And Yaffe calls that a counterexample to the general principle that he says P2 relies on, to wit, that one cannot be motivated to have a preference that one lacks.

But Elster does not need to rely on that general principle. He is drawing our attention to the fact that if we want to be motivated by the long-term then we are already giving long-term prospects some deliberative weight. After all, if I want to be motivated by long-term concerns it is because I think it would be better for me to forego some short-term benefits in favor of long-term ones, and I am trying to motivate myself to follow my opinion. In that case we can expect me to change my first-order preferences. If I do not do so, I am being weak willed. But if I am indifferent to the long-term, we cannot expect me to simply stop satisfying my first-order preferences in favor of future-directed ones, because I have no reason to do so. A preference for the future, like a preference for better health, structures my set of preferences and makes me wish for preferences that I in fact lack. But I will not want those preferences to be in the set if I lack the structuring preference. Indeed, it is hard to see why we should ever expect people to even want to like wheatgrass juice if they are indifferent to their health: that would amount to the expectation that they should be motivated to acquire a preference that has nothing to do with the things that give their life its shape and purpose. It would be inconsistent, in fact, to prefer delaying pleasure while at the same time preferring present benefits to future ones. Although Yaffe is correct that one *can* be motivated to have a preference that one lacks, it seems that the possibility he points to really obtains only in cases where one has a second-order structuring preference that gives a reason for the first-order preference. It does not apply to the structuring preferences themselves. This observation leaves us with another way to characterize the addict as irrational; it is irrational, we can assert, to be an addict because the structural preferences of addicts should be different.

But the grounds for the assertion are unclear. It will not do to argue that a life that centers on one activity is irrational unless one wants to argue that all strongly held vocations are irrational (as, we saw above, Rawls may have thought). Doing so does not single out addicts. The experience of repeated craving, too, is common enough that it does not single out addicts. Nor will it do to insist on the irrationality of short-term thinking, as we have seen. Considerations like these lead the likes of Fingarette (1988) to wonder whether a life based around alcohol is any more irrational than a life devoted to scholarship or athletic achievement (although that does not, in the end, seem to be his view, despite his rejection of a disease model).

The traditional basis for attacking addiction as irrational is that it is incompatible with human flourishing, since the addict is in thrall to the lowest part of his nature. But many things we value very much are low pleasures in that traditional sense of expressing our animality with some urgency—for example, sex. It is apparent though, at this point, that the scrutiny of addiction on rational grounds has become moral scrutiny. That is not an objection in principle: moral scrutiny does not have to be irrational or unconvincing. But it does mean that the project of resting our account of the desirability of addiction on a neutral scientific account of addiction as a disease has disappeared along with the appeal to considerations of instrumental rationality.

The current disease model of addiction is based on two claims: drugs have measurable effects on the brain, and long-term drug use is self-destructive. The skeptical response is that many things are self-destructive but we do not pathologize them. Skeptics are often Szaszians who dismiss the disease model out of hand. The neurological effects of drug consumption are undeniable, however. But even so, it so not clear that the brain science alone justifies a medicalization of drug use, given that any subset of people who are distinguishable psychologically or behaviorally have, on materialist assumptions, different brains. The brain difference may be one that is incompatible with normal brain function, or it may not: some addicts, as we have seen, function well. But even if we identify a neurological deficit we must ask whether that deficit is one that accompanies a mental disorder or merely a self-destructive habit. These are hard questions, and I do not ask whether the biological model or the skeptical response is correct. The point is that we cannot, as far as we can tell, decide between the disease model or the skeptical response based on a mechanistic theory of central systems that would let us decide whether the addict is psychologically dysfunctional in virtue of the computational facts. Further, our account of the rationality of addicts remains folk-psychological, in the broad sense in which decision theory is continuous with folk psychology. And addiction, unlike delusion, is not a case where we should expect widespread agreement about the pathological nature of the phenomena based on purely instrumental views of rationality.

Of course, some may think that even a good mechanistic theory of rationality would not decide the case, since it would not establish whether a syndrome is a mental illness rather than a self-destructive or immoral suite of behaviors. Many people see self-destructiveness as irrational, and a number of philosophers have read enough Kant to think that immorality is irrational. The idea that mental illness often is diagnosed where we should just see deviance or nastiness is common. An examination of pyschopathy highlights this idea and shows that a science of rationality is our best chance of disentangling the conceptual knots.

5.6 Psychopaths

Both moral and cognitive values are implicated in our attitude toward psychopathy, so we may take that as our example of the way attitudes to putative pathologies reflect views of human reason that involve conceptions of flourishing. I have said that delusions appear irrational to almost everyone whereas the irrationality of addiction will remain contested. Psychopaths are similar to addicts in that their behavior, as well as their emotional responses, puts their mental status in question. But in the case of psychopathy destructive effects on the lives of others are what chiefly command our attention. The question is whether the absence, in psychopathy, of normal regard for others warrants calling psychopaths mentally ill. I argue that psychopaths flout both our moral and our rational norms. I do not defend the Kantian idea that moral requirements are just a species of rational requirement, but psychopathy does involve patterns of behavior that suggest a widespread, if partial, collapse in cognitive skills. This may matter to moral cognition as well.

Behavioral and affective abnormalities are central to our understanding of psychopathy, but theorists who study psychopaths regard the behavior as indicating some underlying deficits. And in the sense that there is more than behavior to be concerned with, the very existence of psychopathy as a condition is controversial. The DSM-IV-TR (701) has no entry for psychopathy distinct from antisocial personality disorder, which is "a pervasive pattern of disregard for, and violation of, the rights of others that begins in childhood or early adolescence and continues into adulthood." Antisocial personalities are impulsive, aggressive, and neglectful of their responsibilities. "They are frequently deceitful and manipulative in order to gain personal profit or pleasure (e.g. to obtain money, sex or power)"; typically, they show complete indifference to the harmful consequences of their actions and "believe that everyone is out to 'help number one'" (DSM-IV-TR, 702–703).

Many career criminals merit the diagnosis of antisocial personality, and Raine (1993) has argued that the personality of the career criminal should be considered a distinct form of psychopathology. Researchers in psychopathy, however, insist that the purely behavioral definition of the disease is too inclusive and obscures the genuine psychopathology in psychopathy. Meloy, for example, calls it "too descriptive, inclusive, criminally biased, and socioeconomically skewed to be of much clinical or research use" (1988, 6). Hare distinguished three behaviorally very similar groups on the basis of "personality structure, life history, response to treatment and prognosis" (1970, 9): these being psychopaths, acting-out aggressive neurotics and subcultural delinquents (see also Cleckley 1976, Hare 1999, and Lykken 1995).

Psychopathy is diagnosed using Hare's Revised Psychopathy Checklist (Hare 1991), comprising twenty questions. Answers are scored on a scale from 0–2, meaning no, maybe, or yes. Items 1–8 and items 9–17 can be distinguished using factor analysis (items 18–20 load about equally on both factors).

The Revised Psychopathy Checklist
1. Glibness/superficial charm
2. Grandiose sense of self-worth
3. Pathological lying
4. Conning/manipulative

5. Lack of guilt or remorse
6. Shallow affect
7. Lack of empathy, callous
8. Failure to accept responsibility
9. Need for stimulation/proneness to boredom
10. Parasitic lifestyle
11. Poor behavioral controls
12. Early behavior problems
13. Lack of realistic, long-term goals
14. Impulsivity
15. Irresponsibility
16. Juvenile delinquency
17. Revocation of conditional release
18. Sexual promiscuity
19. Many short-term marital relationships
20. Criminal versatility

Blair (1995, 1997; Blair et al. 1997) explains psychopathic behavior as resulting from the absence or malfunctioning of a module he calls the violence inhibition mechanism (VIM). Blair took the idea of a VIM from ethologists who had suggested the existence of a mechanism that ended fights in response to a display of submission. A well-known example is the canine tendency to bare the throat when attacked by a stronger conspecific. The assailant then ceases the attack, rather than taking advantage of the opportunity to press it home. Blair hypothesizes that a similar mechanism exists in humans, activated by the perception of distress in others. Since engaging in aggressive activity will lead the victim to exhibit distress cues, aggressive activity typically leads to an aversive response brought on by a mental representation of the victim's suffering, which triggers the VIM. Blair maintains that the linkage between others' suffering and our own aversive response is a crucial step in the development of empathy. Since psychopaths do not have a properly functioning VIM, they do not experience the effects of their violence on others as aversive. Blair argues that this explains why psychopaths become more violent at an early age. The most intriguing part of Blair's theory is his argument that people lacking a properly functioning VIM would not be able to recognize the distinction between moral transgressions that cause other people to suffer and other social transgressions that do not. This prediction was confirmed in a study comparing the moral cognition of psychopathic murderers with the moral cognition of murderers who were not diagnosed as psychopaths.

Even if it is correct (for some skepticism, see Nichols 2002a), Blair's theory only explains why psychopaths are deficient in empathy and other moral traits and emotions. It does not explain the more general failures in planning to which psychopaths are prone. Although there is good evidence that psychopaths are on average at least as intelligent as nonpsychopaths, they are impulsive, irresponsible, and very poor at long-term planning. Psychological testing on psychopaths suggests they "have the capacity for genuine judgment and sound affect but that a cognitive deficiency interferes with their ability to integrate the products of these faculties with ongoing behavior" (Newman 1998, 83). Similar deficits are found in patients with frontal lobe injuries (Damasio 1994), and although lesions in the frontal lobe have not been found in psychopaths (Hare 1993), it has been argued that some sort of information processing problems in frontal-lobe systems may account for some of the cognitive shortcomings of psychopaths. These theories try to account for rational shortcomings by positing computational mechanisms, just as Blair tries to explain the abnormal moral development of psychopaths in terms of computational failures.

Psychopaths, then, appear to be both immoral and irrational.[8] In both cases, we are dealing with the transgression of norms. And in both cases, too, we face the issue of how much transgression is enough to call someone wicked or irrational: we all act wrongly and reason badly sometimes. Psychopaths act badly enough for us to call them evil people. Do they make enough mistakes in reasoning to count as irrational? We all fall short of perfect rationality. And the experimental tradition implies that everybody makes not just mistakes in reasoning but *systematic* mistakes in reasoning. If this is correct, we must come up with a theory of rationality that allows human beings with our limited cognitive capacities to count as rational (Cherniak 1986; Harman 1995).

Still, when we assess psychopaths in the light of a less demanding, naturalistic view of reason they do appear to display a number of rational shortcomings without which they would be much better at long-term thinking, acting, and planning (Newman 1998) Their deficits are reminiscent of those

---

8. Indeed, in the thinking of many philosophers, the two are intimately related, since the Kantian tradition in modern philosophy holds that moral requirements just are requirements of reason (Smith 1994). Murphy (1972) and Nichols (2002a) argue that psychopathy is an important challenge to this theory, but both overestimate the degree of rationality that psychopaths exhibit. Kennett (2002) discusses empathy and moral reasoning in autism and psychopathy, but seems to run together empathy and theory of mind. Theory of mind is normal in psychopaths (Blair et al. 1997).

in frontal-lobe patients whose capacity to rationally integrate the product of particular systems elsewhere in the brain is famously poor (Damasio 1994). This certainly looks like a central systems deficit.

Psychopaths are also evil. Is there any connection between the cognitive deficit and the moral one? It depends on how we consider moral cognition's purpose. A number of thinkers with an evolutionary and/or economistic bent have thought of morality as essentially a set of conventions or norms designed to solve the coordination problems that arise in social life (Binmore 1994, 1998; Skyrms 1996). Other traditions, such as the pragmatist one, might see moral cognition as continuous with everyday cognition, viewing cognition in general as a problem-solving capacity designed to predict and control the world. If game-theoretic or pragmatist approaches to morality are correct, we should expect cognitive deficits also to affect the ability to solve problems of social cooperation via moral cognition. In that case, the basic point about psychopaths is not that they lack empathy, but that they suffer from a cognitive deficit that impairs their reasoning in a number of areas of great importance for creatures like us.

To properly defend a theory of psychopathy as a rational (if non-Kantian), moral deficit would take a naturalistic account of the functions of moral judgment rooted in Darwinism or pragmatism and a mechanistic theory of central systems that would apply to moral cognition as a special case of thought in general. This is true too even if the origins of our moral judgments are broadly affective, since affective responses still need to be applied to actions and states of affairs. Indeed, Damasio's frontal-lobe patients are notably bad at decision making because of their lack of affective involvement with the world.

Without a theory to explain the computational deficits that implicate moral cognition, we are left with a theory that appeals to evolved social coordination or other naturalized views of morality. Any naturalized theory of moral cognition is currently even more speculative and less likely to convince than the little theory of delusion that I based on ideas about the difference between abnormal belief formation and our natural epistemic infirmities. The two theories, though, do show that we can approach psychopathology via naturalistic theories of reason even in the absence of a good mechanistic understanding of cognition. But agreement on the naturalistic rationalization project will require equilibrium between folk theory, various sciences, and naturalistic philosophy of mind and epistemology. Normative debate will be central to it, and continue to afflict psychiatric theories that rely on the rationalization project.

5.7 Conclusion

I promised three morals. Here they are. First, the absence of a neurocomputational description does not mean that no reason for calling something irrational exists. It does mean that construction of psychiatric theory often will be guided by normative considerations. This shows the failure in some instances of the two-stage picture, which hopes that such theory construction can be guided only by positive fact. Despite widespread hopes for the two-stage program, much biological psychiatry already works in this way. It takes accepted psychiatric categories for granted and looks for their biological basis. That is, it searches for the biological underpinnings of normatively evaluated conditions.

Second, despite the absence of computational theories of good reasoning, there may be impressive agreement that some thought or behavior is irrational or otherwise pathological. Even if the agreement is based on intentional-level, normative attributions of irrationality it can be defended, and often the defense can appeal to empirically testable naturalistic claims about human nature. Sometimes the ensuing agreement will be impressive enough to make the lack of foundational computational theory seem beside the point. If anything is irrational, florid psychosis is, and we are unlikely to give up on the belief that some things are rational and some are not. However, normative arguments will continue to haunt some diagnoses. Intentional-level theories of irrationality, even if formalized, will not resolve all debates about mental symptoms.

Third, the absence of computational accounts of thought means that some theories will miss out on some levels of explanation: we will have intentional phenomena at the top, characterized in folk-psychological terms, or in terms of formalized renditions of folk-psychology such as rational-choice theory. At the bottom, we will have several levels of physical description of brain mechanisms where the intentional capacities live, but we will not have the computational picture that once was supposed to show how the intentional could be recast in a form that made it physically realizable. This means that there will be gaps in our knowledge of psychology that make the multilevel picture hard to fulfill. We will be left with behavioral neuroanatomy, but not computational neuroanatomy. This will affect some disorders, and indeed some symptoms within some disorders, but leave others untouched and amenable to a more comprehensive multilevel treatment. If central systems are intractable because they are not modular, then the computational picture will be applicable to disorders that reflect dysfunctions in modules. If theory of mind is a modular capacity, then perhaps a computational theory of autism will be possible. If basic emotions are modular, some elements of mood and anxiety disorders may be computationally tractable. I discuss a theory of phobias as implicating evolved modules in chapter 8.

How should we react to my three morals? In the case of perceptual systems, we can identify computational mechanisms and processes objectively. Rationality is not like that. We cannot identify what counts as normal or abnormal function without making normative judgments. This is true even if we develop a satisfactory formal theory, because building the formal theory involves making decisions at the outset about whom to exclude and what counts as rational.[9] What counts as normal vision is not contested in the way that we contest nondemonstrative inference, or the correct level of risk-aversion, or the right amount of self-esteem. But for psychiatry to turn its back on the fuzzy aspects of our functioning would be for it to give up much of what people want from it. The fuzzy aspects are central to our thinking about human nature and they are implicated in the paradigmatic mental illnesses, such as mood disorders and psychoses.

The project of naturalizing reason and the project of mechanizing it are different. We may be able to carry out the former, but the theory will remain a normative one. The mechanization project, at the moment, looks unattainable. And it is notable that the successes of the mechanistic program seem to coincide with the areas where the normative aspects of our thinking about human nature do not occur. Cognitive theories of attention, perception, motor behavior, parsing, working memory, and so on seem not to raise the sorts of normative considerations involved in theories of planning, judgment, emotion, interpersonal relationships, and problem solving.[10] Psychiatric theory construction may be guided by personal-level theories of good human cognitive performance, but these will, in many cases, be contested on nor-

9. Such initial judgments can be revised, and so can the norms which inform them. But these revisions, like the rest of the theory construction, will involve an interplay between empirical investigation and folk psychology.

10. It seems too that the division between central and peripheral systems demarcates areas of our mental life where self-knowledge matters from those where it does not. Nobody cares whether we have introspective access to the mechanisms we use to parse sentences or compute the direction from which a sound is coming. But a failure of self-knowledge with respect to the sources of our decision making, beliefs, and emotional life is much more unsettling. This must be bound up with the other properties of central systems, but I do not examine that here.

mative grounds. This means that the foundation in positive facts that the two-stage picture wanted is not always attainable. But in many cases it may be, and we cannot tell in advance where the mechanization project will work and where it will not. If there are a number of modular capacities in our psychology then the two-stage picture might in fact be quite widely applicable, and we can chip away at cognition piecemeal. If there are very few modular capacities then our ability to decompose and localize human cognition is very limited. The best we can do is to locate aspects of our rational abilities in major brain areas, but without intervening computational levels. That will lessen the scope of the two-stage picture and inhibit the development of multilevel models of disorders, thus reducing the number of points in the system where we might try to intervene. The implications of this for psychiatry are examined in more detail when I discuss explanatory models of schizophrenia.

To conclude this discussion of the potential ineliminability of the normative in wide swathes of psychiatry, I revisit the constructivist position I attacked in chapter 2. I said there that constructivists typically argue that what we call mental illness is just deviance with respect to norms. Have I now vindicated this idea? Not really. The constructivist mistakes remain: their view still is not plausible as conceptual analysis nor as a general picture of mental illness. And we now can see that something else is wrong as well. Constructivists have argued again and again that the mentally ill flout our moral values, and these constructivists often have often seen normative issues as reflecting purely subjective points of view. The discussion in this chapter suggests that both points are wrong. First, often it is rational norms that matter, not moral norms—although they are related. Second, it is too quick to assume that there cannot be objective agreement over at least some norms of human rationality and irrationality. The naturalization project, if it could be carried out, would rebut constructivism. The mechanization project would refute it wholesale. But I have said that as things stand, the mechanization project, which the objectivists about psychiatry bet on a great deal, cannot be viewed optimistically across the board. The success of the two-stage project and the full explanation of mental illness depend on exceptionally difficult issues in human cognitive architecture.

A More or Less Realist Theory of Validation as Causal
      Explanation

A theory stands to its subject matter in exactly the same relation as a map does to
the territory it represents.
—James Fitzjames Stephen, *English Jurisprudence* (1861)

6.1   Introduction

The previous two chapters introduce a perspective on psychiatric explanation that pays special attention to molecular biology and cognitive psychology. These are the dominant theoretical components of biological psychiatry
and cognitive neuroscience, the two disciplines I would like to see merged.
Chapters 4–5 look at the prospects for generalizing the explanatory strategies of computational psychology, and the problems cognitive architecture
puts in the way of that generalization of cognitive neuroscience throughout
the mental disorders. The molecular and the computational are the two areas
in which psychiatry is likely to see the most progress as it synthesizes with
cognitive neuroscience again, after their decades of estrangement.
   This chapter goes into more detail about how they relate. Taking schizophrenia as an explanandum, it examines how cognitive and molecular perspectives might intertwine in the explanation of that disorder. I also go into
more detail about what an explanation ought to look like. The current
picture of explanation in psychiatry puts unattractive philosophical roadblocks in its own path. Biological psychiatrists have taken from physicians
the goal of knowing the causal structure of illnesses. But their understanding of validation is still the neo-empiricist notion of construct validity,
borrowed from mid-twentieth-century behavioral science. The current
understanding of validating a disorder is an unstable blend of two themes:
operationalist beliefs about the validation of concepts in terms of other
concepts serve a broadly realist ambition of relating our concepts of mental
illness to the causal structure of the mind/brain.

The desire for causal understanding should be fostered, and construct validity suppressed. But the causal structure we explain will be, in appropriate ways, an idealization, since the variety across patients in terms of symptoms is enormous, even where those patients share a syndrome. So we need to find, among the variety, an idealized representation of the symptoms and course of a mental illness, and try to uncover the idealized causal relationships that account for that idealized picture. I call the idealized representation of the syndrome and its course an "exemplar" of the disorder, and causal validation is the explanation of why exemplars have the form they do. The result is a model, consisting of the exemplar plus its causal backstory. I understand exemplars as not only the objects of psychiatric explanation, but also the taxa employed in psychiatric classification. Nosology, I stress in chapters 9 and 10, is the sorting of explained exemplars into classes that reflect causal-historical structure, a blend of proximate (neural) and distal (developmental and environmental) causes. I also argue later on that the models we develop by explaining exemplars can be exploited for clinical use, because they admit of articulation in specific ways to fit specific patients.

For now, though, the focus is on explanation. In this chapter and the next two I discuss some ways to explain exemplars. The order of topics is, first the biological, then the social, then the historical. I begin by relating the cognitive and molecular themes sounded so far to the conceptual structure of exemplar explanation. Molecular and computational processes go on inside the skin. But the skin is not a very interesting boundary in the philosophy of mind, and in the next chapter I discuss social processes that can have an effect on psychology when they interact with internal processes. Then in chapter 8 I discuss possible avenues of evolutionary explanation.

I begin with exemplars, which are what psychiatric theories explain. An exemplar is a representation of the clinical features and typical course of a disorder, abstracted away from the detail of individual variation. One way to think of an exemplar is as a description of a typical patient that psychiatrists and clinical psychologists can carry in their minds. Experimental evidence shows that clinicians do in fact think of disorders via theoretical representations that privilege some causal mechanisms over others. Kim and Ahn (2002) have shown this, and their study also suggests that clinicians base diagnosis on symptoms that in their mentally represented theories are considered to be causally central rather than causally peripheral. The exemplar includes a number of features of real disturbed mind/brains, but it is possible that no actual patients embody all the features of an exemplar, although patients who share a diagnosis embody at least some features of the exemplar of that diagnosis, and the typical patient will exhibit the theoretically central features. An explanation in psychiatry aims at modeling the causes and development of psychiatric exemplars.

An explanation might not include information at every level of explanation, since although in general we should identify causal mechanisms at different levels, there may be disorder for which a level is inapplicable. In some cases, for example, there be a characterization in terms of neurological abnormalities and one in terms of functional or affective abnormality but nothing at an intermediate level—it may be just a brute fact, for example, that some forms of brain damage produce a given pattern of cognitive deficits in more or less random ways without an available generalization in terms of subpersonal mechanisms at an intervening level.

Perhaps we should think of exemplars as akin to maps. The newly fashionable analogy between maps and scientific theories (Giere 1999; Kitcher 2001) has attracted philosophers because it lets us say that theories, like maps, can be accurate even though we do not want to say that they are in all respects true, and even though they are produced to serve specific purposes and may selectively exclude or distort aspects of reality. Maps correspond to selected aspects of reality but require additional knowledge to be used successfully. Similarly, we may think of exemplars as representing diseases as discrete entities of a certain type, even though we know that in reality, the match between exemplar and patient will be only partial, and will need to be supplemented by clinical background knowledge (which I talk about in chapter 10).

After I introduce exemplars, my second topic in this chapter is a comparison of the psychiatric ideal of validating disorders with my project of explaining exemplars. Many current theories of validation suffer from unmotivated operationalist commitments. My approach, in contrast, is more or less realist. Exemplars are intended to be a compromise between the needs of a causal explanation of mental illness and the great variation in symptoms that mental disorders present.

We need a way to understand the typical causal profile of typical specimens of disorders. Understanding every last detail of each patient is not a feasible theoretical goal. In fact, it is not a feasible goal from any perspective, although it is a useful regulatory ideal for clinical practice. From a theoretical point of view, the best we can hope for is the development of explanations that trace the major symptom-types of each disorder to the pathological processes that give rise to them.

Because of the variability in symptomatology, similarities are swamped by differences, the statistical effects are small, and the sheer number of components is difficult to track. The idealization can do this work, but there are risks in placing idealized exemplars so centrally in a realist account. The first risk is that idealizations are treated by many philosophers of science as a mere predictive instrument, and so metaphysically eliminable (and so antirealist). I do not want an antirealist account, but it makes sense to allow for a development of theory through pragmatic stages, where predictive adequacy is all we have. As the theory matures, though, we should hope to see real causal relations filled in, with the end result being a realistic model of the disorder. At that point, mature exemplars locate the connections among attributes or factors in the underlying structure of the syndrome rather than in the conceptual structure of our exemplary concepts. We are some way from having mature exemplars. My current stress on idealizations is designed to capture the practices of psychiatry's current, immature state, and admit of development toward a final mature theory in which realist aspirations are fulfilled.

The second risk involved in appealing to idealizations as part of a more-or-less realist account is that although idealizations do play an important role in mature sciences, such as frictionless planes in physics, the legitimacy of appeals to idealization enjoys different presumptive favor in each field. Some idealizations are statistical constructs, and the construct may have some predictive power. But no claim is being made that the predictive accuracy derives from the accurate assignment of, say, weights and directions (both causal powers) to the components of the model. Normally, predictive accuracy is the result of such accurate theoretical assignments. In psychiatry, as I have conceded, there is, at the moment, often no causal story to tell. Ultimately, I expect that causal powers will be accurately assigned to the components of the model, and at that point we will be able, as I mentioned, to think metaphysically, seeing disorders as causally structured slices of nature.

So my approach in this chapter uses much machinery, both exemplars and models, because it is designed to interpret current practice in a way that permits not just a growth of knowledge but a way to make that growth serve a realist agenda. To put a realist spin on an immature science, I aim to establish a set of structures that can be treated differently over time, becoming steadily more metaphysically committed as the science develops and driven by a commitment to uncovering the causal structure of mental illness.

My approach preserves everything attractive in the existing project of validation, and does so because it is a realist theory that looks for unobserved processes—proximate causes—that intervene between etiologies and symptoms and are expressly related to the cognitive neuroscience of the normal mind/brain. The existing approach to validation, on the other hand, is divorced from the science of the normal mind/brain and hampered by its lingering operationalism.

The third topic in this chapter is a concrete treatment of schizophrenia. I set out, on the basis of current theories, how the schizophrenia exemplar looks. Then I sketch an explanation for it: it is intended to be illustrative rather than definitive. I compare explanations that look for a proximal cause of schizophrenia in defective cognitive mechanisms and more distal causes that stress the importance of developmental problems in schizophrenia. We need to combine both approaches, and typically the causal validation of exemplars will involve both a distal and a proximal dimension.

6.2 What Is Explained When We Explain a Mental Illness?

6.2.1 Exemplars and Models

Some histories do repeat themselves, albeit imperfectly. In trying to explain a mental disorder, we prescind from clinical variation across individual patients and treat the explanatory target as a process that unfolds the same way over and over again—a set of phenomena that usually occur together (the signs and symptoms) and that have a natural history (or course)—as a characteristic process that unfolds in a typical, though not wholly determinate, way. On this conception of psychiatric explanation, we think of mental disorders as histories that have an identity apart from the individuals in whom they, as histories, unfold. To explain a mental disorder, then, is to explain an idealized picture of that disorder, to show what causes and sustains it, abstracted away from many of the details of its realization in individual patients. The detailed forms that pathologies take in individuals are the focus of the clinical project, not the scientific one. The scientific project generalizes, whereas the clinical one uses the resources of the science to deal in particulars.

This is not a neutral approach. It uses a picture of mental illness that many reject. One rejection asserts that no shared histories can be found among all the variation. A less dramatic objection concedes that diagnoses aim at discriminating shared histories but claims that exemplars are too remote from the facts and miss most of what needs to be explained. The objection is partially correct, since I assume that explanation in psychiatry targets idealized exemplars that represent ideal types but do not necessarily correspond to any actual case histories. So, in a moment I defend my idea that an exemplar gives the form of the psychiatric explanandum against the objection that it is too removed from clinical reality. For now, I continue to elaborate the basic idea. First, I sketch the concept of an exemplar in a little more detail and then briefly explain the motivation for it, which is that in the absence of deterministic laws in psychiatry we must idealize to arrive at appropriate generalizations.

I use the term *exemplar* to denote the idealized theoretical representation of a disorder that I envisage. I distinguish an exemplar from a model. An exemplar is a representation of the typical course and symptoms of a mental illness, whereas a model is a representation of those symptoms, that course, and the causal determinants of both of them. A model is an exemplar together with an explanation.

Another term that is often used to describe idealized representations of mental disorders is *prototype*. This too has some unfortunate connotations, since for many philosophers and cognitive scientists a prototype is a structured representation that encodes a statistical analysis of the features possessed by the objects the prototype represents. Although statistical analysis is important in distinguishing psychiatric populations from each other and from normal groups, I do not wish to assert that statistical analyses are built into representations of mental disorders, as opposed to simply being one of the tools involved in the construction of those representations.

Actually, all the good words are taken, since Smith and Medin (1981) used *exemplar* to define a kind of mental representation. Although I have no view on the mental representation of categories, one feature of Smith and Medin's exemplars is congenial. In their treatment, all exemplars comprehended by the concept are retrieved from memory and object recognition depends on matching one of the sample exemplars. In my view, exemplars include, along with information about course, all the symptoms that a patient might show, even though most individual patients will not show all of them. Diagnosis works by fitting a patient to a portion of the exemplar, and the exemplar is explained by modeling the process whereby the symptoms in the exemplar express the state of neurobiological system (pathology) that depend in its turn on logically prior causal processes (etiology).

And now let me say something about models. Model-theoretic views of scientific theories now play an important role in philosophy of science as part of a general view about what a scientific theory is (Giere 1988). Not every theory is a model, nor is doing science necessarily a matter of model-building, nor does every scientist mean the same thing by "model." Also, scientific models often are thought of as abstract mathematical descriptions, and although psychiatric models might use mathematics (for example by operationally defining a symptom as present when some result on a test is reached), psychiatric models are not primarily mathematical structures.

A model is an explained exemplar: the exemplar is the typical manifestation of the symptoms and course of a disorder, and a model is the representation of the causal relations that obtain between features of the exemplar and various aspects of the organism. The model has an adjustable relation to the target, which in this case is the disorder. The relation of parts of the model to reality can be adjusted—it will resemble the target (the condition) differently depending on how it is employed.[1] In particular, a model of a disorder that represents its basic causal structure can be supplied with more and more detail, which makes the model steadily less applicable to all the patients who fall under the exemplar, but also steadily more accurate and capacious with respect to some subset of them. At the limit, we might have a model that applies only to one subject.

When we explain a mental disorder, we show that some biological, cognitive or other processes cause the symptoms and course. The exemplar represents the syndrome and course, and the model explains the relations between features of the exemplar in terms of a representation of causal process that occur in a patient. If the model is a good one, its structure resembles the actual processes taking place in subjects who exhibit the symptoms included in the exemplar. This resemblance relation is adjustable in two respects. First, the model presents opportunities for therapeutic actions. The model defines a set of relations that differ from those in a real patient in two ways. One is precision: qua idealization, the model contains a certain looseness, in that the causal relations it represents need to be made more precise when we look at a real patient. The degree to which a symptom is present, for example, might need to be specified precisely in a clinical setting, whereas in the exemplar the symptom can be defined as inhabiting some range of values, any one of which might apply in nature. And not every patient instantiates every feature of an exemplar, and so not every part of a model will apply to a given patient. Once we understand the relations that exist between parts of the model, we can try to manipulate those relations so as to change or forestall selected outcomes. I cover this in a little more detail in chapter 10, where I discuss the ways in which clinicians can employ an exemplar-based nosology.

So when we move from the biological perspective to the clinical one (to adopt Guze's formulation), we move from a more abstract to a more precise model, since we are shifting from a general description of a disease process

---

1. This property of models has been stressed by Peter Godfrey-Smith and Michael Weisberg in unpublished work. I thank them both for helpful conversations, which have influenced this chapter in several places.

to a specific description of the biology of an individual. We do this by censoring some aspects of the original model, making others salient and defining the salient aspects more precisely.

Second, the model can be more or less realistically construed, depending on the information available to the model-builders as well as their general intellectual commitments. The lesion method correlates biological insults with symptoms, but it does not aim to justify the claim that the site of the lesion is responsible for the normal form of the behavior. The correlation is as a predictive instrument rather than a causal hypothesis. Lesions alone "do not authorize cerebral *localization of function*, that is, they do not mean that a function disturbed by the lesion was somehow inscribed in the tissue destroyed by the lesion" (Damasio and Damasio 1989, 16–17).[2] So a model that relates lesions to features of the exemplar is not a realist's model of causal structure, but it does provide predictive utility. In some cases, this might be all we get, or all we want. In other cases, the model may depict the actual causal relations responsible for the symptoms. This is what biomedical model building aspires to in general, and I treat it as the goal of psychiatry. But the utility of the lesion method shows that less realistic models can be of great use.

So the fidelity of the model-target relation is adjustable. It can be manipulated to fit clinical demands as well as scientific constraints, and it can be more or less metaphysically committed, as my programmatic account requires. The picture, then, is one in which a clinical description and an account of the natural history of a disorder give us an exemplar: they tell us what has to be explained. The explanation should proceed by displaying causal relations. The causes are diverse biological and nonbiological factors, often interacting in complicated ways, and typically they raise the probability that something will happen, but they do not make it certain. Modeling exemplars is not a search for laws governing mental phenomena, but a search for the causal relations that explain the presence of exemplary features. The assumption is that those relations in the exemplar mirror relations that

2. Damasio and Damasio argued that representations are distributed across the brain and brought together as needed, rather than being localized, so the lesion could not possibly indicate the cognitive part responsible for the behavioral deficit. That is a controversial theory about cognitive architecture. But on almost any theory a lesion might not indicate the cognitive part responsible for the capacity. The lesion could interrupt a pathway that projects from the cognitive part, or damage some other part that provides a resource for the affected cognitive part, but in a way that lets others function, since they need less of the resource (Glymour 2001).

obtain in actual humans who meet a diagnosis, but an exemplar is a representation of an ideal patient. A disorder is a destructive process in a human being, and an exemplar represents one such process and its typical outcome. I am not claiming this is a general picture of explanation in science, or even in the life sciences more broadly. I do claim that it preserves most of what is attractive in contemporary psychiatry, makes sense of many of the semi-articulated theoretical claims of psychiatrists, and gives the whole business a realist slant.

The slant is realist even though the relation between explanation and reality is indirect, mediated as it is by the exemplar. The relations that hold between bits of the exemplar track relations that obtain between entities in the patient, and these entities include not just observable symptoms but unobserved causal processes. Not every relation in the exemplar exists in the patient, since some processes do not occur in every case and some symptoms are not displayed. An exemplar is a faithful depiction of causal relations in nature, but not all the causal relata exist in any given case: that is, some people instantiate only some aspects of the whole exemplar. People may also instantiate more than one exemplar at a time; they can differ with respect to the aspects of an exemplar they instantiate. I now turn to simple exemplars.

6.2.2 Simple Exemplars

Exemplars represent the symptoms and natural history of disorders. In psychiatry we aim to explain exemplars, and in classification they are the basic taxa. Some exemplars may be very simple, with few symptoms and an uncomplicated course. Symptoms include any measurable or observable characteristic, so a lesion is a symptom, as is a behavioral test score, a brain scan, or a biopsy result. Simple exemplars need few additional assumptions to apply to individuals. Disorders that we presently see as neuropsychological rather than psychiatric often have simple exemplars. Associative agnosia, for example, is defined as the inability to recognize by sight an object that one otherwise has intact knowledge of, and that one can recognize by touch or from a description (Farah and Feinberg 2000, 81). This description, together with a description of the associated neural damage, is a simple exemplar. Often we can associate the symptoms with a lesion, which also counts as apart of the exemplar, since part of the normal course of the disorder is that it is a consequence of the lesion. Most disorders do not begin with physical insults in this way. But even simple exemplars mask individual variation, since some patients are more handicapped than others, and there are differences in the form of the disorder, with some patients being especially poor at face recognition and others doing worse at printed-word recognition. Furthermore, the exemplar for associative agnosia, as with many neurospychological conditions, abstracts away from other cognitive deficits that a patient may have, and offers us a partial picture of the total psychopathology. In any case, the disorders that neuropsychologists seek to explain are typically abstractions away from the additional deficits that almost invariably accompany a fine-grained disorder, since usually brain damage involves not just the pathway leading to the symptom of interest, but other pathways too.

In my example a statement of a neurological deficit plus a description of a lesion together make up a simple exemplar, and the explanatory task is to show how the lesion causes normal relations among cognitive parts to be interrupted or otherwise deformed, so as to produce the symptoms. This is the basic picture of exemplar explanation: describe a symptom, find a physical correlate, and explain their real-world relation in a theoretically significant way. How plausible is it to hope that the much more complicated exemplars we will generally need for psychiatric disorders can work well enough to make this method a useful general approach?

Some psychiatric disorders are very simple. They are defined "monothetically," in terms of just one symptom. The paraphilias, for example, are defined simply as intense, sexually arousing fantasies, urges, or behaviors involving sexually deviant objects or processes: each paraphilia has its customary focus. As I pointed out in chapter 3, there are serious problems with regarding the paraphilias as disorders. I refer to them here because they are such a clear case of monothetic categorization. Frotteurism, for example, is defined as follows (DSM-IV-TR, 570):

A. Over a period of at least 6 months, recurrent, intense sexually arousing fantasies, sexual urges or behaviors involving touching or rubbing against a nonconsenting person.
B. The person has acted on these sexual urges, or the sexual urges or fantasies cause marked distress or interpersonal difficulty.

The paraphilias are not the only monothetic disorders currently recognized, but most DSM diagnoses do recognize diverse symptoms and great variation across subjects. Sometimes one or more symptoms are essential to the diagnosis, but often none are. In Major Depression, for example, there are nine possible symptoms which, in short, are: depressed mood most of the day; diminished pleasure in all activities most of the time; unexplained changes in weight or appetite; abnormal sleep patterns; psychomotor agitation or retardation; fatigue; feelings of worthlessness; inability to concentrate; thoughts of death or suicide. To merit the diagnosis of major depression, at least five of the nine symptoms must present together during a two-week period, and at least one of the presenting symptoms must be depressed mood or loss of pleasure (DSM-IV-TR, 349–356).

DSM categories often group together very different symptom profiles as manifestations of the same disorder. Thus one can qualify as having a Major Depressive Episode even though one does not experience a depressed mood, provided that one does exhibit a markedly diminished interest in daily activities. And by one reckoning there were 56 different ways to satisfy the DSM-III criteria for Borderline Personality Disorder (Clarkin et al. 1983). As one might expect, these heterogeneous categories often are poor predictors of patients' future trajectory or their response to treatment and thus the vast majority of DSM categories remain "unvalidated." I deal with these complications in a moment and revisit them in chapter 9 when I discuss classification. But first, we should look at the other main aspect of an exemplar, its course.

6.2.3 Kraepelin Redux

DSM-style psychiatry often is called "neo-Krapelinian" in honor (or not, depending on who uses the term) of Emil Kraepelin. Kraepelin, born in the same year as Freud, argued that different mental disorders had characteristic histories, and he kept detailed records on the progress of his patients until he felt able to generalize across them, classifying individuals together on the basis of their shared histories. This stress on following the course of a disorder led him to distinguish what we now call schizophrenia. It started, he argued, with a slow decline of general psychological function, accompanied by headaches and physical lassitude. Over time, a progressive dementia set in, at different speeds in different individuals. In some, it did not progress very far. In others, it was swift and calamitous. A stage of excitement and euphoria occurred, followed by a period of calm that appeared to indicate improvement, but soon showed itself to be the onset of severe and profound mental collapse. Kraepelin pointed out that this progressive deterioration characterized a number of patients who had, on the basis of their clinical presentation, been previously thought to exhibit different conditions, including Kraepelin's own diagnosis of dementia paranoides, as well as catatonia and hebephrenia.

Kraepelin applied the term "dementia praecox" to the syndrome whose course he had charted. He argued that only a small minority of hebephrenia patients had dementia praecox, on the grounds that they did not fit Hecker's understanding of the course of hebephrenia. He also argued (1899,

173–175) that observation of patients with dementia praecox revealed that their mental capacities tended to regress over time, and that this contrasted with Hecker's understanding of hebephrenia as the arresting of normal development. The slow onset and initially less severe symptoms he took to distinguish dementia praecox from "periodic insanity" (which we would now call manic-depressive illness or bipolar depression).

Kraepelin (1896) also argued that the sudden onset of dementia praecox showed that it could not just be the result of an inadequate inherited psychological constitution that deteriorated slowly until it attained clinical significance. Rather, he thought, the cause of dementia praecox had to be a distinct neuropathology that occurred in adolescence.

Discrimination among conditions via information about their histories is a characteristic method of immature medicine: Sydenham used information about course to distinguish similar syndromes in the seventeenth century. Contemporary psychiatry still uses this method to disentangle related diagnoses or support a unitary interpretation of a category. Disagreement over whether a diagnosis is unitary or a collection of several subtypes often turns on the extent to which we can distinguish distinct populations of patients based on separate outcomes. Jellinek (1960), for example, argued that there were several classes of alcoholics, each with a characteristic natural history. But Vaillant (1995, 38–39), on the basis of a decades-long study, argued that "when a longitudinal view of alcoholism is substituted for the cross-sectional view, there do not appear to be many different alcoholisms."

So the course of a disease should be represented in the exemplar along with its symptoms. Exemplar explanation works by displaying the causal relations among pathogenic processes that produce the symptoms and cause them to occur when they do and unfold as they do. The issue still on the table is whether this is possible even in principle once we move beyond the monothetic neurological cases. In cognitive neuropsychology, we may hope to associate a pattern of physical trauma with a circumscribed, measurable symptom that can be treated in isolation by the theory even if it seldom shows up in isolation in the data. An exemplar is a set of symptoms that can be abstracted from the data and studied together as the target of explanation. At this stage, I just argue for the possibility in principle of generalizing the exemplar approach beyond the simple cases (and even the simple cases are difficult to explain). The discussion of schizophrenia in section 6.4 is a more detailed defense by example. But now I turn to the difficulties involved in moving to the polythetic diagnoses that concern psychiatrists.

6.2.4   A Simple Example of Variation in a Symptom—Sleep

In some disorders we see simple exemplars in which patients share a key symptom or two. The overwhelming majority of disorders seem to involve several symptoms that vary across patients. In this section I defend the idea that much more complicated exemplars are amenable to the same basic treatment as simple ones. First is a case in which one symptom can be instantiated differently in different people who share the same diagnosis, then cases of greater variation.

As an example of variation in one symptom, consider that major depression often makes one sleep much more or much less than usual. Clearly we cannot expect to see these two forms of the symptom coincide in the same person at the same time. But what depressives can share is a disruption to normal circadian rhythms. To explain this aspect of the major depression exemplar, then, we must find a causal pathway that can lead to either type of circadian disruption—excessive or insufficient sleep. This might involve showing how the neurons in the suprachiasmatic nuclei (an inch or so behind your eyes), which house our circadian mechanism (Hastings 1998), can have their normal autoregulatory function interfered with by cognitive dysfunction in other systems. A model of depression might in that case include a causal pathway running from disrupted central systems, via physiological arousal due to anxiety or despair, through the circadian mechanism, and ending up, depending on the details of the arousal, in either excessive or insufficient sleep.

The crucial point is not the symptoms themselves, but the causal pathway that runs through the circadian mechanism, because it is disruption to that mechanism that causes the abnormal sleep patterns. The symptom—sleep disturbance—can take either of two abnormal forms, so the exemplar should recognize both of them. But even if individuals with major depression do not have the same profile of sleep disruption, they may share a common proximal mechanism that when affected causes the sleep problems. In modeling the etiology of depression, we can try to show how variation in circadian regulation depends on variation in other cognitive mechanisms. We can model the suprachiasmatic nuclei as the final common pathway through which the values assigned to cognitive variables elsewhere in the model can influence sleep.

People who share a diagnosis share an exemplar. They sometimes share symptoms in a precise sense: they will return the same result on a test. More often, they share a disruption to a mechanism that can produce one of a range of test scores, depending on the context. Imagine that the circadian mechanism reacts to stress by either overexpressing or underexpressing the relevant genes (mammalian *per* and *tim*, as it happens) in the suprachiasmatic nuclei. In that case, we can figuratively imagine a mechanism with a switch that reacts to stress by moving up or down the dial. We can say that depressives have a circadian mechanism that is disrupted by stress and which moves either up or down depending on the other processes at work in the mind/brain of the depressed person. The goal, then, is to understand the relation between properties of the circadian neurons and states of the cognitive system, such that, when the cognitive system is working normally the circadian system has an output within normal bounds. But if the cognitive system moves into some abnormal state, its relation to the circadian system ensures that the latter system is pushed out of its normal state, too, with the result that sleep is either excessive or insufficient. We can readily imagine a feedback effect in which disruptions to normal rhythms have an adverse effect on psychological capacities, like reaction times, that degrade in step with sleep deprivation. Such a behavioral outcome could feed back into the cognitive systems, making one more anxious and thus pushing the circadian mechanism even further askew. We know for example that sometimes an interaction between low mood and poor sleep arises from reduction in exposure to daylight, which affects some people very adversely. I am imagining a similar, endogenous relationship between stress or anxiety and sleep mechanisms, including the circadian clock, brought about by nonphotic interventions from the outside world.

The generalization about depressives I envisioned just now would, if we could make it, reflect a robust relation, one that holds as other conditions change. Indeed, I assumed that changes in the values of cognitive variables such as reaction times, and circadian variables such as hours of sleep, vary in step because they reflect a systematic causal relationship. The relationship continues to hold even when one or the other end is pushed out normal function by some pathogenic process. The process could be a chemical one or a cognitive one: either way, it will have an effect on the cognitive processes that can impact the mechanisms underlying normal sleep.

I would not advise betting a pension on the story I just gave. But it does raise an important idea we can exploit when we try to generalize the simple story of modeling an exemplar that I referred to in discussing cognitive neuropsychology. The simple story was one in which a behavioral deficit is associated with a lesion and our job is to explain why a lesion *there* should have a robust relationship with *that* behavioral outcome. On the more complicated story about depression I just made up, we explain the causal process linking some neurocomputational abnormality in one cognitive part to a second cognitive part (the circadian regulator) that mediates between the first part and the behavioral effects, and can generate a range of outcomes.

Two parts of the story I just told are central to the strategy of generalizing the simple neurological exemplar to all psychiatry. The first is the identification of cognitive parts in the brain that produce the symptoms of the exemplar when their normal functioning is disrupted. That is, a systemic breakdown produces a change in the observable properties of the subject's behavior. The explanations of delusion in chapter 4 had this form: they explained delusions as the product of either breakdown in belief fixation systems, the interaction of unusual but normal belief systems with breakdowns in peripheral systems, or the effect on cognitive systems of a brutely organic change in neurotransmitter profiles. If variation in behavioral outcomes reflects the variety of possible outputs of a cognitive part, then we can generalize across patients at the level of cognitive parts, even if we cannot do so at the level of outcomes.

An effect of pathogenic processes that in its turn causes behavioral outcomes admits of many interpretations: cognitive, endocrinological, neuroanatomical, genetic, and so on. The relevant system could be characterized at any level, or at several, and the relationship could cross levels, going from a neurocomputational characterization to a very-low level one, as in my example of computational disruptions in cognition causing dysregulation of *per* and *tim* genes.

The recent enthusiasm for endophenotypes reflects an explanatory bet on a small number of mediating variables rather than a large array of causes and outcomes. The endophenotype (Gottesman and Gould 2003) is an unobserved intermediary between phenotype and genes. Given the typically large number of genes involved in complex traits, and the variation in phenotype across individuals within a diagnosis, Gottesman and Gould advocate a search for endophenotypes as a simplifying heuristic. They suggest that a complicated set of symptoms probably depends on a smaller set of abnormalities at various levels, and those underlying abnormalities depend on the expression of a set of genes. Rather than look for a great variety of genes and a great variety of phenotypes, we should look for endophenotypes, which depend on a set of genes and give rise to smaller sets of symptoms.

Gottesman and Gould see endophenotype as uncovering the genetic causes of disorders. But we can liberalize the concept. Many biological entities can cause systemic failures. And many different outcomes can result. The hope is that by identifying systems that mediate between causes and symptoms we can impose some tractability amid all the variation. If exemplar explanation is to work, it will do so by tracing those systemic relationships.

But the relations we are likely to find between systems, and between systems and behavior, are context dependent. As chapter 4 shows, psychiatric generalization often will fail to be fundamental: the more detail that goes into the model, the less likely they are to hold. A good place to start thinking about the significance of this is to return to the issue of clinical variation in presentation of symptoms. An actual patient is, in a sense, a model of a disorder enriched with a set of real-world facts. So, with the simple picture in hand, we take up the objection that patients have too much diversity to make the project of extracting and explaining an exemplar a viable one.

6.2.5 Theory and Clinical Presentation

I have introduced exemplars as the objects of explanation and classification in psychiatry and explained that they are idealizations intended to capture what is common across patients who share a diagnosis. This approach may seem to give insufficient attention to the important fact of individual variation. Individual patients are all different, and each case of mental illness is a unique tragedy. The stress on individual variation is characteristic of the clinician, who typically sees patients who fit several diagnoses up to a point, but none very well, and who differ from one another even when they share a diagnosis.

But the idea that psychiatric inquiry and explanation should be built around idealized exemplars is not at odds with a proper recognition of individual variation, which may serve different ends. Individual variation raises two problems, in fact. The first is the degree of idealization that is feasible in a model. I discuss this problem in detail in section 6.4, when I deal with the problem of individual variation in the case of schizophrenia. The second problem of individual variation is that of the relation of the exemplar to actual people, who suffer from an actual condition, not from its idealized representation. I address this issue by contrasting the needs of the theorist and the needs of the clinician, who has to treat actual people.

An optimistic view of individual variation concedes that existing categories are too heterogeneous but argues that once we have better categories that reflect basic science more fully, we will sort patients into classes that exhibit greater unity. On this view, individual variation is an artifact caused by a failure to disentangle subtypes within disorders, and insufficient sensitivity to comorbidity, the co-occurrence of disorders in one patient. With greater statistical and causal sophistication, we will delineate patient populations more precisely and the problem of variation will go away.

I agree with this response up to a point. Even so, it is overoptimistic to think that scientific breakthroughs will solve the problem. The amount of similarity displayed by subjects who share a diagnosis always will be limited: too many contingent interactions exist inside the mind/brain and between the mind/brain and an individual's biography and environment. Cognitive systems, when functioning properly, probably enter only a very small number of their total possible states. If they are perturbed, they are likely to shift into some other part of the space of their possible states, and small initial perturbations could produce great differences in the final states. Individuals who differ only a little in their cognitive malfunctions could differ greatly at the level of symptomatology. And many disorders also display great variation in time, and variation across individuals as to when successive episodes strike. Two people could share a diagnosis and exhibit the same symptoms in January but look very different by the summer, even though they have the same condition and have had it for the same length of time.

So some degree of individual variation is inescapable. The limited concession to variation that an explanatory theory built around exemplars can make, however, almost certainly will not be enough to satisfy clinical demands. So clinicians will need to apply the information contained in exemplars in ways that are unconstrained by the demands of idealization. The contrasting needs of clinical and theoretical settings do not mean, though, that explanatory theories are clinically useless. On the contrary: in making explicit the causal relations among features of a patient's psychology and environment, exemplar explanations can suggest all manner of interventions to the clinician who wants to improve the patient's condition. At the limit, patients may show symptoms that partially correspond to several exemplars, none of them very well. If the variation is great enough, the clinicians may be better off without trying to fit a patient to an exemplar and instead thinking of an individual as having a set of problems that are drawn from several exemplars, and viewing the patient's life as a unique biography rather than an exemplary natural history. But even in such a case, the clinician still may exploit the resources gathered by the attempt to explain exemplars. I take this question up again in chapter 10. For now, I continue my examination of the issues raised by exemplar explanation, which is a way to validate syndromes by showing how symptoms depend on underlying causes. That view contrasts with the existing picture of validation, to which I now turn.

6.3   Validity

6.3.1   Construct Validity

Validity in contemporary psychology and psychiatry is usually construct validity. "A construct is some postulated attribute of people assumed to be reflected in test performance" (Cronbach and Meehl 1955, 283). In this section I explain construct validity and show how it is applied in the "neo-Kraepelinian" psychiatry of the DSM-III and its successors. Validation, on my account, should show that what we measure in psychology really exists. The point of it is to gain a genuine causal understanding of the natural processes that produce psychopathology. Construct validity, on the other hand, does not aim to do this: it aims to define a disease operationally, by relating observations to other observations, rather than defining it causally, by relating observed properties of the disease to unobserved underlying properties. The time has come to reject this picture and look for causal relations. I discuss validity in the DSM and defend my realist's distaste for construct validity section in section 6.4.2. There, I also respond to some objections to the idea that validation can be realist. Exemplar explanation is ultimately a realist theory of validation, although, it may pass through other stages on the way.

Demonstrating construct validity requires showing that measurable aspects of psychopathology occur together in diagnoses that presume them to be present and do not occur in diagnoses that presume their absence. A measurable aspect of psychopathology is typically some quantity that is determined by a test. It might be a test of autonomic arousal, or a psychological test designed to indicate the presence of psychopathology. The Minnesota Multiphasic Personality Inventory (MMPI) is probably the most widely used diagnostic test in psychology: it contains 567 statements about oneself with which one must agree or disagree. Some have clear intuitive psychological import, whereas the significance of others is not immediately apparent. It is no surprise that depressives tend to agree more than nondepressives that "at times I think I am no good at all," although it is rather surprising to learn that paranoid individuals are more likely than other people to enjoy reading mechanics magazines.[3] The MMPI is a personality test that scores individuals along ten psychological dimensions. A number of tests specifically dedicated to diagnosing particular disorders have also been created.

As well as scores on test instruments, less formal indicators of pathology exist, but measurement is the currency. Construct validity's purpose is to show that there is a relationship between all the indicators we observe, and the relationship obtains in all and only the situations where a theory says it should obtain. Independent measures are supposed to converge on the construct: that is, independent measurements give the same result. But the

---

3. Dawes 1994 reviews the utility of such instruments. Some—like the famous Rorschach inkblots—appear to be completely useless (Wood et al. 2003).

upshot of all this is not a claim that we have discovered something in nature that has a causal influence on what we observe and measure. Rather, one claims to have shown that a concept introduced as part of a theory has a satisfactory statistical relation to other concepts. As Blashfield (1989, 26) puts it, construct validity "focuses on the extent to which a test measures a particular concept as that concept is explicated within a particular theory." To establish the construct validity of a depression scale, for instance, we could compare people who score highly on the scale with people who do not, to see whether people in the first group share some property that the low-scoring group lacks. The property under investigation is one that the theory behind the scale assumes to be associated with depression, such as personal loss. Construct validity attempts to understand concepts in terms of other concepts in the same theory. It is not a test of concepts against an underlying reality, but is more akin to theoretical definition in the empiricist tradition, comparable to philosophical techniques like Ramsification (Trout 1998, 69–72). The point of construct validity is not to uncover something in nature, but to find agreement across measurements. What in the world that agreement ultimately depends on—the nature of what is measured—is not supposed to be reached by these methods. The methods reflect the operationalist assumptions of mid-twentieth-century behavioral science. Construct validity explains consistency of observation by postulating a coherent relationship between concepts that can serve as sources of measurement (McHugh and Slavney 1998, 301–302; Trout 1998, 70–71). It is important to realize that a construct is a relationship among concepts within a theory of observable behavior, and "the theory specifies in some fashion or other (1) the meaning of the construct, (2) how it is related to other constructs, and (3) how it is related to specific observable behaviors." (Ghiselli et al. 1981, 282). Note that Ghiselli et al. do not say that the construct is related to underlying casual processes. Construct validity betrays its operationalist heritage—it detects a relationship between concepts, not a relationship between measurements and unobserved natural processes.

Now, construct validity is not worthless. Consider self-esteem, which psychologists have studied extensively. Researchers readily propose and study functional relationships between self-esteem and other psychological variables, such as depression or academic achievement. For example, we find that damaged self-esteem is involved in depression and that complex self-representation lowers self-esteem (Woolfolk et al. 1995). This research does not suggest that there is a self-esteem system in our brains whose properties are being measured. Rather, a valid self-esteem construct enables us to make predictions about behavior by relating a measure of self-esteem to measures of other human traits. Suppose we want to boost the self-esteem of ninth grade girls, because we think that increased self-esteem will lead to fewer pregnancies and better test scores. Then we need common agreement on something that measures self-esteem, so we can compare scores before and after the implementation of measures designed to boost it. If the relations between self-esteem and the other measures are what we care about, then the underlying neuropsychology of self-esteem may not matter. All that matters is that we can measure self-esteem and relate it to pregnancies and academic performance. The point is not that there is no such thing as self-esteem. But the causal processes that produce and alter self-esteem are not addressed by the theory; the causal processes are unobserved aspects of the individual's mind/brain. The study of self-esteem in relation to depression and academic achievement investigates relationships between observable properties. That is how construct validity works. It does not mean that construct validity is a nonsensical concept. It is, though, an operationalist concept that fails to relate observed properties to unobserved processes.

"Operational" concepts are defined in terms of techniques of measurement and observation. One role for operationalism in the construction of a taxonomy was outlined in a classic paper by Carl Hempel (1965). Hempel understood the process by which a science acquired a mature taxonomy as follows. The process begins with a "natural history" that describes the phenomena and makes preliminary generalizations. This is followed by theoretical stages, beginning with the introduction of operationalized terms. Bridgman (1927) defined an operational definition of a term as a stipulation that a given operation yields a given outcome. Hempel acknowledged that in psychiatry Bridgman's definition would need to be relaxed somewhat, to include terms defined via observation without any measurement, for example. But Hempel saw operational definitions as only the first stage in the construction of a mature taxonomy on the foundation of a natural history.

As inquiry progresses, Hempel maintained, it is "marked by the introduction of new, 'theoretical' terms, which refer to various theoretically postulated entities, their characteristics, and the processes in which they are involved; all these are more or less removed from the level of directly observable things and events" (1965, 140). According to Hempel, theoretical terms distinguish a mature science, and as psychiatry developed, "classifications of mental disorders will increasingly reflect theoretical considerations" (150). Hempel expected either the then-dominant psychodyamic approach or biological psychiatry to become the theoretical basis for taxonomy.

Hempel was a great philosopher of science but an unreliable prophet. In the last thirty years the old psychoanalytically based paradigm has been abandoned but not replaced by a nosology grounded in a theory. We still await the mature stage. Instead, the classification system has become officially wedded to a view that sees operationalism not as a stage on the way to taxonomy, but as the fruition of the nosological enterprise. The official idea is that disorders are defined simply in terms of what is observed and can be agreed to on that basis, without attention to controversial theoretical hypotheses.

So the makers of the DSM-III deliberately turned away from Hempel's view that a mature taxonomy in any science should be causal theoretical, a view that has gained widespread support in the philosophy of science. Instead, they embraced his second stage—the stage at which operationalized concepts are employed as a move toward theoretical understanding. Diagnostic criteria are typically specified by a list of sufficient conditions (often disjunctive and with an occasional necessary condition thrown in) stated almost exclusively in the language of "clinical phenomenology." This jargon draws heavily on folk psychological concepts and protoscientific clinical concepts (such as self-esteem, delusion, anxiety, and depressed mood).

The concepts and categories of clinical phenomenology are notoriously vague, imprecise, and unquantified. By limiting the data gathered in diagnosis to the salient and easily identifiable signs and symptoms of clinical phenomenology, the DSM-IV-TR scheme ignores a wide range of other data about mental functioning that can be gathered by psychometric techniques and by methods used in cognitive science and neuroscience. The advocates of the DSM approach continue to insist that their concepts are operational. But as several people have pointed out (e.g., Wiggins and Schwartz 1994), most of the concepts of clinical phenomenology are not operationalized even in a loose sense. They often involve assumptions about the inner state of the patient and are not tied to measurements or direct observation but to psychiatric lore. In addition, a number of ostensibly theory-neutral terms actually are drawn from some theory or other. Wrenching a term out of context is not the same as giving it an operational definition. The applicability of these concepts can be quantified, so that a patient can be rated for, say, anxiety, on a scale using standard testing measures. These testing measures are nontheoretical in the sense that they are independent of any one theory and can be used to measure traits that are recognized in different theories. However, the measured traits usually are posited on the basis of psychiatric lore and are qualitative concepts. The extent to which someone is anxious can be measured, but the concept of anxiety is not defined in terms of measurement, nor really in any terms beyond the folk psychological, as modified by clinical training. The difference is between using a test to define whether a trait is present (as for example, the ability to pass the false-belief test has come to operationally define possession of a theory of mind) and using a test to determine the magnitude of a trait that already has been assigned to a subject on impressionistic grounds.

Modern psychiatry relies on four main indicators to validate a diagnosis. These four criteria descend from the seminal work of Robins and Guze and their colleagues in St. Louis, who proposed a fivefold process designed to distinguish disorders from one another (Feighner et al. 1972; Robins and Guze 1970; North et al. 1993). They envisaged a process by which clinical description was supplemented by laboratory studies (perhaps including psychological tests) and exclusion criteria to distinguish diagnoses from each other. Follow-up studies would see whether the initial patient population could not be further distinguished based on the course of the syndrome, and family studies would see if the patients' relatives included other people who met the diagnosis. In theory, a valid syndrome must meet these four conditions, although any of them may be fudged in practice, and most diagnoses in the DSM-IV-TR remain unvalidated.

Natural history The natural history of a syndrome is its characteristic clinical course and outcome. Kraepelin, for instance, distinguished the two major functional psychoses, schizophrenia and manic depression, by noting that schizophrenia (which he called *dementia praecox* or "premature dementia") tends to be chronic and steadily and unremittingly worsen over a lifetime, whereas manic-depressive illness (Bipolar Disorder) is episodic and had, Kraepelin thought, a better prognosis (Goodwin and Jamison 1990). Some disorders are distinguished by the age at which patients tend to fall foul of them. Kraepelin viewed dementia praecox as largely a disorder of adolescence. Panic disorder has a mean age of onset of 25 (Goodwin and Guze 1996).

Family history Most psychiatric conditions run in families, although apparent exceptions include anorexia nervosa. The conditions currently studied by neuropsychologists rather than psychiatrists are also exceptions. Many families must cope with behavioral problems from generation to generation, even though the causes might not be in their genes, and even though the conditions may vary across generations. Generally speaking, though, mental disorders do seem to have a genetic component.

Differential response to treatment   Two seemingly similar sets of symptoms can be distinguished if they respond to different kinds of treatment. When psychiatrists talk of differential response to treatment they usually have medications in mind. Often, however, this "pharmacological dissection" does not work; most disorders respond to more than one drug. For example, both major depression and Obsessive-Compulsive Disorder respond to Prozac, which also is being marketed under another name as a remedy for premenstrual tension. (Just as headaches, sprains, and backache all respond to painkillers, which rather suggests that the utility of pharmacological dissection is limited.) Furthermore, many disorders include symptoms that respond to drugs and other symptoms that do not. In addition, some classes of patients may differ in responsiveness from the general patient population. Prozac is markedly less effective on adolescents, for example, and it seems to increase the risk of suicide in the mildly depressed, although it lowers it for the profoundly depressed.

Laboratory findings   Laboratory findings link symptoms to underlying physical markers. But in many cases no distinctive physical markers can be reliably associated with a given symptomatology: correlations are often very low, and the markers also correlate with nonpsychiatric problems. The following, for example, all have been claimed to be indicators of schizophrenia: increased ventricle-to-brain-tissue ratio; abnormalities in the basal ganglia, corpus callosum, or cerebral hemispheres; blood flow in the left globus pallidus; lowered platelet monoamine-oxidase activity; hearing loss; eye-tracking dysfunction; oligodendroglial dysfunction; elevated levels of Borna virus antibodies; and morphological abnormalities, including high arched palates and low-set ears (Davis et al. 2003; Goodwin and Guze 1996; McHugh and Slavney 1998). Laboratory findings associate a physical marker with a set of symptoms: this is another relation among measurements. The physical markers should occur in the presence of the diagnosis, but that does not mean they explain the symptoms.

6.3.2   What Is the Purpose of Validity?

We have seen what construct validity is supposed to be. But why is it important? The processes by which validity is established are clinical. That is, they are supposed to separate similar sets of symptoms by careful observation. These were the methods that Sydenham, in the seventeenth century, used to separate scarlet fever from measles, and discover that the pox was in fact two diseases, chicken pox and small pox. Sydenham's achievements were profound, and Robins and Guze (1970) claimed that their methods enabled them to distinguish between schizophrenics with a poor prognosis and those with a good prognosis, based upon clinical features. Nobody denies that good clinical observation can sort patients into classes. And construct validity works by measuring properties of members of the classes that result from the sorting process, and relating those properties to measurements of other postulated properties of members of that class.

The motivation behind the concept of validity, and construct validity in particular, emerges in McHugh and Slavney's discussion. McHugh and Slavney (1998) stress the fourth of the hallmarks of validity I mentioned in the last section—the search for a physical marker. A disease, they say, "is a construct that conceptualizes a constellation of signs and symptoms as due to an underlying biological pathology, mechanism and cause" (1998, 302). But McHugh and Slavney also seem to take the construct idea very seriously and deny that diseases exist as natural entities. If they deny that the construct represents real natural processes and properties, then their commitment to discovering what happens in psychopathology becomes harder to interpret.

A valid construct does seem indeed to be a model of something in nature—it shows us how the effects of some underlying destructive process show up in a form that we can observe. Other theorists have been much less equivocal on this score than McHugh and Slavney. Kendell (1975) was explicit that he wanted a classification of disorders that carved nature at the joints, and Robins and Barrett (1989, v) argued that the point of validation is to assure us that our conceptualization of a disorder matches an actual process in nature:

Psychiatric diagnosis enables a wealth of facts regarding the patient's history and current state to be communicated in just a word or two. But we ask more of diagnosis than efficient communication. We want it to be valid, by which we mean that we want it to correspond to what exists in nature—to describe a "real" disorder.

As I have indicated, my sympathies lie in this camp (although I would not have put those scare quotes around "real"). I think that we aim in classification to track genuine structure in nature, on the basis of causal theories that explain why nature has the structures that it does. As Kendell says (2002, 12–13) the strongest source of validation is causal understanding. Ultimately, we want not just a correlation of observable effects with one another, but an understanding of the natural processes that give rise to these effects.

Construct validity treats the mind/brain as a black box and relates one set of outputs to others. But science is maturing and we are increasingly able to look inside the black box. To justify the psychological concepts recognized by our current theories it is no longer enough to simply ignore the inner workings of the mind/brain and validate the concepts by their relation to other behavioral measures. The inputs and outputs that we want to keep must be grounded in knowledge of the mechanisms we see when we open the box.

There are two ways to reject the purely operational version of construct validity. My own preference is simply to deny that the underlying causal relations cannot be measured. If we can measure them, then much of the motivation for relying on constructs is lost. And as psychiatry turns into cognitive neuroscience, we are busily measuring the processes in the brain. A slightly more concessive rejection of simple operationalism admits that underlying processes might be hard to reach, but treats the correlations between measurements that we can make as indicating some underlying relationships, and erects a theory about those relationships and their connection to the measurements. This may be what some contemporary psychiatrists have in mind when they talk about construct validity and causal discrimination in the same breath. But it is important to see that there are more or less realist ways of thinking about what underlies the statistical correlations. And even if, on grounds of pragmatic ease, we adopt the view that the causal dynamics are difficult to access directly, we still need some theory that tells us what the measurements are tracking. And the theory will be better the more it aligns itself with knowledge of causal dynamics where we have it, since then the inferences from statistical and predictive factors will be drawn with more chance of being true.

A theory with considerable predictive power might have to be discarded when we move from the understanding associated with construct validity to that which rests on causal knowledge. Stephen Turner (2002, 6) illustrates this point with the example of a clock. It might be possible, based on the clock's observed behavior, to develop a predictively powerful theory of how clocks work. But once we open up the clock and learn that its inner mechanism includes a wheel, then any earlier theory of the clock that does not mention wheels must be heavily revised. No matter how predictively useful the earlier theory was, its virtues do not make the actual presence of the wheel irrelevant, and the putative reality of the entities mentioned by the earlier theory cannot override the "prosaic reality" of the actual presence of the wheel. Turner assumes that the earlier theory just will be discarded. But it need not be discarded, merely revised in a way that takes the input/output relations it recognizes and connects them with the workings of the clock's real mechanism.

Now, in some cases one might argue that an earlier theory, with its predictive virtues unquestioned, should be preferred, at least for some purposes, to a less useful theory that recognizes the real parts of the system. This position is defensible, but it is not open to defenders of construct validity to adopt it, for in the study of abnormal psychology we simply do not have theories that both ignore the inner workings of the mind/brain and exhibit the predictive power that might tempt one to hang on to them in preference to theories that recognize prosaically, rather than putatively, real phenomena.

Predictively, psychological theories are almost always, in Lykken's terms, "low-powered" (1995, 148). Predictions about what distinguishes patients with a diagnosis from a control group usually do not single out some categorical difference that all and only the patients with the diagnosis will exhibit, such as a lesion or the presence of a particular behavior. More usually, we predict that some naturally continuous quantity will be higher among subjects than among controls, or that one variable will be positively correlated with another. But such statistical effects can seem frighteningly cheap; a prediction that manic depressives are taller than average probably stands a 50 percent chance of being true just because any group of people we single out and measure carefully is likely to be taller or shorter than average. The problem is that in recent editions of the DSM, the model of disease built around construct validity is severed from the idea that measurement is a causal relation between features of the world. So although recent editions of the DSM assume that mental disorders are distinct and identifiable entities that should be validated, it leaves itself without the resources to say what kinds of distinct entities they are. What is objectionable about DSM psychiatry is its simultaneous reliance on intuitive theory and disavowal of causal explanation. Since theory is inevitable, it would be better to base classification on the best explanatory theory we have of how the mind works.

Now, the best theory could be best in the sense that it provides the most ambitious causal explanation, or it could be a different theory that is best in the sense of having the most evidential support. A pair of related objections to the theme of this section exploit the fact that these may be different, so that (1) our best current theory will be overturned as we proceed, and (2) it is not unreasonable to refuse to classify on the basis of causal explanations if all our etiologies are highly speculative.

Of course our current theories will be overturned, but there is a difference between a theory's being damaged in a way that leaves its basic assumptions intact, and a theory's being so mutilated by a falsification that its basic assumptions have to be rejected. My contention is that our knowledge of social, cognitive, and molecular neuroscience is now good enough that we can build a theory of psychopathology that borrows its basic assumptions from neuroscience, and that those basic assumptions will survive the refutation of the detailed explanations and predictions to which the framework gives rise.

So the foreseeable overturning of the next generation of psychiatric theory need not impugn the general framework that the theory will employ. My answer to the second objection is to recapitulate Hempel's expectations. I am not saying that operationalism is never defensible; as I noted, clinical observation can provide a means of distinguishing populations. I agree with Hempel that we should employ operational criteria at an early stage, when we are building a theory in comparative ignorance of causes. But when operationalism is regarded not as a valuable heuristic but as the basis of a general approach to classification, then I do think it is indefensible. Hempel warned us to be alert to the importance of causal information at a later stage; the mistake that the architects of the DSM have made is to have become locked in to a system of classification that has made a virtue of operationalism (of a sort) and thereby isolated itself from theories and discoveries that could produce a causal taxonomy. A system of classification can be devised in a way that makes it open to the replacement of merely operational concepts with concepts defined in terms of causal relations to the world. The current system is designed to exclude causal information in principle because of anti-realist hang-ups; we want a system designed to include causal information in principle, even if the relevant facts are unavailable. A purely clinical classification is acceptable as a pragmatic stopgap; it is not acceptable as a goal.

Construct validity is valuable as a way of increasing our confidence that we have uncovered something real, if it reflects measurable relationships we expect to see between phenomena that our theories recognize. But it does not provide causal understanding that can be exploited in the way that medicine can exploit its knowledge of human biology.

Now I switch gears to a more concrete discussion. The remainder of this chapter shows what explaining an exemplar might look like by discussing schizophrenia in some detail. I set out what I hope represents fairly the contemporary consensus on schizophrenic symptomatology, and then show how to build and explain an exemplar on this basis. Again, the science is largely speculative and designed to illustrate the philosophical points at issue. A toy model along the lines suggested above shows how the explanatory agenda of the cognitive neurosciences might be a good bet for the biological model as I understand it, and shows what causal validation might look like.

6.4   The Schizophrenia Exemplar

6.4.1   Schizophrenia and Other Psychoses

Theorists disagree about whether schizophrenia is one disorder or several and about the boundaries of the core disorder. Indeed, researchers lament the lack of a generally agreed upon, precisely specified clinical description of schizophrenia just as often as they lament our lack of a clear understanding of its etiology. For example, uncertainty about the natural boundaries of the disorder causes uncertainty about what population we should study when we try to locate the genes that, everyone agrees, must be involved: "the *technological* advances in genetics have outpaced the *conceptual* advances." (Green 2001, 65).

Still, here I attempt to sketch an exemplar of schizophrenia that represents a consensus on its main features. This section describes the key symptoms and course of schizophrenia. Section 6.6 asks how that exemplar might be explained. Everything is resolutely tentative. Much of the work I talk about has survived for generations and much of it will last, but many details surely will not hold up. However, I do think that the *strategy* I pursue here is sound.

What I am offering is an illustration of exemplar-based psychiatric explanation. The empirical claims are plausible, but their purpose is illustrative: it tells what to expect in a good causal explanation of schizophrenia that is part of a synthetic theory of the mind/brain. The picture is that the great variety of schizophrenic symptoms can be explained in terms of a smaller number of causes. Specifically, we have an abnormal process of neural development that results in a set of cognitive abnormalities that are the proximate cause that sustains the symptoms. The developmental story and the description of the cognitive abnormality together provide two dimensions of causal explanation. The cognitive abnormality can be seen (as Gottesman and Gould [2003] suggest) as an endophenotype. It vindicates Bleuler's (1911, 8) suggestion that the essential feature of schizophrenia is a broad disruption to normal cognition with psychotic symptoms as accessories to that essential feature, not the main locus of the disorder. If the cognitive deficits are primary, the problem is relating the shortcomings in the machinery of cognition to the symptoms. Here the issues of the last chapter come to the fore again. We can distinguish between explanations in terms of one large cognitive deficit and explanations in terms of several different deficits. I do not try to adjudicate between these approaches definitively: that would be premature (and, indeed, presumptuous of a philosopher). Rather, I try to indicate what sorts of empirical and conceptual issues about cognitive architecture they raise, and suggest some considerations that might tell on each side.

First, I suggest how a schizophrenia exemplar might be built out of the empirical raw material by establishing the stereotypical features of the symptomatology at the appropriate level of generality, along with information about course. Then I show how an explanation of the exemplary features of schizophrenia might proceed and identify the issues that it raises.

Many attempts to define schizophrenia are searches for exemplars or tests of exemplars proposed by others (Mellor [1970], for example, was evaluating Schneider [1958]). My account of the clinical features of the schizophrenia exemplar is intended to capture symptoms that would be regarded as canonical. They are what you would expect to see in a patient if you were a psychiatrist who had been informed that the patient had received a diagnosis of schizophrenia. They are, therefore, mostly assembled from textbooks. I indicate areas of disagreement, since different ways of conceiving of schizophrenia make a difference to the explanandum and impose different constraints on the theory.

The standard way of setting up schizophrenia (e.g., DSM-IV-TR, 297–298) distinguishes it from a variety of related conditions, including other conditions in the schizophrenic spectrum and nonschizophrenic psychotic disorders.[4] The distinction between full-blown schizophrenia and the schizophrenia-like conditions is mostly made based on the duration of the symptoms. Active schizophrenic symptoms must be present for six months to confirm a diagnosis of schizophrenia. If symptoms are present for between one and six months without impaired social functioning, a diagnosis of Schizophreniform Disorder is given—if the symptoms persist the diagnosis is changed to Schizophrenia. If the symptoms remit within one month of their appearance the diagnosis is Brief Psychotic Disorder. If they do not remit, the preferred diagnosis is Psychotic Disorder Not Otherwise Specified, which can also be diagnosed in cases where the cause is unclear.

---

4. Psychosis, by the way, is a slightly slippery term: the DSM-IV-TR uses it to mean "delusional" in the context of some psychotic disorders. In other cases it refers to the presence of "delusions or hallucinations not accompanied by insight" (298); in the case of the family of schizophrenic conditions it refers to "delusions, any prominent hallucinations, disorganized speech, or disorganized or catatonic behavior" (297).

The DSM-IV-TR distinguishes other psychotic disorders on etiological grounds: Shared Psychotic Disorder (*folie à deux*) occurs when an individual develops delusions under the influence of someone else who already is psychotic (332); Psychotic Disorder due to a General Medical Condition (334) is diagnosed if the psychoses are caused by a medical condition such as epilepsy, which is associated with a variety of psychotic states. Psychosis also may be caused by substance abuse (338).

Two other conditions raise conceptual issues of special interest. Schizoaffective Disorder is diagnosed when schizophrenic symptoms occur at the same time as symptoms of a mood disorder. Typically, the psychotic symptoms precede or succeed the mood disorder. Schizoaffective patients, with their combined disorders of affect and cognition, have long been a puzzle: in the nineteenth century Batty Tuke bemoaned the existence of a "large class" of whom "it is impossible to say whether they are melancholic maniacs or maniacal melancholics" (quoted in Radden 1996, 9). Radden argues interestingly that the persistent puzzlement over schizoaffective conditions is a residue of faculty psychology, which divides cognition from mood and makes their joint disability a theoretical embarrassment.

Another interesting case is Delusional Disorder (DSM-IV-TR, 323), which is diagnosed solely on the basis of "nonbizarre" delusions, defined as "situations that can conceivably occur in real life" (324). It is hard to know what to make of this. If we think of the Capgras Delusion, for example, do we suppose that one's wife's being kidnapped and replaced by a robot is or is not "conceivable"? Incidentally, the Capgras Delusion, along with the other delusions of misidentification discussed in chapter 4, is not included in the DSM-IV-TR. However, it certainly seems that a general treatment of psychotic conditions should include delusions of misidentification, which are not uncommon in schizophrenia. It is also the case, as the DSM-IV-TR acknowledges, that many of the symptoms characteristic of the psychoses, such as hallucinations, delusions, and catatonia, can occur in other disorders including the dementias. The schizophrenia exemplar includes symptoms that occur in other disorders too. This is generally the case in psychiatry, as in medicine generally. It means that theory construction cannot employ locally plausible explanations without checking them for consistency with other explanations of the same symptom in other disorders, or explaining away the apparent identity of the symptoms.

6.4.2 The Symptoms of Schizophrenia
The DSM-IV-TR (312) gives five characteristic schizophrenic symptoms, although each may perhaps be better though of as a kind of symptom that

can occur in several ways. They are: delusions, hallucinations, disorganized speech, grossly disorganized or catatonic behavior, and negative symptoms. In going over the clinical picture in more detail, I use textbook accounts that do not usually follow the fivefold DSM picture.

What might we expect to see when introduced to a schizophrenic? To begin with, we might expect to see at least some of Schneider's eight first-rank symptoms (Schneider 1958):

1. Hearing one's own thoughts spoken
2. Hearing voices conversing about oneself
3. Hearing voices keeping up a commentary on one's actions
4. Experiencing bodily sensations caused by external agents
5. Disturbances of thought, especially thought withdrawal and thought insertion
6. Thought broadcasting
7. Delusional perception
8. Feelings, drives, and volitions experienced as the work of others or as being under the influence of others.

The first-rank symptoms have been studied extensively. Despite Schneider's claims, they are not now regarded as diagnostic of schizophrenia on their own. A wider conception of core symptoms, such as McKenna's (1994), encompasses: abnormal ideas; abnormal perceptions; formal thought disorder; motor, volitional, and behavioral disorders; and emotional disorders. I essentially borrow McKenna's definition of schizophrenia in terms of these five types of symptoms to serve as the schizophrenia exemplar, but I also add a sixth category, the neurocognitive deficits that show up in experimental tests as opposed to clinical contexts. I also include in the exemplar information about the course of the disorder.

6.4.3 Abnormal Ideas, Abnormal Perceptions

The main form of abnormal ideas in schizophrenia is the delusion, in all its varieties. One common delusion is "delusional mood" in which familiar surroundings seem uncanny or threatening: "There is some change which envelops everything with a subtle, pervasive and strangely uncertain light" (Jaspers 1959, 98).

Delusions of misidentification also are common. Delusions of misidentification are included in the Capgras and Cotard Syndromes, but schizophrenics tend to have more promiscuous misidentifications than those classic syndromes, although they do sometimes suffer from them (Fish 1962, 76–77). In schizophrenia, almost any neutral object can seem threatening or portentous, and ordinary circumstances can be construed in ways significant to the patients, who may believe that they are being followed or that events in the news refer to them or have a special significance only they can fathom. McKenna (1994, 3) also includes grandiose delusions, typically that one is divine or divinely inspired, sexual delusions, and delusions about the diseased or transmuted state of one's body.

Schneider's eighth cardinal sign is the experience that one's feelings, drives, and volitions are under the control of others. This is now normally referred to as "delusions of control" (McKenna 1994, 27), such as the experience of being forced to pace the ward or urinate on oneself that was complained of by "Laura" (McKay et al. 1996).

Although it is unclear how much confidence we can invest in the idea of a separate category of bizarre or fantastic delusions, McKenna does give them separate mention: "giving birth to thousands of children, walking all over the moon, having hundreds of people inside their body, and so on" (1994, 4).

I have mentioned hallucinations. These are the central category of abnormal experiences in schizophrenia. Like delusions they can be broken down into types, and I follow McKenna's typology, which draws on earlier surveys (1994, 7). Although schizophrenic hallucinations have been reported in every sense modality, auditory hallucination is the most common, so common that Fish (1962, 56) argued that auditory hallucination was often diagnostic of schizophrenia. The experience is typically that of hearing voices, and Schneider's first cardinal symptom, audible thoughts, belongs here. In audible thoughts one's own thoughts are experienced as spoken out loud. Voices also typically issue commands to the patients or comment on their actions. Often the commentary involves several voices discussing the patient in the third person.

So far my tour of the cognitive abnormalities in schizophrenia has looked at delusions and hallucinations. However, the aspect of schizophrenia that most evades ordinary understanding is surely thought disorder.

6.4.4 Thought Disorder

Thought disorder covers various phenomena, including formal thought disorder, which is not the most common schizophrenic trait and the hardest to understand. Fish (1962, 38) makes a fourfold distinction between

1. Disorders of the form of thought
2. Disorders of the stream of thought
3. Disorders of the possession of thought
4. Disorders of the content of thought (Delusions).

We have already looked at delusions.

Formal thought disorder is diagnostic of schizophrenia in the absence of obvious brain trauma. It usually is diagnosed on the basis of linguistic abnormalities that nearly all schizophrenics eventually exhibit. This has led to a tendency among researchers to lump breakdowns in conceptual mastery together with different types of speech deficits, including problems in speech production. There is, in consequence, much disagreement about the precise forms that formal thought disorder takes.

I continue with McKenna's (1994) categories. The first is *derailment of thought* (Kraepelin 1919, 72), a tendency among schizophrenics to slip further and further from the point as they follow seemingly random associations. McKenna (13) quotes this example:

A donkey was carrying salt and he went through a river, and he decided to go for a swim. And his salt started dissolving off him into the water, and it did, it left him hanging there, so he crawled out on the other side and became a mastodon . . . it gets unfrozen, it's up in the arctic right now; it's a block of ice, the block of ice gets planted.

Bleuler attributed this tendency to slide from the point into neighboring concepts to the loosening of customary associations: "Associations which used to be made regularly are omitted, while material is associated which is not normally connected with the initial idea" (1911, 355). Loosening of associations is central to Andreasen's recent theory of "cognitive dysmetria" as the core deficit in schizophrenia, which I discuss in section 6.7. Bleuler's associationism, is, I assume, entirely optional, but the idea of a general, nonspecific breakdown in the computational processes underlying central systems might have something to recommend it. Cameron (1944), for example, argued that derailment was a breakdown in the causal relations among thoughts.

Other schizophrenic speech is just *incoherent*: "lettuce is a transformation of a dead cougar that suffered a relapse on the lion's toe" (quoted in McKenna 1994, 13). Still more exhibits *poverty of content*, in which statements are uttered at a length that would normally suffice to answer a question or state an opinion, but no information is given. Instead the patient just produces highly abstract, cliché-ridden or vague speech. McKenna (1994, 15) calls this type of speech *empty philosophizing* (which caused me some momentary embarrassment).

Although some schizophrenic speech is poor in content because of excessive abstraction, theorists also have claimed to find an obsession with concreteness, showing itself as an inability to generalize or attend to context. Goldstein (1944) argued that this accounted for the presence of perseveration in schizophrenics and a tendency for their attention to be captured by one momentary stimulus after another. It resulted, he argued, from an isolation of some aspects of the cognitive nervous system from others.

Another interesting feature of schizophrenic speech is Ganserism (*vorbeireden*, or talking past the point), called so because it is the core feature of the Ganser syndrome (Enoch and Trethowan 1979; Whitlock 1967). In Ganserism the patient gives incorrect but approximate answers to questions; answers that betray the fact that the patient actually knows the correct answer (McKenna 1994, 16; Fish 1962, 167). One of Whitlock's patients, for example, when asked to give his name, said "It may be the same as yours" and said that the street he lived on "May be the same street we are trying to find tonight" (Whitlock 1967, 23). One of McKenna's patients said Spain was the capital of Italy, and that the average height of an English woman was three-foot-six, but that his mother, who was over five foot tall, was about average height (McKenna 1994, 24).

The Ganser syndrome features a variety of psychotic or hysterical states as well as *vorbeireden*, although the diagnosis seems to be made often just on the basis of talking past the point. In addition, *vorbeireden* has been found in a number of organic conditions, and it often is viewed suspiciously, as it can be found among malingerers.

Ganserism raises some issues about what counts as an approximate answer (many undergraduates utter approximations to the right answer, though not always in a form that suggests they really know it). It also illustrates the way in which polythetic exemplars can contain symptoms that appear in many other exemplars. *Vorbeireden* crops up with a small number of additional symptoms in the Ganser syndrome, as well as occurring in schizophrenia and various organic psychoses. This means that the explanation of an exemplar must be constrained by consistency with other, related, exemplars—an explanation of the set of symptoms in the schizophrenia exemplar must not be inconsistent with the explanation given to Ganserism elsewhere, or it must show that the similarity is just apparent.

I have been discussing formal thought disorder, the first of Fish's four types of thought disorder, in which language that is hard to understand reflects underlying distortions in thinking. I have previously discussed delusions, which Fish classified as disorders in thought content, so his last two categories of thought disorder remain.

The first category comprises abnormalities in the stream of thought: patients stop following one train of thought and abruptly pick up another. Fish (1962, 47) regarded this as diagnostic of schizophrenia if it is severe or if the patient complains of it spontaneously.

Disorders of the possession of thought make up Fish's last category. These disorders are perhaps the most bizarre and incomprehensible symptoms of schizophrenia. Unlike formal thought disorder, the language the patients use to express these abnormalities is, in one sense, perfectly comprehensible. It is the content of their complaint that defies understanding. Take, for instance, thought insertion. One of Mellor's (1970, 17) patients, for example, famously stated that "the thoughts of Eamonn Andrews come into my mind. There are no other thoughts there, only his . . ."

Eamonn Andrews was a British TV celebrity. What can it mean to have his thoughts in one's own mind? The patient is not complaining that her own thoughts have been manipulated by Eamonn Andrews so that he is causing her to think things she would not otherwise have thought. Her complaint is that she is introspectively aware of someone else's thoughts; an episode in her psychological history belongs to another. Whatever makes thoughts seem to be one's own is missing, but it is hard to see what that could be. Normally we are not called upon to sort out, among the contents of our minds, those that belong to us and those that belong to others. We just experience our thoughts as ours.

The strain put on our ordinary conceptions of mental life by formal thought disorder seems to suggest a role for the philosophy of mind. A good descriptive analysis of what alienation from one's own thoughts might be is required, and the explanation must be faithful to experience that the patient undergoes. That is hard to achieve for formal thought disorder because the experience seems so weird. This is not like providing an explanation of how thought can become derailed: in formal thought disorder the thoughts themselves can be perfectly well-formed but they are experienced as not one's own. But before I get onto the details, let me deal with the last two of McKenna's five core symptoms, motor disturbances and emotional disturbances.

6.4.5 Motor Symptoms

Chapter 2 discusses some of the motor symptoms of schizophrenia when setting out Frith's model of motor disturbances. Other disorders of movement in schizophrenia do not fit the model, however. Jaspers (1959, 180) distinguished between neurological impairments of motor activity, psychologically motivated impairments, and a residual category of psychotic motor disturbances. This last category encompasses most of the catatonic symptoms, which are a varied bunch. Bleuler (1911, 442) thought they were held together only by their appearance at a severe stage of the disease and lacked any common underpinning. Bleuler also noted (444–445) the phenomenon of negativism, in which patients passively let their bodies be adjusted without helping, or move in a way opposite to that desired by a caregiver.

Catatonia includes not just the stupefied, motionless behavior that the term suggests, but also types of impulsive and vigorous, albeit aimless, activity (McKenna 1994, 18–21). Schizophrenia also may include mannered or stereotypical motion, posture and facial expression, and a variety of non-linguistic abnormalities in speech production, such as sudden changes in the volume of speech, and "clicking and smacking sounds, sniffing and snorting" (Kraepelin 1919, 66).

6.4.6 Emotional Problems
Affect is often flat in schizophrenics (McKenna 1994, 25). Kraepelin in particular stressed this emotional diminution as characteristic of the progress of schizophrenia. Schneider (1958) thought that anhedonia, the inability to experience pleasure, was a first-rank symptom. As the disorder progresses emotional responses and expressions become attenuated, stiff, and shallow. The emotional expressions of schizophrenics, especially in chronic patients, can also be inappropriate to the context. Patients are also socially withdrawn or aversive.

6.4.7 Neurocognitive Deficits
Schizophrenics (and, interestingly, their nonschizophrenic relatives in some cases) exhibit a variety of measurable cognitive deficits. These include sensory gating, prepulse inhibition, signal detection, and aspects of memory and social cognition. Prepulse inhibition is the phenomenon of prior inhibition of the startle response (see the discussion of Patterson's mice in section 4.5.2). Sensory gating, the phenomenon by which our response to the second of an identical pair of stimuli is suppressed, usually is assessed by measuring the P50 component of the auditory event-related potential (ERP). The second of a pair of clicks usually elicits a lower P50 than the first one did. This suppression is impaired in patients with schizophrenia, as well as sufferers from some forms of head injury and dementia (Marshall et al. 2004). Ringel et al. (2004) found the effect in chronic schizophrenics with advanced negative symptoms, but not in acute psychotics. Sensory gating hypotheses explain these perceptual anomalies as the effects of faulty "gating" or filtering of sensory input. Many theorists assume this is involved in the poor attention that schizophrenics display.

The symptoms presented so far are the textbook collection. I follow Bleuler and various moderns in thinking that they are expressions of some underlying disruption rather than forming the essence of the illness themselves. Having gone over them, though, it is worth asking what this collection of linguistic, cognitive, sensory, motor, and emotional phenomena have in common. The answer appears to be, nothing: "schizophrenia has a special kind of qualitative heterogeneity" (Meehl 1990, 5). We have what looks like a very diverse set of phenomena, with almost nothing in common. Meehl inferred from this diversity that there was no point expecting only one kind of causal mechanism to explain schizophrenia, but that "the strategy should be one that permits us to move up and down in the conceptual hierarchy of the life sciences" (1990, 6). I already have defended that as a general methodological point, but it does seem to resonate especially for schizophrenia. Most theorists now agree that myriad genetic and environmental factors play a role in the genesis of schizophrenia, and that the causal influences may differ from one patient to another. The strategy that appears to be gaining favor is one in which schizophrenia is seen as a neurodevelopmental disorder that has as its endpoint a state of the mature organism in which all manner of psychological capacities are disrupted. But the disease is not progressive once the nature stage is reached. Although schizophrenics may be more vulnerable than most people to the psychological stress of aging, there does not appear to be a further deterioration once the basic disruption is reached (Lewis and Levitt [2002, 425] call this a "steady-state outcome"). So the explanatory model has two main components. One tracks the process by which the steady state is reached, and the other represents the steady state as the proximal cause of the exemplary symptoms.

6.4.8 Course

The symptoms of schizophrenia typically make their first appearance in the late teens or twenties, and almost never after age forty (Lewis and Lieberman 2000), but it does not follow that schizophrenia begins in adolescence. If we define schizophrenia in terms of its characteristic etiology, then it may begin prior to adolescence but not show itself until later. Our decision on this topic will determine whether we regard some early indicators of schizophrenia, such as poor motor skills, as symptoms or not. Physicians can retrospectively identify children who grew up to be schizophrenic based on their abnormal movements in old home movies (Lewis and Levitt 2002, 416–417). Once full-blooded schizophrenia sets in during early adulthood, the prognosis is poor. Most patients deteriorate, although if they take their medication they may bounce back from a first attack. Nearly everyone relapses, however (many patients stop taking the drugs, for one thing) and the typical pattern is of a cycle of remission and relapse that causes a steady deterioration for the first 5–10 years of the illness. At that point a plateau is usually reached, although the most severe cases may continue to worsen (Lewis and Lieberman 2000, 327).

This tendency to reach an endpoint rather than continually deteriorate leads researchers to conclude that schizophrenia is not a progressive disease that strikes in early adulthood but a neurodevelopmental one. That is, there is a process of abnormal development that ensures that a normal cognitive state is not reached, rather than a degeneration from a normal cognitive state to an abnormal one, as happens in many dementias.

So, the schizophrenia exemplar includes the signs, symptoms, and course I have just summarized. This is an idealization, as I stressed. Not every patient suffers from all these symptoms, and there is no attempt to represent the course of the disease very precisely, since the stages through which the disease progresses will vary in timing, intensity, and occurrence across sufferers.[5] Now we turn to the explanation, which has two components. We need both a developmental component and a component that represents the cognitive and other disruptions that make up the final state of the schizophrenic.

6.5   Explaining the Schizophrenia Exemplar—Development

The etiology of schizophrenia appears to have two dimensions. The challenges to modeling successfully the formation and sustaining of the schizophrenia exemplar are twofold: to link the developmental story with the cognitive, and then to understand the proximal, cognitive disorder. This second component includes both the accurate characterization of the cognitive deficits and the story of how the cognitive deficits give rise to the accessory symptoms.

In schizophrenia, all hell breaks loose in the mind after about the middle of the second decade of life. Three main proposals attempt to account for this timing (Lewis and Levitt 2002). On one account, a lesion that occurs early in life remains silent until the damaged tissue is supposed to come on line. At that point the tissue, because of the early damage, is unable to play its normal role. A variant of this view says that the crucial event occurs much later on, with a correspondingly short latency period. Most of the evidence, however, appears to indicate a much longer period of abnormal development rather than a crucial insult. Lewis and Levitt (2002, 424–425) offer a four-part neurodevelopmental model.

---

5. I am setting aside here some questions to do with the processes whereby symptoms are selected for an exemplar and defined. I discuss them in chapter 10.

The first two components in their model (although this is my language, not theirs) are the genetic and environmental inputs, both of which appear to be very diverse. The general consensus is that a genetic susceptibility to schizophrenia exists, but the search for particular genes has been fruitless. In fact, combinations of two sorts of genes may be important. There could be gene mutations that are fairly directly responsible for schizophrenia but which have different effects depending on variations in other genes that interact to produce brain structures. Even so, the combination of these genes appears only to predispose the brain to schizophrenia in some way. For the disease to actually manifest, gene expression must interact with environmental causes at particular stages of development. As I note in chapter 4, maternal influenza infection at some stage of pregnancy may be one such environmental factor, but there are many more. Even if we stick to prenatal factors, we find increased risk of schizophrenia associated with poor maternal nutrition, birth at certain times of year, greater urbanization of the place of birth, small head size, and others. These risk factors presumably track some physical mechanisms that effect gene expression, but no one knows what they are or how many of them exist. In so far as they can be isolated, they have low predictive value, since many of them occur in the biography of nonschizophrenics, and their absence does not prevent schizophrenia from occurring.

Given the number and poor predictive power of the genetic and environmental contingencies, and the complications of their interaction, there is probably no chance of finding any lawlike connections between genes, environment, and final state in schizophrenia. We may be able to express some statistical relations one day. At the moment, we must imagine a model, or a family of model types, representing very many interactions of genes and environments, that permits us to hold some interactions fixed while others are examined and manipulated in model organisms or other experimental setups.

The third part of the model represents the effects of gene-environment interactions on development. A developmental cascade seems to be involved, as the shape and timing of normal brain development is altered. A number of brain abnormalities have been found in schizophrenics. Lewis and Levitt (2002, 421) note that attempts to tie specific abnormalities to aspects of full-blown schizophrenia resemble searches for the effects of individual genes, in that the whole system is complex and poorly understood, and there has been no success in tying developmental processes to brain circuits dedicated to any psychological process.

The final component of the model is the steady state, as the effects of all the interactions between genes, environment, and development are stabilized to produce the adult brain, with all its abnormalities, that is the proximal cause of the symptoms. I now turn to the question of whether we can explain this final state in standard cognitive-science terms. The natural tendency is to reason that the physical bases appear to be so various that interesting generalizations are only possible at the cognitive level. This raises two problems. The first is whether the cognitive bases will be similarly various. If there is great physical variation and great variation is symptoms, perhaps there is no clean cognitive endophenotype at all. The second problem is the issue of cognitive architecture. Schizophrenia affects our psychology in ways that look to implicate both central and peripheral systems. And if central systems can only be understood in terms of folk psychology and neuroscience, then there will no useful cognitive level anyway.

I turn now to explanation of proximal causes.

6.6 Explaining the Schizophrenia Exemplar—The Final State

6.6.1 Introduction

If underlying cognitive deficits are central to schizophrenia, then its symptoms must receive explanations in terms of those deficits. Two ways of doing that each make possible a family of simple models. The first tries to locate the source of psychotic experience in some cognitive part. The extent of the symptoms that the faulty part explains varies across models—some aim at considerable generality, others are more specific. I call these *modular* models, not because every exponent is committed to a strong view about psychological modularity, but because the models aim at identifying one component of the mind/brain. Modular models are usually *piecemeal*. A piecemeal model explains schizophrenia as failure in a number of different systems, which together comprise the cognitive deficit. That is, one theorist could parcel out the symptoms among different systems. One sort of breakdown could explain derailment, for example, while another explains delusions of control. But most of the core cognitive symptoms could be explained by the failure of one cognitive part. For example, some theories try to explain psychotic symptoms in terms of defects in a system that monitors one's own mental states.

Global models, on the other hand, explain a wide variety of deficits in terms of a breakdown in a generic, rather than specific, mental processes. Typically such models appeal to high-level, general mental capacities that are implicated in a wide variety of cognitive tasks. If we think of the problem as specifying an endophenotype, we can see the modular and global approaches as affirming and denying the possibility of decomposing the endophenotype into psychologically more basic components. A modular approach decomposes the mind and maps symptoms onto functional abnormalities. A global approach blames the suite of symptoms on some very general property of cognition. Modular theories understand the proximal cause of schizophrenia as a set of computational disorders that prevent normal cognitive processes from occurring, one at a time. The approach updates Bleuler's idea of loosened associations in terms of an array of disruptions in information processing. Global models, on the other hand, look not for particular deficits but for general deficits that have a common effect on different capacities, which is truer to the spirit of Bleuler's original proposal.

To bring the issues into clearer focus, I approach them through the problems of thought disorder and delusions of control, or passivity experiences. I start by borrowing a general approach to the problem of how these experiences are to be understood on the surface, in broadly folk-psychological and phenomenological terms, and suggest a general philosophical framework for understanding that conception of the psychotic symptoms. Then I will illustrate the properties of the models I just distinguished by testing them out in this case. Then having separated them for heuristic purposes, I suggest that they may need to be combined.

6.6.2 How Can We Understand Thought Disorder?

To begin, we must get a handle on the experience of thought disorder. Stephens and Graham (2000, 172–176) suggest that in thought disorder one experiences parts of one's mental life as intelligent intentional performance, but not as expressive of one's own intentions or beliefs, which is the experience of the psychologically healthy. In other words, although thoughts seem disconnected from one's mental economy, they do not have the random character of items that usually are experienced as disconnected, such as transient thoughts or snatches of song running through the head. In thought disorder, Stephens and Graham suggest, mental activity seems purposive, which is characteristic of a mind. However, the subject does not experience it as her own, because it does not seem expressive of her mind: her occurrent beliefs and desires do not accord with what she takes her psychological dispositions to be.

Stephens and Graham do not mean that we can never surprise ourselves. After all, I might, in response to someone's question, find myself arguing a position I had never entertained before, or suddenly demonstrating some previously undreamed of political extremism. But in that situation I would experience my thoughts as my own, because they express my current state of mind. Even if, on further reflection, I disavow what I said, I do not disavow that I was the one who expressed it at the time nor that the thoughts arose from a process of rational activity on my part, even if that process subsequently strikes me as untrue to my conception of myself. In the case of retrospective disavowal, I fail to identify with my first-order mental states. In thought disorder, I fail to experience those first-order states as my own in the first place. Stephens and Graham liken the experience of thought disorder to finding myself writing a coherent letter and having no idea to whom it is addressed nor any recognition of the thoughts it contains. In that case, they say, I must either change my conception of myself to acknowledge that the thoughts in the letter really are my own, or, if they remain unfathomable to me, I must conclude that someone else is using me to write a letter.

The letter analogy suggests the experience of someone else using me to perform their action. Yet schizophrenics sometimes complain that someone else's thoughts (those of Eammon Andrews, as we saw in Mellor's [1970] example) are not just controlling their own thoughts, but are actually present in their mind. Stephens and Graham's picture of thought disorder, then, seems insufficiently fine-grained: it cannot distinguish between (1) someone else forcing my mind or body to engage in intelligent performance via means unknown to me, and (2) part of someone else's mind or body intruding into my mind or body and there performing an intelligent action. Schizophrenics complain of (1) with respect to both thought and action (the latter being delusions of control) and (2) with respect to thought. Stephens and Graham (2000, 142) note the importance of this distinction, but they go astray by characterizing the alien experience in terms of experiencing the agency of another person (172). That seems to miss the relevant distinction, because the effects of another's agency might be felt in one's own thoughts. What is wanted is an account of how we can experience someone else's thoughts. They do not manage to show how such an experience is possible, although that is hardly a criticism, since, as they convincingly show, neither has anybody else.

So the explanation Stephens and Graham offer is partially successful. But it does have an interesting property, as they note (2000, 175); in so far as it as applicable to passivity experiences across the board, it offers a general treatment of both motor experiences and cognitive ones, in the spirit of Frith (1992). I looked earlier at Frith's recent attempt to unify a cluster of motor abnormalities in terms of the forward and inverse models. The model of Frith's in chapter 2 was restricted to motor behavior, but it could be supplemented by analogous structures to explain the self-monitoring of thought.

Frith (1992) has tried to explain schizophrenia in terms of a more general monitoring system that keeps track of one's willed action (as opposed to stimulus-driven ones). Frith argued (1992, 81–83) that breakdowns in the general monitoring system cause schizophrenic delusions of control. He based this claim on experimental evidence from video games. Like normal subjects, schizophrenics could correct their errors in response to visual feedback, but unlike the normal controls, they could not adjust their performance when the visual feedback was removed. Frith concluded that in the absence of visual feedback normal subjects could respond in the light of their knowledge of their own intentions, but schizophrenics could not. This suggested their intentions were opaque to them. (Schizophrenics can tickle themselves, for instance, which suggests a sense in which their actions might as well be someone else's.) As for the cause of this self-monitoring failure, Frith suggested that deficits in theory of mind were ultimately responsible for it, as well as underpinning many types of delusion (1992, ch. 7). Stephens and Graham do not propose a cognitive architecture. But we can read them as positing some system that, when working normally, is responsible for the self-ownership of one's integrated psychological life and maintaining the boundaries between the phenomena that are part of that integrated structure, and the other phenomena that enter the mind unbidden.

Ever since Kant, it has been a recurrent philosophical theme that in cognition the self is best seen as a formal or structural requirement on interpreting and integrating experience. There must be some way of relating my behavior to my plans, assessing the effects my plans have on the world, and distinguishing changes in my experiences brought about by my movement through the world from the world's movement relative to me. We can think of the monitoring capacities that include the forward and inverse models as part of this overall capacity to integrate movement and experience, along with introspection, perceptual feedback, and motor feedback. All are monitoring capacities that are necessary for the self as a structural or organizing principle of experience, so that monitoring capacities of various sorts are integral to there being a self at all.

In a general way selfhood of this Kantian sort must involve the capacity to delimit boundaries between the mental phenomena one generates oneself and those of others. And this capacity involves not only introspection in the classic sense of being aware of one's thoughts, it also involves, as we saw when looking at Frith's models of passive motor experiences in chapter 2, the capacity to keep track of one's bodily movements and respond to feedback about them. Indeed, on the broadly Kantian conception of self-monitoring, perception and action are intimately linked, since at a minimum I must be able to perceptually track my own path through space. But how is the loss of the boundaries around the self to be understood?

6.6.3 The Kant Machine

Despite my reservations, there is something very appealing about the connection Stephens and Graham draw between experience of thought disorder and delusions of control on the one hand and, on the other hand, the blurring of boundaries around one's experience and the subsequent inability to retain the structural integrity of one's mental life. Notice, too, that this is a philosophical position that has a scientific upshot, since we now are in a position to argue that cognitive models of thought disorder and movement disorder in schizophrenia need to be related.

Call it the Kant machine. The result is the computational basis of a Kantian cognitive self, supplemented by other systems. Imagine, then, that the theory-of-mind system supplies knowledge of one's own intentions and that this knowledge serves as input to the forward and inverse models, since they have to be able to keep track of my intentions to move my limbs one way or another. We could also imagine that we have computational bases for other monitoring capacities. To explain delusions of control, we might appeal to a deficit in general ability to distinguish self-generated from externally generated beliefs and desires. This might be needed in order to keep track of contents that enter the mind via the testimony of others, and to index them so that they are not confused with my own beliefs and desires— I want to know what my enemies believe about me, but I should not treat it as on a par with my own self-assessments.

Call that mechanism the *testimony calibrator*. Problems with the testimony calibrator lead us to mislabel our own thoughts as generated by others. To explain thought insertion, we might imagine a system that ensures that random mental events like snatches of song or momentary speculation are not treated by our reasoning systems as if they were on a par with beliefs and desires that are structurally related to each other and to perception and action systems. Letting too many random elements in might place arbitrary constraints on reasoning, so it would be good to have a *gatekeeper system* to keep them out.

My imaginary embellishments of Frith's model of motor monitoring indicate the sorts of things that would need to be added to the elementary architecture to generalize the self-monitoring model of delusion from motor behavior to thought. It just assumes the architecture needed to explain one set of the symptoms that go into the schizophrenia exemplar, other things being equal. Because of this limited explanatory ambition, it is also modest and leaves much unexplained.

The Kant machine can be interpreted as containing distinct cognitive parts or just as a representation of the functional components of self-monitoring and self-other differentiation. That is, we could interpret it as a set of distinct computers at different physical locations in the brain, or merely as a representation of a set of capacities that cognitive parts must discharge if experience and perception are to be organized properly. It decomposes the mind into functional components, and ties schizophrenic symptoms to failures in those components. But it can be understood in more than one way.

Here is a quick look at a simpler modular machine. Nichols and Stich (2003, 190–191) object to Frith's claim that theory-of-mind deficits explain many delusions. They demur on the grounds that schizophrenic patients appear to have intact theory-of-mind capacities. On the other hand, individuals with Asperger's Syndrome, whose theory of mind is deficient, are able to report their inner experiences and do not suffer from thought disorders. This combination of results leads Nichols and Stich to conclude that Frith is correct to see deficits in self-monitoring as the source of delusions of control, but that we should distinguish separate mindreading and self-monitoring mechanisms. That distinction, they argue, explains the double dissociation between schizophrenia and Asperger's. Nichols and Stich (2003, 160–163) hypothesize the existence of a separate monitoring mechanism, which takes representations tokened in the "belief box" as input and yields *I believe that p* as output. They suggest this as a general solution to the problem of detecting one's own propositional attitudes, and argue that in schizophrenia the monitoring mechanism is impaired but theory of mind is intact. There are some problems with this view. The actual picture may not present quite the smooth dissociation that Nichols and Stich assume, since some theory-of-mind deficits do seem to exist in schizophrenia, and indeed in nonclinical patients with high schizotypal ratings (Langdon and Coltheart 1999). Such patients do not merit a diagnosis of schizophrenia but have psychological properties that suggest they may be prone to psychosis. Langdon and Coltheart, contrary to Nichols and Stich, argue that poor mentalizing is a primary cause of psychotic symptoms.

Second, Nichols and Stich need to say more about the deficit they attribute to their monitoring mechanism. As we saw above, delusions may exist in schizophrenia alongside veridical beliefs that are integrated into one's mental economy in the usual way. Since schizophrenics and other psychotics have some intact rational capacities, the monitoring mechanism is dealing with some representations from the belief box in a normal manner. The system cannot be completely broken.

In fact, the data seem to show that the monitoring mechanism, if one exists, is probably not the problem. When schizophrenics report weird beliefs, the monitoring mechanism is intact. If *I am God* gets tokened in the belief box, it should be copied to the monitoring mechanism, detected normally, and available for report. In this case, then, the fact that the system is working just right explains why schizophrenics can report being God. Of course, it is a delusion, but a monitoring failure does not explain the weirdness of the belief. Why, then, should the oddness of schizophrenic experience more generally be explained by a monitoring failure? Nichols and Stich argue that many first-rank symptoms are explained by a monitoring deficit. They support this hypothesis by pointing out that the inner life of many schizophrenics, as judged by their introspective reports, is bizarre or impoverished. Note that this means that their monitoring mechanism cannot just be a belief and desire detector. First-rank symptoms often involve reports of a strange experiential inner life, as when thoughts are experienced as being sucked out of one's mind. The greater generality in the monitoring mechanism need not be a problem. It just means that the system is more widely connected to other mental processes, not just the belief and desire boxes. However, there is an obvious objection to the theory that monitoring deficits account for the first-rank symptoms. Why should we not say, rather, that the monitoring system is working as designed, but the input it receives from other systems is corrupt or abnormal, and these pathologies earlier in the flow of information explain the strange results of introspection? We can explain the oddness of the introspective reports that schizophrenics make as the accurate (or at least, best available) report of abnormal experiences. This interpretation seems preferable. As we just saw, schizophrenics can accurately report their delusional beliefs as well as some veridical ones. The monitoring mechanism is working properly in these cases but dealing with odd data. It does not explain the strange contents, which are generated elsewhere. The monitoring system is doing the best it can. It is working normally, but lesions elsewhere in the cognitive architecture have left it struggling to make sense of the abnormal output of the systems it monitors. The appeal to broken introspective capacities may be beside the point, then. And, to return to the first point, even if it is not we lack a theory of why the broken monitoring system sometimes permits correct reports of both veridical and delusional beliefs. Why is the nature of the fault such that it only hampers some self-reports and not others?

The need to explain symptoms as the partial, rather than total, failure of a system is a problem for Frith's original model of the failure of behavioral monitoring in schizophrenia too, since schizophrenics do not lose all ability to monitor their bodily movements. And a form of the second objection also applies to the Kant machine. If symptoms coexist with normal forms of the symptomatic phenomenon, a model that attributes a symptom to a breakdown in a system must either supplement the model with connections to additional systems which operate so as to partly correct for the output of the compromised system, or show how the inner working of the system accounts for the preservation of some normal behavior. We could imagine, for example, a psychodynamic interpretation of the model in terms of some trauma that causes the testimony calibrator to disavow only some contents as self-generated, perhaps those that are punitive self-assessments, which now become attributed to some other presence. Or, we could imagine the system as realized in a neural net that displays varying types or degrees of degradation depending on specific patterns of damage.

These modular models also fail to explain all the symptoms of schizophrenia. Nichols and Stich, for example, aim to account for Schneider's first-rank symptoms. They do not speak to linguistic or emotional problems, nor do they account for derailment or other problems in thought that do not involve the self. While I leave these problems aside and move on to some global models, I hope that I have shown what modular models look like, and what some of the problems they face are.

Modular models endorse a specific deficit or a family of specific deficits that explain symptoms of schizophrenia one at a time in terms of damage to underlying systems with the job of producing the normal form of the behavior. Global models, on the other hand, are the heirs of explanations like Bleuler's "general loosening of associations." They exemplify a search for a property of cognition, rather than a component of it, which can explain the variety of schizophrenic symptoms all at once. Hoffman's "synaptic pruning" hypothesis and Andreasen's idea of "cognitive dysmmetria" are examples.

6.6.4 Global Models

Modular models in psychiatry operate on the assumption that a profitable way to proceed in addressing an exemplar is to go symptom-by-symptom (Frith and Johnstone 2003), looking in the first instance for local explanations that relate symptoms to causes. The final step is thus the attempt to integrate all these partial explanations. Global models, on the other hand, try to identify some general causal agent that can explain as much of the exemplar as possible, with the final step being the demonstration that changes in some cognitive or neurobiological value bring about corresponding changes in behavioral measures. Bleuler argued that the crucial causal agent was a loosening of normal associations. More recently, Ralph Hoffman (McGlashan and Hoffman 2000; Siekmeier and Hoffmann 2002) has modeled this process in terms of the computational properties of excessive synaptic pruning in associative cortex.

Hoffman devised a model to connect two findings about schizophrenics. First, they show (in postmortem and imaging studies) an abnormal degree of synaptic pruning, especially a dramatic loss in corticocortical connections, which may indicate a failure to arrest the normal reduction in synaptic connectivity that occurs in adolescence. Second, schizophrenics show unusually high levels of semantic memory priming. An example of semantic priming is that subjects are more quickly able to distinguish words from nonwords if the words are semantically related to words they have previously seen. Schizophrenics show very high levels of priming, although this lessens as the disease progresses and all cognitive capacities slow down (Siekmeier and Hoffman 2002, 345). Indeed, one of the features of derailment of thought is the way in which schizophrenics become stuck following random associations rather than following through a train of thought. In Siekmeier and Hoffman's model (2002), a neural net was trained to display normal levels of priming and then degraded by eliminating nodes whose activation weight fell below a stipulated level in order to simulate neural pruning. After pruning, the model became worse at retrieving memories and displayed a higher degree of semantic priming.

Siekmeier and Hoffman concluded that their model captured the psychological significance of excessive synaptic pruning, since they took themselves to be simulating associative memory and showing how it is hijacked by random associations when functional ones are lost. Even if we grant this, their claim (345) that their experiments suggest that schizophrenia arises from excessive loss of local cortical connections seems excessive. They do not show anything that connects with any symptoms other than derailment, and even there the connection is left to the reader to discern. Nonetheless, it is true that they modeled an empirical finding in neurally plausible terms. The question is how to connect the cognitive property they discuss to the symptoms of schizophrenia.

A more general global model is Nancy Andreasen's appeal to "cognitive dysmetria." Cognitive dysmetria is hypothesized by analogy with "dysmetria," a neurological term of art referring to "disruption in the fluid coordination of motor activity." Cognitive dysmetria, therefore, is disruption to "the fluid coordination of mental activity that is the hallmark of normal cognition" (Andreasen et al. 1999, 911). Andreasen et al. argue that normal cognitive coordination is subserved by the cortico-cerebellar-thalamic-cortical circuit (CCTCC). They argue that this circuit is abnormal in schizophrenia and appeal to imaging studies that show relative inactivity in key areas of CCTCC during word-recall and face-recognition tasks.

This model is too vague about what cognitive dysmetria actually is. Andreasen et al. suggest that poor cognitive coordination might lead one to "incorrectly connect perceptions and associations and misinterpret both external and internal processes" (1999, 916). But the details are unclear, and we are left with the impression that cognitive dysmetria is just general cognitive failure, or another term for Bleuler's "loosening of associations."

One possibility, though, is that the unclarity of Andreasen's formulation is inevitable, since all that can be said about the cognitive level in schizophrenia is that cognition generally is wrecked. Next I contrast the two approaches a bit more, say why this might be so, and then suggest a way to avoid this dead end by combining the two approaches.

6.6.5 Joint Models

Modular models try to understand schizophrenic symptoms in terms of breakdowns in particular cognitive mechanisms, or failures of normal cognitive functions, that produce abnormal forms of those functions. The crucial supposition is that all the diverse genetic, developmental, and environmental factors involved in schizophrenia conspire, in the final state, to produce a set of computational failures in cognitive parts, with different parts explaining different aspects of the disorder. Global models look for a generic deficit and work bottom-up, identifying a widespread flaw that explains many problems. Global models may cite the computational properties of neurons (Friston 1996) but have not connected mathematical models of neurocomputation with psychological processes or symptoms. Modular approaches tend to go symptom-by-symptom, tying parts of the syndrome to distinct processes and global models. Modular approaches seldom model the properties of cognitive parts mathematically, but they provide a much clearer sense of the various tasks that the mind/brain performs, and how the symptoms might relate to the lack of performance of those tasks.

The search for specific deficits, although it makes intuitive sense, may be misguided. Without an understanding of central systems, we may not be able to develop good computational models that can connect intentional descriptions of cognitive function with the computational properties of neurons.

And given the great diversity and heterogeneity of problems in schizophrenia, it may be that a view like Bleuler's is the best we can do. Although we can come up with descriptions at the intentional level of the various deficits that schizophrenics display, we may not be able to explain them other than as the different manifestations of an underlying failure of communication between neurons. In a nutshell, this view holds that an interaction of genetic susceptibility and environmental factors leads to a failure of normal communication among brain cells, with the consequence that, cognitively, everything falls apart.

Something similar obviously is part of the truth about schizophrenia. As Meehl (1990, 14) puts it, the big fact about schizophrenia is that things are haywire pretty much everywhere, in all systems, at all levels. Perhaps when the system slips out the space of functional states, stochastic factors play the largest role in determining where in the space of nonfunctional states it ends up, and there are no useful causal generalizations apart from the very broad and uninformative. Perhaps if we could take a schizophrenic mind, rewind the tape back to birth, and let it play again, tiny differences in the various environmental factors would cause a very different type of schizophrenic mind.

But models of intermediate scope could exist somewhere between the big picture and the individual history. A set of explanatory stories could capture what happens in a way that connects a decomposition of the schizophrenia syndrome, in terms of description of defective cognitive functions, to a particular pattern of cortical disconnection at the neuroanatomic level.

Frith's latest model, for instance, retains a general approach to schizophrenia as a failure of self-monitoring and a symptom-by-symptom approach (Frith and Johnstone 2003, 124). But now "the self-monitoring model assumes that the damping down of responses to sensations that we cause by our actions results from signals coming from brain regions involved in generating the actions" (135). Normally, corollary discharge enables us to track whether we are responsible for the changes in our view of the world. The shift in view is predictable on the basis of the commands sent to eye muscles, so that command is copied to a "self-monitoring" mechanism. Frith assumes that similar corollary discharge is necessary to monitor the connections among epistemic relations to the world, which fits with the picture of a Kant machine that organizes and updates the mind in response to experience. Frith blames the failure of corollary discharge in the motor control system on a disconnect between frontal lobes and posterior regions. And this disconnect is explainable in terms of disruptions to the normal computational and anatomical properties of brain cells. If we interpret the Kant-machine model not as a representation of modules but as a representation of functional relations within cognition, it may be possible to deform the model as Frith deforms the model of motor awareness. We can take a model of normal functional relations, sever some of the relations, and then try to trace the failure of those relations to hold to a systemic breakdown in the brain. The systemic breakdowns probably will be a fairly heterogenous set, physically speaking, with many different fine-grained stories to tell about disrupted cellular relations. The final picture, though, is of functional disconnections throughout cognition, each mapped onto disjunctions of functional disconnections between brain systems.

Such model takes the symptoms of the schizophrenia exemplar, divides them up into families that reflect failures of normal relations to hold among parts of cognition, and then looks to connect those failures with the developmental story by specifying the end state of abnormal brain development in a way that makes it clear how the cognitive disconnections map onto systemic brain malfunctions. The big model comprises many small ones. At the intentional level, the smaller models each represent a part of the exemplar in terms of functional failure among parts of the cognitive decomposition. These failures track specific forms of that the global neurocomputational breakdown. Most of the explanatory burden is likely to be borne by the specification of the neurocomputational problem, so in that sense that model is a version of a global approach to explaining schizophrenia. But modular models at an upper level are still necessary. The modular approach at the cognitive level can sort disjunctions of fine-grained specifications of neurological breakdown. Figures as high as one quadrillion are bandied about as the measure of the number of connections between brain circuits. Detailed differences among disconnections of these circuits might make a difference to how schizophrenia manifests. So although the neurocomputational story in terms of cellular disconnection probably tells us what causes schizophrenia, cognitive characterizations of failures can have a vital role to play in tracking types of pathology. Rather than looking for generalizations across cellular relationships, we can sort them into classes based on the systemic effects they have. Just as talking about beliefs allows us to group physically different setups together in terms of the difference they make to behavior, so talking about failures of normal intentional relations allows us to group patterns of neural connections together in terms of the difference they make to which aspects of the schizophrenia exemplar are displayed.

So we can envisage a hierarchy of models. To explain the exemplar we construct a collection of model types, each representing a set of disruptions to normal functional relations in cognition, which correspond to different ways in which a token symptom can appear. Because the cognitive models have a box-and-arrow form, they permit different interpretations across different research programs, and different reductive strategies within one research program. Some boxes may be localizable in terms of brain anatomy, for instance, and some may not be. But each box and arrow represents a disjunction of numerous fine-grained physical setups. Therefore, a further family of physical models can correspond to each token of a cognitive model, with these physical models being different forms of one neurological anomaly, or different neurological anomalies, depending on how the neuroscience turns out. Causal validation works by taking symptoms defined as measurements of functional disconnections among psychological states (cognitive, affective, etc.) and showing how these track causal relations at lower levels.

We can represent the relation between symptoms and course either as the stochastic effects on the interactions between genetic, environmental, and developmental variables, or as predictable causal relations. In the former case, we will have a set of models of the interaction between distal causal factors but little way to link them up with models of the final state. In the latter case, we might be able to understand distinct causal histories in terms of the production of symptoms by particular distal interactions. That would let us subtype schizophrenia into several different syndromes.

Psychiatric explanation within the synthesis of existing psychiatry and cognitive neuroscience will take this form. Symptoms described at an intentional level will group neurobiological processes into relevant classes. That does not mean there is no role for computational etiologies. A good computational story will serve to make the intentional story much more precise, and let us track physical causes more perspicuously. (And, as I say in chapter 5, various formal and philosophical ways of naturalizing the intentional let us tell the intentional story more precisely). But, good computational stories may not always be forthcoming.

The models can be more or less precise. One way of making them precise is to fit them to classes of individual patients. This might be one patient or a group. In these cases, the symptoms will take forms that depend on other facts about the patient. And some facts may not be captured in these models at all. A class of patients may suffer from a delusion that the CIA is after them, for instance. We can sort patients into this class at the intentional level, thereby lumping together people who may have little in common at a detailed physical level. But that sorting will not explain why the symptom has that particular form, even if we can explain delusions of persecution as a class. The question of why it should be the CIA as the persecutor, rather than Al Qaeda, the Vatican, or General Motors, is another story, and it requires an answer in terms of cultural factors interacting with the psychology of the patient. Indeed, a symptom of mental illness may be comprehensible only in relation to some norm or other cultural factor. My next topic is how to make room for such social and cultural differences within the overall framework.

Human beings in society have no properties but those which are derived from, and may be resolved into, the laws of nature of individual man.
—J. S. Mill, *A System of Logic*

I knew that nought was lasting, but now even Change grows too changeable, without being new.
—Byron, *Don Juan*

7.1 Introduction

Mental illness often is called socially constructed, but it is not always clear what social construction is, nor how we should view the status of its products.[1] In this chapter, I look at social constructionist explanations of mental illness. On my view, some aspects of psychopathology are caused by cultural forces, and to explain them I should add the causal powers of mental representations of the social world to the explanatory resources of psychiatry. Rather than carving up the domain of psychopathology into two subsets, and consigning one of the ensuing groups of disorders to biological psychiatry and the other to the social constructionists, I recommend integrating the biological and social approaches. Instead of explaining condition A in terms

---

1. Philosophical work on social construction has not kept pace with the flood of work in the social sciences, but the last decade has seen some stirrings of interest. Haslanger (1995, forthcoming) and Mallon (forthcoming) are general philosophical discussions of social construction. Other philosophers have discussed particular candidates for social constructions. Hacking (1999) lists many conditions that have been claimed to be socially constructed and distinguishes several different constructionist projects. As well as the papers by Hacking on mental illness cited in the text, see Andreasen (1998), Appiah (1996), and Kitcher (2003) on race, Griffiths (1997) and Mallon and Stich (2000) on emotion, and Stein (1998) on homosexuality.

of brain chemistry and condition B in terms of social construction, we ask which aspects of both A and B can be explained by which method. But to understand how social and cultural forces shape behavior, we need a mechanism, and the mechanism needs to be material, otherwise social and cultural causation will remain mysterious. The obvious place to turn for a mechanism of social causation is mental representation. The cognitive sciences suggest that aspects of a social environment could be represented by the mind/brain. Yet because mental representations are physical entities they can stand in cause-effect relations within the organism.[2] Of course, this is not a new idea, but taking it seriously allows us to integrate the social constructionist approach into scientific psychiatry.

Opposition to this integration might have several sources. I start, in the next section, by presenting three general objections that, if well founded, would undercut the rationale for integrating social constructionist accounts with neuroscientific ones. These three problems all have been raised at some point as objections to the use of social construction as an explanatory notion. I rebut the objections and then, to develop a positive account, I look to the fullest philosophical discussion of social construction, Ian Hacking's.

I use Hacking's approach as a foil against which my own can be spelled out. Hacking's treatment is unsatisfactory for three reasons: first, it is a semantic rather than a metaphysical proposal; second, although it makes room for both biological psychiatry and social construction it still leaves them in separate spheres, and does not tell us how to integrate them; third, Hacking leaves us unsure how generally he thinks his theory applies, because of his stress on transient and unstable pathologies.

I will argue against all three of these features of Hacking's account, and use my objections to motivate my own case, which preserves much of Hacking's machinery. His approach to the social dynamics underlying the expression of mental illness is enormously promising, but we need to understand how the social dynamics he discusses can be mediated by neuropsychological structures, providing a causal understanding of the relation between social structures and psychological ones, and not merely, as Hacking prefers, a semantic one. Taking cognitive structures seriously might help us to see how mental representations of the social world interact with neural structures to produce abnormal behavior. I distinguish three ways of using this type of explanation: first, social pressures on individuals can push them

2. There are many different proposals about what mental representations are. I mean my broad view to be consistent with all of them, subject only to the constraint that some must be unconscious.

into developing a pathology, such as eating disorders; second, a failure to acquire the right sort of information can lead someone to misrepresent the social order, which may be the problem in dysthymia; third, the idea of representation of the social helps resolve some issues in the understanding of cultural variation in psychopathology, such as the Malay condition, *latah*.

7.2 Three Worries about Social Construction

7.2.1 The Antirealist Objection

The first objection to social construction as an explanatory notion is the anti-realist objection that denies that human kinds are natural kinds and argues that because they are not natural kinds they cannot be scientific explananda. I discuss natural kinds in more detail in chapter 9, but examples will have to do for now. Stein (1998) distinguishes natural human kinds from social human kinds. I use the less cumbrous "human kinds" (Hacking's term) instead of Stein's "social human kinds." By "natural kinds" I mean Stein's "natural human kinds" as well as other natural kinds. Natural kinds are biological kinds such as blood groups. Human kinds include, in Stein's examples, registered Democrats, Ivy League students and convicted felons. Membership of human kinds depends on the rules of social practices. As Griffiths (1997, 145) points out, categories like "felon" or "member of parliament" exist because of sociolinguistic processes, so it is trivial that such categories are social constructions. The nontrivial claim is that social processes have a causal role over and above their role of categorization: in virtue of being socially sorted, people acquire new causal powers. Stein seems to deny the nontrivial claim about causal powers when he says that social human kinds "may play a social role, but they do not play an explanatory role in scientific explanation" (Stein 1998, 429).

Social scientists can measure and predict behavior based on political affiliation. So Stein's claim looks odd in one way—social kinds do sort people into classes that have effects that need to be explained. But you may think that social-kind membership is not useful for explanations; being a Democrat may track a number of other properties that one has, such as secularism, but without explaining why one has those other properties—and this is presumably because being a Democrat does not cause those other properties. Stein's point is that social kinds lack the causal essences that natural kinds possess, and that in consequence social kinds do not enter into explanations in the way that natural kinds do. What makes Democrats share various attributes is not the fact that they are Democrats, which is just a matter of membership in a socially defined group. Presumably there is some shared psychology underlying both registration as a Democrat and above-average secularism.

The contrast is between *natural explanations* that understand human kinds as natural and play a role in causal explanations of behavior, and *social explanations* that understand human kinds as sociolinguistic artifacts. The way to meet the antirealist challenge is to find a social-causal process that makes social explanations genuinely causal and not just shorthand for sociolinguistic categorization: there have to be effects on members of a group that are explained by their being put in that group by social forces, and not just their sharing some psychology that is tracked by membership of the group.

The idea of social construction is designed to capture such a causal role, but there is a catch. Many theorists claim that mental illnesses, for example, are social constructions, but as we have seen social explanations of mental illness often insist that people are not ill, being instead socially deviant or rationally adapted to intolerable circumstances. So both Stein's treatment and the notion of social construction display the thought behind the antirealist objection: social explanations of a category show that the category is not a real one.

7.2.2 The Instability Objection

The second objection to social explanations is that social kinds are unstable. If a human kind is social, the properties that determine the kind stem from conceptualizations within a given society. This conceptualization changes over time, so the properties that determine the kind are unstable and the kind itself changes with the social context. Omi and Winant (1994, 54–55) argue that races are unstable in this way. They are not biological kinds but products of social and historical contingencies that make race an "unstable and 'decentred' complex of social meanings." The instability of social kinds is widely advertised as grounds for doubting their inductive utility. The thought is that social explanations have no stable *explananda*, since social construction makes a kind unstable and hence inductively useless. This implies that social constructionist explanations cannot be integrated into a psychiatry with types of disorder as its subject matter and scientific explanation as its goal, because the types of disorder that are socially explained are too labile: they cannot accurately be characterized in terms of symptoms and course, as both exhibit too much variation. Disagreeing with the thought behind the first objection, notice, does not preclude agreeing with the second. That is, socially caused mental illnesses could be real but scientifically intractable because of their instability.

7.2.3 The Crowding-Out Objection

So the first two objections to social construction are on the grounds of unreality and instability. The third objection is "crowding out": this applies explicitly to attempts to integrate social and natural explanations. The idea is that the provision of one type of objection leaves no room for the other. Social-construction discussions invite this objection in a particular way. As Griffiths (1997) argues, social-constructions can be overt or covert, and many social-constructionist projects are attempts to show that what purportedly has a natural explanation in fact has a covert social one.

Griffiths's idea of a covert social construction relies on the idea that although we would all happily admit that being a registered voter is just a matter of being treated a certain way and not a matter of intrinsic biological processes, we might resist being told that, say, being in love is just a matter of the sociolinguistic practices we have internalized and responded to. Overt social construction produces, in John Greenwood's terms, a social group: people in the group "share the fundamental dimension of sociality: they are oriented to the cognition, emotion, and behavior of other members of the group" (2003, 100). Registered Democrats make up an overtly constructed social group, since they share a conception of themselves and know that others in the group share it. But suppose being in love is covertly socially constructed. The existence of romantic love, if it is a covert social construction, depends on everyone denying that it is socially constructed and thinking instead that it is natural. Lovers do not form a social group; their amorous behaviors and understandings of themselves *qua* lovers do not just reflect their shared beliefs about membership of the class of lovers. Although they may have expectations about each other, lovers form these expectations because they think amorous behavior reflects a natural state of being in love. In this case a natural explanation is necessary to vindicate our understanding of a human phenomenon and the natural explanation is at loggerheads with the social one. The social explanation says that love is something that we just thought was natural. Substituting a social explanation for a natural one does not provide a new explanation for why something is the way it is, but an explanation for why we mistakenly thought it existed in the first place. The crowding-out objection obviously has affinities with the antirealist objection, but it is worth distinguishing for two reasons. First, the crowding-out objection comes from within the social constructionist camp, whereas the antirealist objection is a naturalistic worry about the propriety of social explanation in general. Second, the crowding-out objection may admit that there is a genuine causal process that a social constructionist story can narrate, whereas the antirealist objection holds that social explanations are, in fact, fake: when they do not confuse use and mention, they tell a story that is merely correlated with the actual underlying natural causal facts.

I suggest an explanation of mental illness that blocks all three objections, but I am especially interested in the crowding-out objection, because theorists use it to undermine natural explanations of mental illness. Natural kinds and social kinds, on this line, are mutually exclusive. The implication is that social explanations and natural explanations cannot be brought together to explain the same kinds. In that case, mental illness cannot be explained by integrating the natural and social. Yet, given the apparent interplay of culture and biology in many mental illnesses, an integrationist perspective that employs both social and natural explanations seems very attractive.

Integrationist explanations of mental illness, then, face three challenges. First, natural explanations—say in terms of brain chemistry—cannot be all that validate a diagnosis. Second, even if social explanations apply to real conditions, the fluidity of social practices may limit them to unstable psychopathological phenomena. Third, social explanations must not be excluded by natural explanations, nor vice versa.

To rebut these challenges, I first distinguish realist social constructionist projects in psychopathology from skeptical ones then discuss the vexed question of stability and argue that the conceptual repertoire Hacking offers us can apply to stable conditions as well as unstable ones. Then I criticize Hacking's view of the relation between brains and cultures, develop a simple competing picture, and, last, extend my discussion to conditions that occur in specific cultures.

7.3 Social Constructionist Projects in Psychopathology

7.3.1 Constructivist and Realist Projects

This section tries to nail down the slippery concept of social construction. Chapter 2 introduced the constructivists, who regard the concept of mental illness as unmoored in facts about human pathology. In the terms of this chapter, constructivists can be said to see mental illnesses as covert social constructions. This often leads constructivists to ally themselves with antipsychiatry movements, which they do because they think that acknowledging the socially constructed nature of mental illness impugns its reality qua mental illness. Scheff's (1999) idea of "residual deviance," for example, is trying to substitute a covert social explanation for a natural one by showing that it is our unacknowledged social practices that cause us to treat people as mentally ill. But Scheff does not defend his crucial assumptions that natural and social explanations are mutually exclusive, and that socially explained conditions are not genuine disorders.

Now, on any view of psychiatry as a whole, there might certainly be disorders that are crowded out by social explanations—exposing their lack of natural explanatory undermines them. Drapetomania is a plausible case. Once we learn that there is no natural explanation for Drapetomania and we learn the political reasons for its conceptualization, it is undermined. It takes, then, a conjunction of covert social explanation and the absence of a natural explanation to undermine a diagnosis. This combination often is taken to define social construction, in psychiatry as elsewhere. But other social-constructionist projects are not subversive. They try to explain why mental illnesses take particular forms. In these cases, we want to use social explanation to answer the questions that natural explanation cannot. I classify these realist projects as social-contructionist projects because they argue that a group of people share a diagnosis because of their psychological adaptation to social demands; social pressures, not anything biological, bear the explanatory burden. Moreover, these theories meet the realist challenge in the sense that they aim to identify a causal process that leads to a series of psychological effects. They do not just track properties correlated via some process independent of the process that forms the category.

I discuss these realist projects in a moment. If they can be defended then the first objection is rebutted. In pointing to the important role of social pressures I am, of course, relying on the very wide interpretation of the medical model that I advocated against Kandel's more reductionist view in chapter 4. In practice, the effect on psychiatry of the medical model has been to shift the study of mental illness toward molecular neurobiology and genetics. But, as I have argued, the basic conceptual structure of the medical model is compatible with great explanatory diversity. That diversity is present in the ensuing discussion in two main ways. First, important questions about the epidemiology of mental illness can only be answered at the social level. Second, social forces figure in the production and shaping of many symptoms.

We are unthreatened, indeed unsurprised, by the claim that sociocultural processes are causally important in bulimia. Being bulimic is not like being a registered voter, but it is not fully covert either. After discussing bulimia, I treat Hacking's work as covert constructionism that intends, unlike Scheff's, to be compatible with natural explanations. I then contest the idea that social explanations, even if they are not subversive, apply only to unstable phenomena that are of no use in scientific explanation.

7.3.2 Bulimia

Bulimia shows how social forces play an important role in explaining psychiatric epidemiology. For example, young white American females feel the pressure to be thin more than their African-American counterparts, and have correspondingly higher rates of eating disorders (Wilson et al. 1996). The pressure on women to be thin is not new. Doctors noticed fear of fat among female patients in the seventeenth century, and Victorian medical journals discussed "morbid dread of fat" (van Deth and Vandereycken 1995; Parry-Jones and Parry-Jones 1995). But the pressure to be thin seems to be growing: in 1978 the average Playboy centerfold weighed only 83 percent of the average female, compared to 91 percent in 1960 (Garner et al. 1980). Eating disorders surely have something to do with this idealization of skinny women.

We also can trace the effects of Western ideas of female beauty as they spread through the Westernizing classes of other nations—there is, for example, a correlation between body dissatisfaction and the use of English in the home among Chinese girls in Singapore (Ung 2003). In Pakistan, a study of girls at English-speaking high schools found only one case of bulimia nervosa, but when the same investigators studied over 200 British schoolgirls from Indian and Pakistani immigrant families, they found a prevalence rate for bulimia of 3.4 percent (Mumford et al. 1991, 1992). Social constructionist accounts of mental illness seem promising for eating disorders. Their epidemiology appears to track exposure to Western norms of beauty, and we should look to the psychological effects of these norms for the proximate causes of eating disorders.

Of course, these psychological effects occur in brains. So we also must acknowledge the importance of natural explanations. Social factors may explain differences in eating disorders across populations without telling us about differences between individuals exposed to the same social pressures. Not all immigrants develop bulimia. To explain why some do, we might adduce differences in brain chemistry. Steiger et al. (2001) found that platelet paroxetine-binding in bulimic women is lower than it is in nonbulimic women. Paroxetine is a powerful antidepressant, and these results suggest reduced uptake of serotonin at central synaptic sites. Further evidence for abnormal serotonin function in bulimia came in the same study from the lower levels in bulimics of the protein prolactin, which is produced by the anterior pituitary gland. This is believed to indicate down-regulation of post-synaptic serotonin receptors.

Steiger et al. could not decide between "state" and "trait" interpretations of their findings. A "state" interpretation explains the serotonin abnormal-ities as the product of active bulimia, whereas a "trait" interpretation views them as a stable aspect of some women, making them susceptible to bulimia. If the trait interpretation is correct, then we have a fact about some women that might explain why they are pushed into bulimia by the surrounding culture.

Now, explanatory networks can stretch a long way, and the existence in those women of a chemical susceptibility to bulimia might itself be explained in other terms. Perhaps something in their genes or something in their child-hood, explains their chemical susceptibility to bulimia. If, as seems likely, bulimia is a product of a complicated interaction of factors, then the ques-tions will determine which causes we cite.

The representation in the minds of young women of the culture they live in answers some questions about eating disorders. It is not, *pace* Scheff, just a matter of how we treat people. Rather, it is a matter of how the cultural information represented in their brains interacts with other aspects of their biologies and psychologies. Since the social can have a causal role in behav-ior via its representation in psychological processes, social explanations can be incorporated into the cognitive neuroscience of humans qua social animals. In some cases, a narrow biological approach will not answer ques-tions about why a disorder has its distinctive symptomatology, nor why its epidemiology is sensitive to cultural transformations. But biology might explain variation in personal susceptibility to cultural forces: understanding the shaping of epidemiology, symptoms and course requires attending to the social and the biomedical at the same time.

Since it is part of our biological nature to have a culture, there is a strong presumption in favor of integrating social and biological explanations. A full understanding of mental illness requires integrating the insights of the life sciences and the human sciences. I develop my approach indirectly, taking over elements of Hacking's important work on the social dynamics of mental illness (1992, 1995a, 1995b, 1998, 1999).

7.3.3 Interactive and Indifferent Kinds
Much changes as a result of being classified or treated in certain ways by human beings, but only human beings understand their treatment as a

consequence of being sorted into one kind rather than another. This understanding affects how we see ourselves and how we behave as a result of our revised self-understanding. Natural kinds are, in Hacking's words, "indifferent": their falling under one classification rather than another does not cause them to change behavior. "The classification 'quark' is indifferent in the sense that calling a quark a quark makes no difference to a quark" (Hacking 1999, 105). But social kinds are *interactive*: being put into one category rather than another makes a difference to people.

Author and brother are kinds of people, as are child viewer and Zulu. People of these kinds can become aware that they are classified as such. They can make tacit or even explicit choices, adapt or adopt ways of living so as to fit or get away from the very classification that may be applied to them. . . . What was known about people of a kind may become false because people of that kind have changed in virtue of what they believe about themselves. (Hacking 1999, 34)

Furthermore, the inquiry itself responds to the evolving behavior, producing new knowledge that leads to new forms of behavior, and so on. Hacking calls this "the looping effect." His fullest treatment is of Multiple Personality Disorder (Hacking 1995a). Popular dissemination of the lore associated with the diagnosis, he suggests, causes people who fall under it to behave differently. Clinicians and researchers notice the changing behavior and respond by changing their conception of the disorder.

So we have intentionally mediated interaction between theory, practice, and the behavior of the people who are studied. This mediation stems from social mechanisms that enable subjects to understand that they are being classified in one way rather than another. Expert theory and lay opinion reflect and change one another, so that what may appear as a new discovery about a diagnosis does not reflect an accumulation of knowledge about a stable condition. It reflects a change in the properties of an unstable kind because of changes in the self-understanding of its members. The result is that social kinds are labile. They change over time, which means that they are not stable objects about which we can accumulate knowledge.

This looping effect resembles Appiah's (1996) concept of identification, by which he means something like the process of conforming our plans and goals for our lives to concepts available in the culture. Those concepts include prescriptions or expectations about what counts as appropriate behavior for people who fall under them. Identification may lead us to behave in ways that conform to these socially available concepts and the norms internal to them.

Appiah's notion is wider than Hacking's looping effect or, at least, it draws on everyday thought rather than scientific procedures of classification. But the thrust is clearly the same: behavior is shaped by expectations about how members of a human kind should behave. Appiah (1996, 78) argues that his "identification" is central to Hacking's "dynamic nominalism," which sees categories of thought about people as creating those types of people. The complex of ideas centered on the looping effect is one way for kinds of people to be created by social practices.

I must note two things about the looping effect: first, unlike Greenwood's (2003) social groups, groups produced by the looping effect can be held together by "external" rather than "internal" transmission. That is, people may make up a group because of representations of how that group is perceived by theorists, media figures, and other members of the wider culture, not just because they share a conception of the group and are aware that others share it. They do not even have to know that they are members of a given category as long as they respond to their categorization interactively rather than indifferently. The group has a distinct social existence, but not via the awareness of shared norms. This is an important point, since people with mental illness are very unlikely to be behaving as they do because they wish to act like people with mental illness; their motivations are almost certainly different. (I must, however, make room for an exception in the case of a condition that is romanticized or otherwise positively viewed within a subculture.)

Second, the kinds produced by the looping effects are unstable, in just the way that Omi and Winant (1994) stress. Before I address instability, I complete my discussion of Hacking by looking at his second social mechanism for creating kinds—niches.

7.3.4 Niches

The second of the social mechanisms that Hacking sees as productive of mental illness is one that he presents as a less general social phenomenon than the looping effect. Hacking wonders why "transient mental illnesses" flourish for a brief time in a particular cultural context. His example is dissociative fugue, first diagnosed in France in 1887 (Hacking 1998). Sufferers wandered the roads in a trance-like state for months. But by 1909 the diagnosis had all but disappeared, and it never really caught on outside France. This impermanence and regionalism raises the question whether dissociative fugue was a real disorder at all. However, the symptoms exhibited by *fugeurs* were real enough, so even if fugue was not a genuine mental illness, we are left with the puzzle of explaining its sudden flowering and disappearance.

Hacking does not deny that fugue was a genuine mental illness.[3] He thinks that disorders may suddenly flourish because an appropriate niche appears for them to occupy. Then they disappear as the niche disintegrates.

Hacking does not look for special, transitory neurochemical conditions to account for the sudden arrivals and departures of transient mental illnesses. Rather, a condition like fugue depends for its brief existence on an equilibrium occurring among relevant aspects of its social and cultural surroundings. Hacking calls these contextual features "vectors" and says that specific forms of (at least) four vectors need to come together to make a niche for a transient mental illness. Hacking names the vectors "medical taxonomy, cultural polarity, observability and release" (1998, 80–81).

Taxonomy For an illness to arise there must be a suitable vacant position in a psychiatric system of classification (a "nosology"). Fugue, Hacking claims, fitted into just such a ready-made conceptual space, between epilepsy and hysteria. But when the orthodox taxonomic picture changed under Kraepelin's influence, a place for fugue no longer existed.

Cultural Polarity A deviant behavior must oscillate between two behaviors—one condemned and the other admired—and partake of both. Hacking asserts that fugue partook of both vagrancy and a newly popular and widely esteemed behavior, tourism. What seems correct about this is the idea that a new way to go mad existed after traveling to distant parts became fashionable: as new social roles and behaviors become widespread, they, or simulacra of them, enlarge the scope of deviant behavior.

Deviance The illness must cause behavior that can be detected as a departure from approved customs. This was straightforward in late-nineteenth-century France, where everyone carried papers and the police vigilantly sought itinerants.

Release One has no responsibilities on the road, so the disorder frees one from the unbearable stresses of everyday life.

Let me summarize Hacking's proposals: there are two social mechanisms that cause a mental illness to appear and develop in the absence of new biological causes. The first is the looping effect of interactive kinds, which transforms human behavior in the process of studying it. The second is the accumulation of social vectors to form a niche within which a transient

3. Or rather, he does not say it was unreal in the nineteenth century. But he does think, about both fugue and Dissociative Identity Disorder, that they are not real by Peirce's test of reality as that which we will agree on at the end of inquiry (Hacking 1998, 96–99).

mental illness can flourish. One of Hacking's vectors is that of medical taxonomy, so we might wonder whether his work on the looping effect is best seen as a detailed examination of one vector, in which case we have only one social mechanism, with some aspects spelled out at great length and the others sketched in more programmatically.

7.3.5 Hacking: The Issues

Hacking writes about particulars. Despite his metaphor of niche and vector, he has done more natural history than ecology. Although he has given more details in his work than most theorists of social construction provide, we remain unclear whether it is intended to apply only to transient or unstable diagnoses. If its application is restricted to transient and unstable conditions, then even if Hacking's account rebuts the antirealist charge it cannot rebut the other objections, to wit, that social explananda are crowded out by natural explanations and are inductively useless. Not all mental illnesses are transient, so we need to know if something like Hacking's account can cover the stable as well as the unstable cases.

I used bulimia to argue that social construction explains aspects of genuine mental illnesses by appeal not just to social processes but to the interaction of culture and psychological vulnerability. But I want to defend social kinds in psychopathology against the other two objections. First, Hacking's social explanations admit of generalization across stable conditions. Second, a coupling of social causes with biological psychiatry's stress on the underlying physical nature of the patient can rebut Stein's worry that social explanation produces kinds of little scientific use. This allows for integration of social and natural explanations in the same structure.

In the next section I apply Hacking's social dynamics to stable conditions. Then I set out and criticize Hacking's own version of the reconciliation between social and natural explanation in psychiatry. I offer my own amended version that preserves Hacking's social dynamics but meets Stein's challenge more effectively by supposing that cognitive structures make up the crucial link between the social world and proximate, biological causation of behavior.

7.4 Integrating Social and Natural Explanations

7.4.1 Stability

Hacking's looping effect and vectors are designed to show how mental illnesses can flourish, change, and subside. But disorders may persist relatively unchanged across the generations. Is this persistence a sign of pathology that is immune to social and cultural forces? Stability may seem to be evidence that the disorder depends on organic abnormalities to the exclusion of social pressures.

However, the persistence of a disorder in relatively unchanged form across the generations does not show that it is not partly a cultural product. If a niche comes into existence because of the coalescence of vectors, why should it not persist? If the vectors fall into a stable configuration robust enough to withstand cultural changes then the niche may endure indefinitely. And although a looping effect may cause behavior to evolve, theory and behavior may mesh well enough to stabilize behavior over the long term. In the former case we would have an enduring niche, and in the latter a looping effect that causes behavior to repeat rather than evolve. At the other end of the temporal spectrum, very short-lived and unstable effects of social factors may exist, as when social contagion spreads through a crowd. Social effects persist on timescales ranging from minutes to millennia.

There might be some surprising instances of stable social causation in psychopathology. For instance, some historians claim that schizophrenia was virtually unknown before the late eighteenth century, which leads them to wonder if there may have been something pathogenic about the Industrial Revolution that caused schizophrenia to flourish (Shorter 1997). If schizophrenia had disappeared among the Victorians we would have a clear case of a transient mental illness. Although it did not subsequently fade away, perhaps the rise of schizophrenia might nonetheless be explained in the way Hacking explains fugue, as a set of vectors coming together to form a niche conducive to the condition.

The looping effect could be working to keep a pattern of behavior stable. Vectors, such as medical opinion, could stay in place and preserve the form of the illness. Social pressures and stereotypes can freeze a pattern of behavior in place as well as cause it to evolve. Hacking has had little to say about stability, but we can distinguish two general ways in which his social dynamics may work—sometimes transforming behavior, and sometimes keeping it stable. We might use Hacking's ideas to account for the coalesence of several symptoms into one persistent form of madness by the coming into existence of a stable niche. As any ecologist can tell you, a niche, once created, can last a very long time indeed.[4] This endurance would rebut the assumption that social kinds are bound to be unstable and hence of scant inductive

4. See Mallon (forthcoming) for a different treatment of stability. Both of us are indebted to Griffiths (1997).

utility: if a social kind is stable then it may be perfectly projectible (inductively useful) in the long run.

But it may seem that stable mental illness is a poor candidate for social explanation. The worry is that what little we know about cultural sources of long-term behavioral stability might not apply to mental illnesses. One way for behavior to stabilize, for example, is via the stable norms. For example, although Macintyre (1982) is mostly interested in tracing changing conceptions of virtuous conduct between antiquity and modernity, he also says that the fortitude of Penelope as she awaits the return of Odysseus would not be out of place (and would be equally praiseworthy) in one of Jane Austen's heroines. But if the virtues are always changing, why does Anne Wentworth behave like Penelope? Macintyre does not say, but the obvious answer is that feminine virtues did not change because the norms governing female behavior remained unchanged.

A stable norm, like a stable medical taxonomy or theory, might seem an appropriate vector for a Hackingesque treatment. But stabilization by norms might seem a poor model for mental illness, since surely one thing Scheff has correct is that many mentally ill people violate norms. How can we appeal to the stability of norms to help explain the replication of counter-normative behavior? Although social forces can interact with psychology in other ways, the place of norms is significant enough to make this a worthy issue to address. We are creatures who are naturally adapted to constitute and follow norms (Gibbard 1990), and they shape much of our behavior.

7.4.2 The Role of Norms

Norms offer cultural forms that can interact with psychology. Norms of deviance need not exist, but if individuals differ psychologically in ways that make norms hard for them to follow, and if some such individuals exist in each generation, then in each generation there will be people who cannot grasp the norm and are seen as deviant.

This might happen in several ways. We can distinguish, to begin with, between a failure to live up to a norm and a failure to grasp it. Bulimics may endorse cultural representations of womanhood that they are unable to live up to, either because their physiology does not permit them to lose enough weight, or because of motivational conflicts: bulimics may be torn between a fairly stable desire to have a certain shape and a succession of more powerful short-term appetites. Stable counter-normative behavior that is characteristic of a social kind might occur in the long-run if norms are replicated across the generations and each generation includes individuals for whom some norms are both very attractive and very hard to attain. (All of us violate norms sometimes—the question is what in their psychology makes some individuals prone to persistent, compulsive violation.)

Another possibility is that some individuals grow up unable to grasp norms, rather than unable to conform to norms they do grasp. As discussed in chapter 3, some mental disorders might be explained in terms of failure to notice and understand norms. I noted that McGuire and his colleagues (McGuire et al. 1994; McGuire and Troisi 1998) explain dysthymia in terms of an inability to engage in reciprocal altruism, and I said that they had missed the point that dysthymics might be capable of mastering the relevant social norms but still fail to do so. Perhaps some norms are not grasped during childhood socialization because of peculiarities in a family or sub-culture, leading to a looping effect based not on classification but on how people treat dysthymics. If they are regarded as socially deviant, they might be treated as such and not afforded opportunities for normal interaction, which would further disable their ability to learn the proper norms. This is reminiscent of Scheff's view, but applied as a genuine causal proposal. People may suffer because of an inability to understand the norms of the surrounding community.

An account such as Hacking's could make room for norms as one vector. If the norms are stable, then in each generation opportunities will exist for an interaction between norms and individual psychologies in ways that stably reproduce forms of mental illness, because some of the possessors of the individual psychologies are unable to follow or grasp the norms. Therefore, counter-normative behavior could be reproduced in each generation by processes that are parasitic on norms. And norms do not only persist, they can spread in a culture, becoming more attractive or viewed by more people as compelling. (For a review of, and contribution to, this research, see Nichols 2002b.) Norms and other cultural resources differ in their spread and in their appeal. If norms spread then more people will learn of them, and if they become more appealing then more people will endorse them and try to conform to them. (A norm could be widespread yet regarded with comparative disinterest.) So a successful norm, both appealing and wide-spread, will enlist more and more people, and hence more and more people will experience the psychic strain of trying to adapt ill-suited motivational structures, or other psychological structures, to comport with the norms they endorse. In that case, the epidemiology of aspects of mental illness that are parasitic on norms should track the spread of the norms themselves and reflect intrinsic differences between people. As I noted, this seems to have happened with bulimia.

This is not the only way for social explanations of stable mental illness to work. Indeed, psychological disorder might be caused by successfully following norms; some people who successfully socialize into a subculture that prizes excessive drinking suffer grave long-term damage. But the stable co-construction in successive generations of norms and psychologies that have trouble living up to them does seem one avenue to explore.

These suggestions motivate an important amendment to Hacking's account. Hacking thinks in terms of cultural vectors in social explanation, but we might prefer to think of a mix of material and cultural vectors. Some vectors might be purely material, such as genes or neurophysiology, and others, such as diet, might be material but shaped in important ways by culture. Still others, for example patterns of alcohol consumption, might be more purely cultural. I have suggested one possibility for the mix: culture may make demands that some individuals are ill-placed to meet because of their psychology. If vectors mix the material with the cultural or conceptual, that suggests that the causal dynamics are not exclusively social or natural, and likewise that kinds are not exclusively social or natural.

So the opposition we began with is too stark. Although registered voters are a social kind, and people with O-positive blood are a natural kind, many psychopathologies (and, doubtless, other interesting human kinds) are both social and natural. In the philosophical sense they can be called natural kinds. But the explanatory mechanisms that keep them intact integrate natural and sociocultural causes. To understand bulimia, we need to know about individual variation in psychology and physiology, and cultural (and subcultural) variation in representations of female bodies. Therefore, some questions we ask about them will be answered by the human sciences rather than the life sciences, and a full understanding will integrate social and natural explanations. Hacking has a proposal for understanding that integration, but it is one that threatens to enforce, rather than dissolve, the distinction between social and natural explanations.

7.4.3 Putting Culture and Biology Together

Suppose that we have a social explanation of why disorders can be stable across time as well as transient or prone to looping. If we could fit these social dynamics together with the medical model we would carry out a successful reconciliation of the constructionist tradition and the medical model. The reconciliation would explain what in the mind/brain causes schizophrenia, and also explain which differences between West Indian and English society cause Afro-Caribbean migrants to Britain to contract schizophrenia between four and twelve times as often as either white Britons or people living in the Caribbean (Rutter et al. 2001).

Hacking's reconciliation consists of "putting a theory of reference alongside social construction" (1999, 122). He begins with Kripke-Putnam semantics for natural-kind terms (Kripke 1980; Putnam 1975) and reminds us of Putnam's notion that the semantics of a natural-kind term incorporates both the extension of the term and an associated stereotype. To apply this picture to the semantics of terms in psychiatry, Hacking assumes that the referents of psychiatric terms are neurological states. So the extension of "schizophrenia" would be something like "cognitive dysmetria" or "abnormal cortical migration." Hacking introduces social construction by understanding the stereotypes associated with psychiatric kind terms as social constructions reflecting the thinking of psychiatrists, psychologists, and perhaps patients themselves. So an account of autism, insists Hacking (1999, 121),

could perfectly well maintain (a) there is probably a definite unknown neuropathology P that is the cause of the prototypical and most other examples of what we now call childhood autism; (b) the idea of childhood autism is a social construct that interacts not only with therapists and psychiatrists in their treatments, but also interacts with autistic children themselves, who find the current mode of being autistic a way for themselves to be.

So on Hacking's picture "autism" refers to a dysfunctional cognitive part. He adds to it the claim that the symptoms shown by people with the broken part will come to depend on the dissemination of the story. As more and more people who are professionally concerned with autism accept the story as a guide to thinking about the condition, autistic children themselves will interact with the beliefs of those professionals about them.[5] A looping effect results. "Autism" refers to the underlying neuropathology, which is in Hacking's terminology an indifferent kind, but childhood autistics will be an interactive kind, with a way of being that is a product of the autistic stereotype and which will in turn cause that stereotype to evolve.

---

5. Their poor social cognition raises the question whether autistic people could understand their treatment by others well enough for the looping effect to occur. Hacking's first answer (1999, 115) is that their behavior may be shaped by the wider understanding of the disease even if autists do not self-consciously react to the classification. This bypasses the intentional route through the mind/brain of autistic subjects and threatens to collapse the distinction between interactive kinds and kinds that are shaped only by human practices, such as domestic animals. (Hacking has another response to this worry about autistics, which is that the patient's whole family may be affected by changes in theory.)

The reconciliation marries the looping effect with the biological tradition of inquiry into mental illness. Neuropathologies are indifferent kinds and they are the underlying physical basis of mental illness: Hacking says that a "follower of Kripke might call [a neuropathology] the essence of autism" (1999, 121). But the behavioral stereotypes associated with those neuropathologies take particular forms because of cultural and social forces. Hacking discusses only the looping effect when he offers his "semantic resolution," but it is not hard to see how the idea of a niche might figure in this context. The pressures shaping the expression of the underlying neuropathology in behavior might be considered vectors, so that the same brain disorder, put in a different culture with different vectors, will lead to different ways of being mad.

Hacking's picture of underlying pathology being shaped by culture into specific forms of illness is an attractive one. In broad outlines it is the approach I favor as the most promising means of reconciling neurological and social constructionist accounts. But to motivate my amendments to Hacking's account, first I take issue with it. My objections all stem from Hacking's preference for a semantic rather than a causal approach to dealing with the relation of biology and behavior.

7.4.4 Problems with the Semantic Approach

The semantic approach does not answer the questions about mental illness we want answered. It deals with the circumstances under which kind terms are revised, not with the causes of mental illness. In Hacking's view, whether disorders are real or constructed "turns out to be a relatively minor technical matter. How to develop a plausible semantics for natural kind terms?" (1999, 123). But what is worrying about the claimed tension between reality and social construction is not just the semantics of natural-kind terms. At stake are several substantive issues: the plausibility of a medical model as a generalized strategy in psychopathology; the politics of normality and deviance; the relative efficacy of drugs and psychotherapy; the nature and experience of mental illness. Hacking's semantic resolution bypasses these important issues. He gives us instead a collection of

kind terms that exhibit a looping effect, that is, they have to be revised because the people classified in a certain way change in response to being classified. On the other hand, some of these interactive kinds pick out genuine causal properties, biological kinds, which, like all indifferent kinds, are unaffected, as kinds, by what we know about them. (1999, 123)

Note Hacking's contrast in this passage between the looping effect and "genuine causal properties," as if only biology can cause behavior. This begs the question of the relation of social construction to biological psychiatry. Knowing that our kind terms refer to indifferent kinds with associated interactive kind stereotypes does not help us to see how the indifferent and the interactive are related. Just knowing that the semantics incorporates both socially constructed properties and biological ones is no help: we want to know how the two sets of properties relate to one another.

In Hacking's picture, biology produces the impairment and society its manifestation. His view is not one in which the impairment has a causal effect on the manifestation of the disorder. This does not fit either the psychiatric facts or the ideas of Kripke and Putnam, which stress the causal link between stereotype and essence.

In order for the social dynamics Hacking writes about to have their distinctive effects, they must be mediated by the organ that the medical model takes to be its special province—the brain. This means that we cannot separate out the social and the biological and address them separately in Hacking's fashion. To understand how behavior might be socially constructed, we need to understand the neuropsychological structures that mediate between society and behavior. After all, it is their awareness of being studied or treated in one way rather than another that causes human beings to form interactive kinds and not indifferent ones: and the brain is the organ of awareness. The social has its effects on behavior—such as looping—via the way people think about it, i.e. via their brains.

Furthermore, the account of essences Hacking takes from Kripke and Putnam does not seem to fit the neurology well. The form of a brain pathology will vary depending on the nature of the social forces, since the brain, being the organ of cognition, changes in response to changes in the social and cultural environment. If similar changes mediate the alterations in behavior associated with the looping effect, then the neural "essence" of the disorder is itself labile, rather as if the essence of water were H_2O when water is liquid but something else when frozen. Such unstable essences are not essences at all: they are merely a long disjunction of sufficient conditions.[6]

---

6. Hacking (1999, 123–124) does discuss what he calls "biolooping," which is behavior modification that works in opposition to the looping effects of classification. He admits that it can change the brain. He regards biolooping as one sort of feedback, and I am not sure how to place it in his account. As far as I can tell, Hacking thinks that there is nothing of general interest to say about the dynamics of classification except for the semantics. Everything apart from semantics is just the details of a given case, and biolooping is one case.

What attracts Hacking is the idea of a contingent relationship between brain pathology and behavior, and the form of the relationship depends on the diagnosis and the surrounding social context. But this point, which is surely correct, can be made without bringing in the semantic machinery, which adds nothing to what Hacking has already established. That might not matter if, as Hacking suggests, his semantic resolution sidesteps the "real vs. constructed" tension in order to study the dynamics of classification. However, any analysis of the dynamics of classification raises just those epistemic, metaphysical, and political issues I referred to earlier, and which Hacking himself recognizes as central. These issues do not disappear when we investigate the dynamics of classification; they reappear, because the dynamics of classification depend on causal processes operating on what is classified.

The dynamics of classification reflect the dynamics of causation rather than the semantics of kind terms. They involve the social and the biological, since classification causes its effects in virtue of operating upon embodied and socially situated people. Hacking's semantic approach is unhelpful in analyzing the causal dynamics that underlie classification, and we must look elsewhere to develop a socially informed biological perspective on psychopathology. The right approach departs from Hacking in noting the dependence of neuropathology on environmental (including social) factors as well as intrinsic factors, and the dependence of the behavioral manifestation on the underlying neuropathology. Whereas Hacking sees neurology as generated by endogenous factors and behavior generated by social ones, we should rather see both neuropathology and behavior as generated by a complicated array of causes that allow us to ask and answer different questions, and offer many different opportunities for therapeutic intervention.

7.5 A Broad View of Social Construction

Hacking's account of the looping effect and the establishment of niches is the fullest theory of social construction of mental illness, and it can explain stable and persistent conditions as well as transient or unstable ones. But Hacking's proposal for reconciling the biological and the social is semantic rather than causal, partitioning the biological and the social into separate causal realms rather than reconciling them. The cognitive neurosciences explain the causal dynamics in terms of mediation of the social on behavior in terms of the representation of social properties by the brain. These representations, in turn, cause behavior. The brain responds to culture, and culture is built by the ideas in people's brains: the developing mind/brain is itself affected by the culture to which it contributes. The opposition between natural and social is in fact much too stark: it requires a brain that is the product of natural factors exclusively and a set of behaviors that are the product exclusively of culture. The true picture sees psychopathology, and no doubt much other human behavior, as the product of interacting natural and social explanations, perhaps on a spectrum from the completely natural, like blood groups, to the completely social, like political allegiance.

Development is an obvious place to look for the incorporation of the social in a model of the mind/brain drawn from the cognitive sciences in (at least) two ways. In the first place, we can explain psychological differences between people in terms of different social and cultural influences on the developing organism.

Second, social properties are represented by the information in the system, and that has to be learned. Understanding mental illness might require understanding how this information interacts with other (external and internal) structures. This point gives us a way of understanding how mental illness can arise even in the absence of obvious brain abnormalities: an important development if we are to reconcile the social constructionist and biological accounts across the board.

The sciences of development often seem to neglect individual differences, dismissing them as irrelevant noise in the search for regularities across individuals. Yet it is these differences in individuals' developmental pathways that are crucial to understanding how similar biology can produce different psychological results depending on a variety of developmental inputs including the sociocultural (Bateson & Martin 1999; Oyama 2000). Not every poor Frenchman became a *fugueur*, so perhaps one way to explain those who did is to search for distinctive patterns in their development. A "trait" rather than "state" interpretation of the neurological differences that Steiger et al. found in bulimics might also refer those neurological differences back to the environments, including the social environments, in which some of the girls grew up. Yet of course those differences might depend in detail on the kind of organism that a particular young girl was at the moment the social factors began to exert their effect.

The complexity of development blurs the hard-and-fast line between natural and social explanation. This reinforces the moral I drew from Hacking's semantic resolution of the social/natural opposition. The true picture is one that sees psychopathology as the product of interacting natural and social explanations, perhaps on a spectrum from the almost completely natural to the almost completely social, with most forms of illness located somewhere toward the middle. This opposition is much too stark in requiring a brain that is the product of natural factors exclusively and a set of behaviors that are the product exclusively of culture.

The culture/biology opposition would make sense if our psychology consisted without exception of informationally encapsulated modules with innate databases, because that would allow no way for new information to make a difference to the information in the databases. But, as chapter 5 shows, that is implausible, at least as a claim across the board. In normal development we learn much about the world, including much about the social world. The brain responds to culture, and culture is built by the ideas in people's brains.

Social explanations still can be of use even without a sharp break between social and natural kinds. To call something socially constructed can be a claim that it exists because social factors produce and reinforce it via cognitive mechanisms. In the unfortunate sense of natural in which "nature" is contrasted with "culture" it may be true that what is socially constructed is unnatural, and in some cases that may be genuinely informative. But being socially constructed does not make something unreal.[7] Social construction is ubiquitous, and social constructionist explanations help us understand psychological phenomena as responses to cultural models. These models can be more or less explicit, more or less persistent, and more or less successful in shaping behavior.

This might seem a long way from customary social constructionist theorizing. Not much turns on the labels (ironically, considering the subject matter) but I prefer to think of social constructionism very broadly, as the use of social rather than natural explanations. This sense is continuous with anthropological theories about cultural variation. "Culture-bound" behaviors are paradigmatic social constructions. But anthropologists are not skeptical about the behaviors they describe. They regard them as real, but not

---

7. Although it may undermine some ways of thinking about it. There was no such thing as witchcraft, if "witches" were social constructions rather than brides of Satan. But, equally, there was no such thing as phlogiston—even though the phenomena were real enough in both cases. The bite of social construction comes when it says not just that we were wrong but that we projected pernicious politics into the world and saw them vindicated there. There seems no reason why the process could not act the other way round—we might discover that what we thought was just a social role is in fact a natural way to be human. Some gay activists welcome the idea that homosexuality will turn out to be "naturally constructed."

naturally explained. The picture I sketched earlier applies to culture-bound mental illnesses.

7.6 Culture-Bound Syndromes

Many syndromes, though they may not be unstable or transient, are "culture bound." They occur only in one culture and there is dispute over whether they are unique conditions or local forms of a common, transcultural condition. One example is *latah*. According to Ronald Simons, "*latah* is the playing out in a Malay cultural context of the behavioral and experiential potential inherent in the neurophysiology of the startle reflex" (1996, 199). Simons regards *latah* as one of a number of "hyperstartling" conditions that exist around the world and that feature a startle response to minimal stimuli that leave most people unaffected, together with matching (echoing what was said and done around one at the time of the startle) and automatic obedience (in which a hyperstartling person obeys an instruction he or she usually would ignore). The other "startle-matching syndromes" may include Tourette's as well as lesser-known conditions in Asia, the Middle East, and Lapland (Gilles de la Tourette 1885; Simons 1996). Simons views all these conditions "as local culture-specific variants of a single syndrome whose form is determined by the properties of an exaggerated startle-reflex and the capture of attention it may engender" (222). Capture of attention occurs when a stimulus crowds out all other awareness. Simons argues that startling induces attention capture, which causes matching and obedience. Different cultures will feature different commands and different matchable behaviors, which explains cultural variation.

However, most anthropologists regard the resemblances across cultures as incidental, and explain culture-bound syndromes as adaptations to specific cultures. Winzeler (1995), who notes that women, the poor, and the old are disproportionately prone to *latah*, gives it a symbolic role in Malay culture as a way for lower-status individuals to subvert hierarchies, typically with humor that transgresses established norms. Winzeler makes *latah* sound like a form of satire and says that "cultural learning, expectation, and individual need and social use appear to be an adequate explanation" of it (136). For most anthropologists, cross-cultural physical similarities in startle-matching syndromes are of no more consequence than the fact that a priest raising a chalice and an athlete raising a trophy employ similar arm movements. What matters is the symbolic role the action plays in the surrounding cultural context. Simons acknowledges cultural variation, but only in the input to the startle reflex.

In discussing disputes between cognitive scientists and anthropologists about whether emotions are culture bound or universal, Mallon and Stich (2000) argue that the competing views depend on different theories of reference. They see the issue as whether emotion terms denote universal psychological mechanisms or "thick descriptions" of culturally specific practices. But that diagnosis does not fit this case, which is a civil war in anthropology. Indeed, Simons says that semantic arguments about the reference of *latah* are "rather pointless" (222). Both Simons and Winzeler strive for ethnographic precision and richness when discussing *latah* in Malay culture, but they disagree about what explains the behavior: is it cultural forces or neuropyschology?

One argument Simons gives for his position can be discarded. When Malays are asked about *latah*, they deny that it has any cultural significance. But this is what one would expect if the social construction of *latah* is, in Griffiths's terms, covert. And Winzeler seems happy to think it is.

So how do we decide who is right? My answer (and this *is* in the spirit of Mallon and Stich) is that we do not have to; we can read Simons as providing a neuropsychological theory of the mechanisms underlying *latah*, and Winzeler as giving a social explanation of the form that the output of the startle mechanism takes when it is shaped by Malay culture. Other explanations are required for the other startle-matching syndromes, because each is shaped by culture. But in each case what is shaped is the startle reflex. The strategy, then, is the same as the one I recommended in place of Hacking's semantic resolution to the social/natural split. We should aim for the integration of culture and neuropsychology via the mediating role of mental representations.

7.7 Conclusion

Merely adverting to social construction or social forces is unsatisfactory in the human sciences. For social explanations to play a genuine explanatory role they must work via a material mechanism. In psychiatry, sociocultural forces can be mediated by mental representations. This can explain the content of particular symptoms (the particular form that delusions take is another example) and the epidemiological patterns exhibited by mental illnesses. Social construction, understood in my broad sense, is not limited to the explanation of unstable phenomena, nor is it antirealist or incompatible with natural explanations. Both social and natural explanations can be integrated in the same neuropsychological structure, which suggests that debates about whether mental disorders are natural or social kinds are beside the point: they can be natural kinds even if part of their explanations are social. But what is a natural kind, and how important is it whether psychiatric conditions are or are not natural kinds? I turn to those questions in chapter 9, after looking at another form of explanation that might fit well into the overall pattern of exemplar explanation: evolutionary explanations, which are the subject of the next chapter.

Homer: That baby-proofing crook wanted to sell us safety covers for the electrical
outlets. But I'll just draw bunny faces on them to scare Maggie away.
Marge: Maggie's not afraid of bunnies.
Homer: She will be.
—*The Simpsons*

8.1 Introduction

In chapters 4 and 6 I mention the search for genetic influences on the developing brain that cause disorders or establish a susceptibility to them. But there is another way to involve genes. Whereas there is no suggestion that the conditions we try to explain by employing psychiatric genetics are anything other than failures of our natural design, evolutionary explanations of psychopathology ask about the psychiatric significance of natural selection. It might have none, of course, but evolutionary accounts are increasingly visible in psychiatry, as in psychology generally. In this chapter I look at theories that focus on two issues: first, whether natural selection might have favored a putative disorder and second, whether it might have favored a behavioral trait that is linked closely enough to a disorder, either genetically or developmentally, that the presence of the trait in the population inevitably causes the disorder to occur as a side effect. Evolutionary explanations often assume that some response to the environment has been implanted by natural selection. The psychological basis of learned responses may also be involved in mental illnesses and often compete with evolutionary explanations.

The evolutionary approach permits three different explanations of psychopathology, which are not mutually exclusive. In fact, one theory could make use of all of them (e.g., Murphy and Stich 2000). First, the overall picture of the normal mind should be that of a mind that has been shaped by natural selection. This is a very minimal concession to evolution, which all rational parties accept. But, as we saw in chapters 2 and 3, accepting the platitude that we have an evolved psychology is consistent with a range of theses about psychological organization and facture.

The second explanation says that disorder results from a mechanism that once was adaptive but, because of changes in the environment, is no longer adaptive. This view does not identify a system within the person as malfunctioning, but locates the pathology in a mismatch between the environment the system was originally adapted for and the current environment.

The third type of evolutionary explanation appeals to the theoretical possibility that some putative disorders are adaptive even in the current environment. It could be that some behaviors or psychological systems evolved in response to properties of the environment that have not changed in a way that renders them currently dysfunctional. This is uncontroversial for large amounts of our psychology. No one doubts that we still need sensory systems, a theory of mind, or an understanding of social reciprocation: they are still adaptive. The controversy concerns whether putative mental illnesses are in fact adaptive in the current environment. The claim is that as well as large tracts of what makes us normal, some abnormal and disturbing psychological traits might be currently adaptive.

My second and third explanations often are taken to distinguish evolutionary psychiatry from other approaches. However, an evolutionary psychiatry that only employed these sorts of explanations would be in the unfortunate position of arguing that psychopathology never should be explained in terms of something having gone wrong with a mind. No thinker, no matter how Darwinian, should want to say that. So the two strategies I discuss—adaptation/environment mismatch and the persistence of apparently pathological but actually adaptive behavior—should be regarded as possible supplements to the main thrust of even an evolutionary psychiatry. The main orientation of psychiatry should remain medical even if psychiatry sees the mind as a culturally embedded product of evolution, as it surely should.

So we have three types of evolutionary explanation for psychopathology—breakdown, mismatch, and adaptive persistence. I discuss breakdown at length in the preceding chapters. In this chapter I discuss the prospects and problems of explaining mental disorder in terms of mismatch or persistence. In fact, there are four possibilities. Either mismatch or persistence could be the explanantia, and the explananda could be, as noted, either disorders or traits closely linked to disorders. An explanation of a disorder in evolutionary terms uses a direct strategy, and an explanation of a disorder that is allegedly a by-product of an adaptive trait uses an indirect strategy.

Hence, mismatches and persistence explanations have direct and indirect applications.

The first of the two variations on the mismatch hypothesis is represented in this chapter by the work of (chiefly) Isaac Marks on the anxiety disorders (Marks 1987; Marks and Nesse 1994) and several recent approaches to depression (Hagen 1999; Nesse and Williams 1995; Price et al. 1994; Watson and Andrews 2002; see Greenspan 2001 for a skeptical philosophical discussion). Before turning to these works, I begin in section 8.2 with a discussion of anxiety disorders, especially phobias, as expressions of evolved predispositions to develop a fear of stimuli that was dangerous in ancestral environments. In the cases I discuss, evolutionary explanations oppose explanations in terms of conditioning, which brings up the idea that learning can be a source of mental illness. Then in section 8.3 I discuss the social competition hypothesis of depression, which is based on the idea that depression is an evolved response to loss of status, or to an unsuccessful attempt to gain status (Price et al. 1994; Nesse and Williams 1995), but I argue that the symptoms of depression do not fit the theory.

Mismatch hypotheses assume that a psychological trait (in these cases, a pathological one) is well understood, and ask what function the trait had in an ancestral environment. The indirect strategy typically takes a disorder and looks for a related behavior that might have been adaptive. While section 8.4 I considers a theory of schizophrenia in this light, it is purely for expository purposes: I do not believe a word of it.

Then I move on to persistence. The main candidates here are personality disorders, which are reinterpreted as traits of personality that are good for their possessor but bad for everyone else. I also discuss persistence theories of antisocial personality.

8.2 Anxiety Disorders as Expressions of Adaptive Reflexes

8.2.1 The Anxiety Spectrum

Some anxiety disorders provide a possible example of disorders that result from a mismatch between the contemporary environment and the environment in which our minds evolved. Many diverse conditions fall under the DSM-IV's umbrella of Anxiety Disorders, including Panic Disorder, Obsessive Compulsive Disorder, and Post-Traumatic Stress Disorder. The relevant conditions for our current purposes are Agoraphobia and the other specific Phobias. Agoraphobia is not, as many people think, the opposite of claustrophobia; it is not a fear of open spaces. In fact, Agoraphobia is anxiety about being in situations in which help may not be forthcoming if trouble occurs or from which escape may be difficult. Typically, this includes an anxiety about being in crowds, or being alone in public places, although it may also include a fear of being left alone at home. Specific phobias are persistent fears of a specific object or situation. DSM-IV-TR lists five subtypes of specific phobias: animal, natural environment (such as heights, storms, or water), blood-injection-injury, situational (such as airplanes, enclosed places, or elevators), and "other" (450). The situational and other types are very hard to distinguish based on the text, which says only that the situational type is cued by a specific situation whereas the other type is cued by other stimuli. The examples are not particularly helpful—fear of elevators is a situational subtype but fear of falling down if one moves away from a wall is a subtype of other phobia (DSM-IV-TR, 445). The difference seems to boil down to that between stereotypical phobias, such as fear of heights, and more unusual ones. If we understand situational phobias as stereotypical ones and adopt the evolutionary explanation for phobias discussed in section 8.2.2, the "other"-type phobias will be those acquired by conditioning to features of the modern world, and not those for which an evolutionary explanation is credible. Not all specific phobias are plausibly adaptations.

8.2.2 Phobias and Preparedness

Marks and Nesse argue that the situations that typically cause phobias are the same as those we would expect to have been dangerous in ancestral environments. They claim that in the ancestral environment, a fear of public places and a fear of being far from home might well have been adaptive responses "that guard against the many dangers encountered outside the home range of any territorial species" (1994, 251). Similarly, they argue that a fear of heights accompanied by "freezing instead of wild flight" would have had obvious adaptive value to our hunter-gatherer forebears (1994, 251). In a modern urban environment, however, people who become extremely anxious when away from home or in public places find it all but impossible to lead a normal life. And people who become extremely anxious in high places find it difficult or impossible to travel in airplanes, ride in glass-enclosed elevators, or work in the higher floors of modern buildings. Thus, because the modern environment is so different from the ancestral one, people who are toward the sensitive end of the distribution of phenotypic variation may be incapable of coping with many ordinary situations despite the fact that all their mental mechanisms are functioning in exactly the way that natural selection designed them to function.

My presentation of the argument is not quite true to what Marks and Nesse hypothesize. I develop the argument in what I take to be its strongest form, in which the idea of an environment/selection mismatch plays an important role, then I look at the evidence.

Marks and Nesse certainly do argue that the nervous system "has been shaped so that anxiety arises in response to potential threats." In effect, we are biologically predisposed to find certain stimuli scary on very little basis, so that fear "develops quickly to minimal cues that reflect ancient dangers" (1994, 254). This is an important point because it aligns the hypothesis with a well-attested body of research in a variety of species documenting the speed with which certain associations are learned. I review this evidence after I finish the presentation of Marks and Nesse's view.

The basic idea is that we possess psychological mechanisms adapted to react to certain stimuli in ways that prepare us for the appropriate response (freezing in some cases, flight in others, for example). As Nesse (1999, 262) notes, it is an important source of support for the theory that the form taken by different phobias match their purported evolved function. He argues that almost every aspect of the flight/fright response matches what is needed physiologically to respond to danger, and that the increase in heart rate associated with anxiety is the result of the sympathetic nervous system increasing blood flow to the muscles. Interestingly, the only phobia associated with fainting is the fear of blood. Marks (1987) argues that fainting is an adaptive response to the sight of one's own blood because the drop in blood pressure that causes the fainting would minimize further blood loss. So the argument is that fears were once adaptive even if they are often not of use in the current environment. To explain the pathological fears labelled as phobias, the mismatch hypothesis is conjoined with another assumption.

The problem Marks and Nesse attribute to phobics is that their prepared mechanisms produce extreme amounts of anxiety. (As they note, converse reasoning suggests that some people suffer from too little anxiety. But such people do not show up in treatment, although they might show up when medals are being handed out for bravery.) But as Marks and Nesse set it out, their suggestion that anxiety-producing mechanisms are overactive in phobias is ambiguous between two readings. Sometimes they appear to be arguing that phobics suffer from broken mechanisms, with the result that they become unduly anxious. The other reading is that forms of anxiety, like most traits, could be expected to show considerable phenotypic variation. Individuals who are toward the sensitive end of these distributions—those who become anxious more readily when far from home or when in high places—might have functioned quite normally in ancestral environments but find themselves at a disadvantage in contemporary environments that have changed in ways that make the stimuli no longer a reliable guide to danger: most of our modern experiences of being considerably above the ground involve looking out of windows, not peering over clifftops.

So my second reading of the claim that phobics suffer from an excess of anxiety, which is that phobics are at the sensitive end of the normal distribution of anxiety response, clearly is correct at least sometimes. Marks says that blood-injury phobics "may be at the extreme end of a normal continuum of cardiovascular responsiveness to tissue-damage stimuli that is genetically influenced" (1987, 46). The second reading does seem preferable. Nearly everybody suffers an aversive reaction to the stimuli that cause specific phobias, even though few of us are phobic about them. It is not unreasonable to expect considerable interpersonal variation in the output of anxiety-producing mechanisms, and the mismatch reading of the Marks-Nesse hypothesis can exploit this expectation without having to make additional theoretical commitments.

Clark's influential theory of panic disorder exploits similar reasoning (1986, 1988, 1997). The theory holds that panic attacks are triggered and sustained by catastrophic misinterpretations of bodily signals, chiefly associated with anxiety but including some others. A feedback loop occurs in which "the crucial event is a misinterpretation of certain bodily sensations" (Clark and Ehlers 1993, 132). This causes heightened anxiety that in its turn has physiological effects that are detected by the sufferer, thus increasing the sense that something is badly wrong and inducing further panicking (see figure 8.1). For example, palpitations might be perceived as evidence of a heart attack, or the pulsing of blood in the temples as a sign of imminent brain haemorrhage (Clark 1988, 149). Up to a quarter of the population may experience an occasional panic attack, but persistent panic attacks are rare. Clark suggests that only some people have a tendency to perceive autonomic-nervous-system events as indications of impending disaster but does not offer a considered view about the nature of this distinctive tendency to panic. He does, though, cite some evidence that the tendency can be seen as abnormal sensitivity in an evolved monitor of autonomic activity: for example, panickers are more accurate than controls at keeping track of their own heartbeats without taking their pulses (Ehlers and Breuer 1992). If it is true that the susceptibility to panic is a matter of an overly sensitive cognitive system, it would be a suggestive extension of the basic principles I have found to underlie the most plausible approach to phobia. It would also deal with the objection to Clark's model raised by Gray and McNaughton (2000, 303). They suggest that if the tendency to panic is a matter of personality

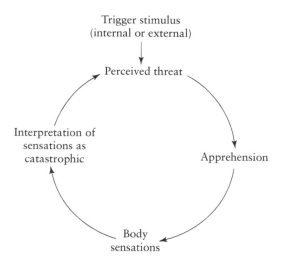

Figure 8.1.
Clark's model of a panic attack (Clark 1986, 463)

type, then people with generalized anxiety disorder, who are prone to panic, ought to develop full-blown panic disorder as a matter of course, but they do not. If the tendency to panic is not a matter of a general tendency to anxiety but a specific oversensitivity, then Gray and McNaughton's prediction is less plausible.

Clark's theory has considerable experimental support (see Clark 1997 for review). In one study, panic patients and normal controls were asked to read out pairs of words written on cards. In the crucial condition the pairs consisted of a word denoting a sensation followed by one denoting a horrible outcome (e.g., "palpitations" plus "dying"). The combinations were designed to mimic the thoughts attributed to panic sufferers by Clark's model, and based on ratings of anxiety symptoms it appears that during the read-aloud study, panic attacks struck 10 out of 12 panic patients but no controls or ex-panickers (who had been treated with cognitive therapy). Other experiments have shown that panic patients who inhale a mixture of oxygen and carbon dioxide are much less likely to panic if it is explained to them that the sensations they will undergo are caused by the gas, or if they are duped into thinking they can manipulate the flow of gas (by turning a dial that in fact has no effect). The alternative explanation for their sensations and the illusion of control each provide a cognitive asset that overrides the natural cognitive process of misinterpretation and reduces the chance of an attack.

The phobias and panic disorder might need to be understood as extreme manifestations of underlying systems that evolved to respond to dangerous situations, and not as expression of systemic malfunction. If, instead, we decided to look for an explanation of phobias and panic disorder in terms of a systemic malfunction, it would require developing an extra hypothesis to account for the nature of the breakdown and to explain why the consequent clinical picture is different from what we would expect from just phenotypic variation. That is, we would have to explain why a malfunction is needed to posit the sort of variation that we would expect. So the mismatch strategy is simpler to defend, and it can be integrated with other well-attested aspects of our psychology.

8.2.3 "Lurking Fears"

In a well-known series of experiments, Mineka et al. (1980, 1984, 1989) showed that young rhesus monkeys develop an enduring fear of snakes after seeing film of a conspecific reacting fearfully to a snake. But juvenile monkeys did not develop fear of flowers when the film was edited to show another rhesus reacting fearfully to flowers. This suggests that rhesus monkeys have an evolved preparedness to respond fearfully to some stimuli but not to others. It is entirely possible that humans have a similar disposition, with, I will assume, an evolved cognitive mechanism underwriting it. Marks reports how his two-and-a-half-year-old son, who had never seen a snake, was terrified by his first sight of "dried skeins of brown, black seaweed looking like myriad dead eels or tiny snakes" (1987, 40). We can assume that a system designed to respond to visual representations of snakes also will respond to snake-like entities.

Like Mineka's research, Marks's story of the sudden terror his son expressed on first confronting snake-like strands of seaweed is an example of the speed with which certain anxious states can be induced by stimuli in a wide variety of organisms. This effect has been noted in many lineages since Garcia first showed how easy it was to condition rats to associate nausea with foodstuffs, and pain with electric shocks (Garcia and Koelling 1966). However, not all stimuli are equally prone to produce such immediate reactions, and not all associations are as easy to learn: Garcia's rats did not learn to associate shock with nausea, nor foodstuffs with nonabdominal pain. His explanation was that the rats were quick to form associations that made evolutionary sense. Mineka's monkeys did not quickly acquire a fear of flowers, even though the videotapes were edited to differ only in the stimulus to which the adult monkeys reacted. If the juveniles saw adults show fear of flowers, they were unmoved, but they did acquire a fearful response quickly when snakes caused the adults' reactions. Again, the suggestion is that the difference lies in the evolutionary heritage of the response; for monkeys, it makes adaptive sense to fear snakes.

A consensus has built up on two points: first, many different lineages including our own display a readiness to respond to some stimuli rather than to others, and, second, the stimuli are ones that are or in the past have been adaptive to avoid. For example, we often avoid certain substances after they make us ill just once, as anyone who drank too much Southern Comfort as a teenager can testify. Experiences like the painful fits of vomiting that accompany too much alcohol indicate danger. Some researchers disagree with Marks and Nesse's claim that phobias typically are acquired in response to situations or objects that would have been dangerous in ancestral environments. Goodwin and Guze, for example, argue that phobias "can develop toward almost any object or situation" and mention a "tennis player who wore gloves because he was afraid of fuzz, and tennis balls are fuzzy" (1996, 161). However, everyone, including Goodwin and Guze, note that the most common phobias are those cued by animals, heights, crowds, and so on. These are the classes of stimuli enshrined in the first three (arguably the first four) of the DSM-IV-TR's subtypes for specific phobia. On the face of it, they make evolutionary sense in terms of the dangers in an ancestral environment. Fear of insects is much more common than fear of cars, even though these days being bitten by a bug is, on average, much safer than being struck by an automobile.

Furthermore, the view of phobias as expressions of evolutionary leftovers is not obliged to predict that no phobias would develop toward other stimuli. There are several ways to explain apparently deviant phobias. First, a claim that some associations are preprogrammed by evolution to be learned with terrific speed does not imply that no other associations can be learned. It is possible that, given enough conditioning, practically anything can become a source of phobia. The evolutionary hypothesis need only argue that the common, readily acquired fears and phobias can be explained in terms of prepared responses to features of the ancestral environment. Other phobias could perhaps be acquired by conditioning. In the heyday of behaviorism, J. B. Watson (1924) argued that classical conditioning could transform any stimuli into one that elicited fear, but C. W. Valentine (1930) rebutted him with studies suggesting that conditioning a fear response in human infants was much easier with some stimuli than with others. Valentine called the quickly conditioned fears "lurking fears." Pictures of stimuli such as snakes and spiders that are assumed to make adaptive sense elicit a much greater galvanic skin response when paired with electric shock than do pictures of unthreatening stimuli like flowers, but they also cause a greater response than pictures of stimuli like guns that are dangerous in the contemporary environment but did not exist in the ancestral environment (Ohman 1979). The distinction between conditioned fear responses and lurking fears is now widely accepted (Gray 1987).

So substantial support exists for the claim that there are two routes to fear response, a prepared evolutionary one and classical conditioning. And it may be that the DSM-IV-TR's category of "other" phobias comprises the residual class of responses to conditioned stimuli rather than the lurking fears produced by evolution. If so, this is perhaps the clearest case of the absence of theory in the DSM producing a lack of diagnostic clarity, since the distinction between specific phobias and other phobias is exceptionally hard to understand.

Another, more speculative, explanation for the occurrence of phobias directed at adaptively irrelevant stimuli is the possibility that an evolved system could be triggered by an input it was not designed for. Sperber (1994) draws a distinction between a module's actual domain and its proper domain (in chapter 5's sense of "module"). The proper domain of a module is all the information that the module has the biological function of processing. Its actual domain, on the other hand, is all the information in the current environment, that having been processed by other modules, satisfies the input conditions of the module. This means that features of the modern, rather than ancestral, environment may nonetheless trigger systems designed to produce anxious responses. In Sperber's example, an "org" has a module with the function of enabling flight, to avoid being trampled when faced with an elephant, but the module also is triggered by oncoming trains. The trains are not part of the proper domain of the module, but they are part of the module's actual domain. Something like this may have happened to "Mr. Michaels," who could not visit his terminally ill father in the hospital because of his fear of anything associated with bodily injury or illness (Spitzer et al. 1994, 245). In his case perhaps a "Specific Phobia, Blood-Injection-Injury Type" (the diagnosis) has come to be generalized to stimuli that were not part of the ancestral environment but share relevant properties with it—the possibility of exposure to physical trauma.

The idea that a phobia is (1) an expression of an oversensitive prepared evolutionary response toward stimuli that (2) it once was adaptive to fear but that no longer brings reasonable benefits in the current environment does rest on a solid comparative and experimental basis. A similar story, which borrows much of the same machinery, seems plausible for the case of panic disorder, but perhaps not for other anxiety disorders.[1]

The pattern of the most common phobias is consistent with the preparedness model. But the environment has changed sufficiently that our encounters with those substances now are largely harmless, and a hair-trigger response is no longer useful. Of course, not all such experiences create phobias, but the claim is that the normal mechanisms underlying lurking fears are sensitive enough in some people to account for the symptoms of many anxiety disorders. In the next section, I turn to a more speculative class of theories, which attempt to apply to depression the idea, that a mismatch exists between the environment we inhabit and the one for which we are adapted.

8.3 Adaptive Theories of Depression

8.3.1 The Social Competition Theory

The most plausible reading of the adaptive construal of anxiety disorders sees phobics as owners of ultrasensitive or overactive threat-detection and anxiety-production systems. In a modern society, according to the theory, these are a hindrance rather than a help, leading to a selection/environment mismatch. Recent work on depression offers a much more speculative mismatch explanation, but one that fails to capture the psychological phenomena adequately.

Mismatch theories of depression view it as an adaptative response to the problems of living in a small, status-obsessed social group. One version assumes that in the small bands in which we evolved, it was adaptive, after having lost status, to accept lowered rank (Price et al. 1994; Stevens and Price 1996). Another story (Nesse and Williams 1995) is that upon being outcompeted, one should change previous behaviors and look for a new way to shine and attract mates, or at least ruminate upon one's problems (Watson and Andrews 2002). The theories concur that depressed mood provides an introspectable marker indicating that self-evaluation in the light of social failure is in order.

---

1. Social phobia remains the outstanding exception. Although infants develop a fear of strangers, it typically evaporates, and there seems little evidence for the idea that social phobia could be an extreme form of a prepared social fear. Clark (1997) has developed a theory similar to his theory of panic to explain social phobia, based on the idea that social phobics use feedback from social situations to construct a negative impression of themselves.

To evaluate these theories of major depression,[2] I begin with the comparative neurological evidence that the theory often stresses, which relates amine levels to social rank. Then I discuss whether the theory extracted from those findings can account for the symptomatology of major depression.

8.3.2 Get Thee behind Me, Crustacean

Many animals change behavior depending on their place in a dominance hierarchy. The shift is accompanied by changes in serotonin levels. This pattern occurs in lobsters (Kravitz 2000) and crayfish (Drummond et al. 2002), as well as in animals closely related to humans. Vervet monkeys show a connection between serotonin, aggression, and rank. Lowered serotonin is seen in monkeys of low rank or who have lost status recently. Higher ranking Vervets show increased serotonin levels, correlated with the amount of submission they receive (McGuire and Troisi 1998, 172–74). Manipulation of serotonin can affect behavior, and, indirectly, rank. Injecting a subordinate crayfish with serotonin, for example, makes it more aggressive (Drummond et al. 2002).

Social-competition theories of depression exploit these links between serotonin, rank, and depression. Various measures suggest that serotonin, at least at presynaptic sites, is abnormally low in at least a large subpopulation of depressives, although attempts to correlate serotonin levels with established clinical subtypes have been unsuccessful (Thase and Howland 1995). But it does seem that among suicidal and gravely depressed patients there is a low level of serotonin release, binding, and downstream processing that is correlated with excessively negative self-assessment and extreme pessimism (Meyer et al. 2003).

The idea behind the social models of the function of depression is that depressed mood is nature's way of telling you to accept that your current behavior will not improve your reproductive lot and motivating you to try behaving differently.[3] In a small group, switching strategies might result in

2. Note the restriction to major depression. More local theories try to explain particular kinds of depression in evolutionary terms. The best worked out is the defection hypothesis of postpartum depression (Hagen 1999), which maintains that ancestral women limited their investment in the new child when, because of social, biological, or environmental factors, a major investment in a particular infant would likely reduce their total number of offspring.

3. Nesse (2000, 17) has more recently argued that depressive cognition might have the function of making one stay in a bad situation, rather than looking for a new one (one which might be worse). He presents no supporting evidence, and no reason to

greater social success. The suggestion is that in ancestral environments, communities were small enough to leave some niches unfilled in which an organism might prosper. Nesse and Williams (1995) think this is now a mismatch: depression is no longer adaptive, since the social setting has changed in modern society, making an affect-lowering response to change in status profoundly dysfunctional. Other theorists seem to think that in so far as depression focuses one's mind on social problems, it might still be adaptive (Watson and Andrews 2002). That is a persistence version.

The mismatch theory of depression is that we have inherited a mechanism that is triggered when we believe ourselves to be outcompeted. The mechanism will fire frequently in the modern world, as we are inundated with information about people who are better than we are. But, of course, in the modern world it is far more likely that the mechanism will fail to achieve the goal it was selected to attain. If the mechanism is set off by the realization that one is not even close to being the best at anything in the global village of the information age, then depression is not likely to be an effective reaction. For it is typically the case that there is no other strategy to adopt—no other niche one could fill—in which one would do much better in that global competition.

If this view is correct, shouldn't everyone be depressed? Nesse (1999) notes that depression is on the rise worldwide and is in fact the third-leading cause of disability-adjusted lost years, according to the World Health Organization. However, given the large groups in which we now live, it is unclear why depression is not even more prevalent, if humans compare themselves to everyone rather than just to those in their immediate social group. Whether depressives do compare themselves to global rather than local populations ought to be testable.

A more general objection, on either a persistence or a mismatch reading, is that the evolutionary theories of depression suggest that it should cause a reevaluation of one's life and a change in behavior. But although depressives do seem to review their own behavior more than nondepressives do (Pyszczynski and Greenberg 1987), and they have a larger repertoire of negative than positive ways of evaluating themselves (Woolfolk et al. 1995), it

think that determination and optimism would not be better aids to seeing things through. More generally, Nesse suggests in this paper that depression is an adaptation for coping with unpropitious circumstances. This is very vague, but (as I note later on) depressive coping styles do not seem to be effective. Furthermore, most people who are in adverse circumstances do not become clinically depressed (as opposed to just ordinarily miserable).

is unclear whether their self-evaluation does what the alleged social function of depression requires of it. The social function demands that brooding be functional, or at least might have been functional once, but this remains to be shown. For example, there are no studies suggesting that people who have suffered a depressive episode subsequently rearrange their lives. Furthermore, the non-affective symptoms of depression seem hard to explain on the social function view.

8.3.3 Depression and Motivation

If the function of depressed cognition is to force a reconsideration & options and to instigate trying something new, then it seems that the best combination would be negative (or at least objective) self-evaluation coupled with positive affect, or at least some effect on motivational structures that causes the changes in behavior that the theory predicts. This effect on behavior *is what depression is for*, according to the theory. But we do not find it.

If we examine the psychology of depressives, we do not find features that take them from introspection to self-reforming behaviors. Depressives do not take up new activities. One frequent symptom is an inability to think, make a decision, or concentrate (DSM IV-TR, 350), which also fits poorly with the idea of depression as a source of cognitive regeneration. A number of studies also have found that depressives engage in wishful thinking, avoidance, and escapism, and do not cope with their difficulties by trying to solve problems (Cronkite and Moos 1995, 579–580). Indeed, depressives do not use cognitive self-control as a coping strategy, whereas the use of escapism as a coping strategy is associated with current and future depression (Rohde et al. 1990). Watson and Andrews (2002) argue that the hopelessness that characterizes depressive cognition is really desperation, indicating that an important problem needs to be solved. They argue that if depression is an adaptation, it should "abate when a problem is perceived to be truly insoluble (2002, 5)" since there no longer would be any function for it. But this prediction is false. Depression does not abate under these circumstances, which is a main reason for its associated increased risk of suicide. Rumination on social shortcomings is indeed a hallmark of depression, but without the motivational effects or cognitive properties that the social function theory requires of it.

If, as on another version of the theory, depression is an adaptation to make lowered rank palatable, it is unclear why brooding over failures is necessary—the actual symptoms of depression do not appear to be designed to reconcile us to loss of status: depressives are not *content* with lower status. It is hard to know what a clear prediction might be, based on this view.

However, we might expect that either that people might suffer depression, or a distinct form of depression, following a loss in status, or that once depressed, they would be reconciled to their lot (since, on this variant of the theory, such reconciliation is the purpose of depression). Neither prediction appears to be confirmed.

The stressors that might cause reactive depression are not obviously status-related, since they include just about anything that might be stressful. And it is well known that people who have been depressed once are at a greater risk of future depression than the general population. Keller et al. (1992) found that 70 percent of patients who have had a major depressive episode will have at least one subsequent episode, and about 20 percent of patients with affective disorders develop a chronic condition. We must assume then, to save the theory, that people who have lost status once are more likely to do so again: if depression is an adaptation designed to make them function better in society, it is not working.

8.3.4 Other Problems

Many symptoms of depression are left unexplained by the social-function view: rapid alterations in weight, physical pain, sexual dysfunction, changes in sleeping patterns, psychomotor agitation or retardation, and the greatly increased risk of suicide (DSM-IV-TR, 349–352). Nesse (2000) for example, only discusses mood. There is some evidence (Cheung et al. 2004) that loss of social rank, degree of rumination, and amount of shame are correlated, and shame may have a social function and some causal connection to depression. But this is a finding about emotions, rather than about depression. The connection between depression and negative self-evaluative emotions like shame and guilt is apparent but poorly understood. And depression involves more than lowered mood or feelings of shame. The other symptoms are central to our clinical understanding of depression, and the evolutionary theorists have neglected them.[4]

The social function theories fail the test that, as Nesse stresses, the phobias seem to pass, which is the matching of putative evolved function with the actual form of the psychological phenomena. In effect, the theory takes one aspect of depression—its pharmacology, and perhaps some cognitive properties—and relies on comparative evidence to find an adaptive role for that one aspect without explaining the other features of depression. This is

---

4. Watson and Andrews (2002) are the exception: they argue that the depressive syndrome as a whole has a signaling function, designed to extract care from others (either honestly or as a form of extortion).

a recurrent problem with evolutionary and sociobiological approaches: explaining only a caricature of a trait, or just a few properties that may seem to have an evolutionary rationale. Some of the data that the social-competition theory relies on is actually quite rigorous, but the data are attached to only one of the phenomena the theory needs to explain.

In light of these problems, we should question the motivation for approaching depression in the light of evolution. We know that many mental disorders are quite highly heritable, and Nesse (1999) thinks that low mood is so common that it must somehow increase reproductive success, which suggests that the burden of proof is on the side of the antievolutionary theorist. But it is equally possible that human beings are just fragile and prone to breakdown, and that brain development in particular offers numerous opportunities for problems. We have reason to believe that many disorders are the result of psychological malfunction, so a theory that looks for an explanation in nonmalfunctioning terms is surely the one that bears the burden of proof.

With that in mind, I turn now to the adaptive theory of schizophrenia offered by Stevens and Price (1996), which is an example of the second, indirect evolutionary strategy. The indirect strategy explains the persistence of genes for a disorder by stating not that the disorder itself was ever adaptive, but that it is a severe form an adaptive trait, caused by a greater genetic load than needed to express the adaptive trait itself.

8.4   The Group-Splitting Hypothesis for Schizophrenia

8.4.1   Introduction

Stevens and Price argue that the incidence rates for schizophrenia are high enough to imply that the predisposition to schizophrenia must be adaptive: "natural selection has fixed it as an enduring component of the human genome" (1996, 142). They do not argue that schizophrenia itself, the full-blown psychosis, was ever adaptive, but they do argue that an adaptive trait must exist somewhere on the schizophreniform continuum, which includes schizotypal and paranoid personality types.

My main interest in the hypothesis is the idea that certain disorders may be inevitable by-products of adaptive traits, rather than being adaptive traits themselves. On this view, it is the rotten luck of schizophrenics to have inherited a greater genetic load than others, and so to express the underlying disposition to schizotypal behavior in a very strong form. Some researchers believe that mood disorders are affected by numerous genes of different penetrance, which can be expressed varyingly (Barondes 1998). It is not impossible that a combination of genes may produce mild symptoms in some and much more severe symptoms in others.

So one of the assumptions that Stevens and Price make, that there could be a genetic basis for the varying severity of schizoid symptoms, is not biologically implausible. Indeed, many researchers agree, but they claim that the relevant genes are defective, not adaptive. The acceptability of Stevens and Price's account depends, then, on its second assumption, which is that there is an adaptive basis to the milder schizophrenic symptoms.

8.4.2 Schizophrenia and Group Selection

Stevens and Price offer what they call a group-splitting account of schizophrenia. They argue that our ancestors lived in groups that would have grown to a point at which they outran available resources, whereupon the group would have to split up. The adaptive value of mild schizoid symptoms, they assert, comes at that point; they contend that mildly schizoid individuals would have been regarded as charismatic, and hence apt to draw followers away with them to start a new group; this would have been good for the group. They claim that "the schizoid genotype is an adaptation whose function is to facilitate group splitting" (1996, 152).

However, it is the charismatic individual they envisage who appears to be the beneficiary of the adaptation, rather than the group. They review anthropological evidence suggesting that charisma increases authority, but this is an individual trait, and the leader of the subgroup must be the only individual with the trait, otherwise we would have to imagine a group of schizoids cooperating, which is implausible if the function is group splitting. Since the leader must be the only person with the trait, we should think of the adaptation Stevens and Price envisage as one designed to secure greater prestige and power for individuals, which has nothing to do with group selection at all.

But the real faults in the theory are social and psychological. The only evidence they adduce is a claim that charismatic leaders who inspire loyalty often seem mentally unbalanced to outsiders, and are given to cognitive dissonance and visionary utterance. But many people exhibit cognitive dissonance, and schizoid personality includes more than shamanistic pronouncements. Stevens and Price make no effort to show that people diagnosed with schizoid conditions are regarded as charismatic—even though one could easily test this hypothesis by asking independent raters to evaluate a group of schizotypal individuals for their leadership qualities. My guess is that if the experiment were done, the patients would just be seen as disturbed, since that is how people typically react to schizoids. Even if we no longer need to split human groups, the genes that cause people to respond to charisma still should be present. If schizoids were charismatic once, then the theory surely should predict that they would continue to be seen as charismatic.

So there is no attempt to show that schizoid individuals are charismatic. Nor is any attempt made to argue that any real individuals have the symptomatology that Stevens and Price consider adaptive—the fact that some people are both odd and charismatic hardly suffices to make the point. They appeal to anthropological studies of leadership in hunter-gatherer societies, but they make no attempt to show that the leaders in these cases have any schizotypal tendencies, rather than mere stereotypic leadership qualities.

So Stevens and Price's hypothesis does not stand up. Its interest lies in its illustration of the second of the two strategies I distinguish for explaining psychopathology in terms of a selection/environment mismatch. Here it is not the trait itself that is considered to have been adaptive in the past, but related traits. Even if the evolutionary support for the view is reinterpreted in terms of individual selection, the adaptive theory of the schizophrenic spectrum has no psychological plausibility. As with the theory of depression reviewed in the previous section, no attempt is made to show that the full array of symptoms can be explained, nor any genuine fit between the symptoms and their alleged function. Both these theories seize on just one aspect of the disorder and construct an evolutionary edifice on that basis.

8.4.3 Mismatch Theories in Retrospect

It may seem that the point of using a broadly adaptationist take on cognitive science to explain psychopathology is an attempt to argue that mental disorder are adaptations. As section 8.1 shows, that is much too simple. An evolutionary psychology can make use of explanations in terms of psychological breakdowns just as readily as any other orientation on the problem. But the idea that some psychopathologies are or were adaptations is certainly a position in logical space, and so far in this chapter I have looked at one way to fill it. The most promising current idea is that anxiety disorders are expressions of evolutionarily prepared tendencies to find certain stimuli threatening. I also have discussed work on depression. Both these research programs share the view that the pathologies they explain were adaptive in the ancestral environment but are no longer adaptive given environmental changes. I noted that this idea, which I call the selection/environment mismatch, might apply to disorders that are by-products of once-adaptive traits; although the only proposal I identify as occupying that position barely stands up long enough to be knocked down, the position itself is a possibility.

As well as traits that are selection/environments mismatches, though, we might also ask about adaptations to features of the ancestral environment that still are adaptive today. That is, we can wonder about persistence as well as about mismatches.

8.5 Persistence Explanations

The mismatch hypothesis assumes that conditions have changed so that a once-adaptive trait is now pathological. However, we need not suppose that the environment has changed in every respect during human evolution. The hypothesis of adaptive persistence argues that psychological adaptations may have evolved to respond to properties of the environment that have not changed in a way that makes those mechanisms currently dysfunctional. The persistence hypothesis says that forms of what we currently take to be pathology may be adaptive in the current environment, just as they were in the ancestral environment.

Specific pathologies for which a persistence hypothesis has been proposed include Antisocial Personality Disorder and Histrionic Personality Disorder. (Harpending and Sobus 1987; McGuire et al. 1994; McGuire and Troisi 1998; Mealey 1995). I outline the persistence hypothesis in 8.5.2, then move on in 8.5.3 to a general assessment of this strategy, which has its roots in the techniques of game theory, as applied to the evolution of behavior.

8.5.1 Two Personality Disorders

Personality disorders are culturally extremely deviant patterns of experience and behavior. They are persistent and inflexible, they arise in adolescent or early adulthood, and they lead to distress, personal difficulties, or impairment for the person who exhibits them. However, although subjects with Antisocial Personality Disorder usually get into trouble with the authorities, it is not clear that antisocial acts are always bad for the individual who commits them, rather than just the people on the receiving end. McGuire and Troisi (1998) suggest that two personality disorders in particular may represent adaptive deviant behavioral strategies: Antisocial Personality Disorder and Histrionic Personality Disorder. Antisocial Personality Disorder is characterized by a disregard for the wishes, rights, or feelings of others. Subjects with this disorder are impulsive, aggressive, and neglectful of their responsibilities. "They are frequently deceitful and manipulative in order to gain personal profit or pleasure (e.g., to obtain money, sex or power)."

Typically, they show complete indifference to the harmful consequences of their actions and "believe that everyone is out to 'help number one'" (DSM-IV-TR, 702–703).

I have adopted the DSM terminology in speaking of Antisocial Personality Disorder, but as I noted in chapter 5, Cleckley (1976), Hare (1970, 1993), and Lykken (1995) regard psychopaths as a distinct psychiatric population, to be distinguished from antisocial personalities, who are typically career criminals.

On Mealey's (1995) sociobiological account, "primary sociopaths" are more or less equivalent to "psychopaths." The traditional distinction between primary and secondary sociopathy is that primary sociopaths (sometimes called "simple" sociopaths) lack normal social emotions; secondary (or "hostile") sociopaths exhibit antisocial behavior but do display the social emotions. Mealey sees the difference between primary and secondary sociopathy as follows: primary sociopaths are designed by evolution to be cheaters in social-exchange situations, devoid of empathy and seeing others in purely instrumental terms. Secondary sociopaths, on the other hand, respond to a difficult early environment by adopting a strategy that involves frequent cheating, but they are not emotionally deficient and might not have grown up to be sociopaths if they had not spent their early years in a pathogenic environment. Mealey argues that there should be a cross-culturally constant, if small, percentage of primary sociopaths whereas the proportion of secondary sociopaths should vary. I assume that the persistence hypothesis is designed to explain only that fraction of antisocial personalities who are genuine psychopaths. This fits Mealey's treatment, and I will assume that it applies to McGuire and Troisi's claims too.

The second disorder McGuire and his colleagues discuss is Histrionic Personality Disorder. Subjects diagnosed with this disorder are attention-seeking prima donnas. Often lively and dramatic, they do whatever is necessary to draw attention to themselves. Their behavior is often sexually provocative or seductive in a wide variety of inappropriate situations or relationships (DSM-IV-TR, 711–712). They demand immediate satisfaction and are intolerant of or frustrated by situations that delay gratification. They may resort to threats of suicide to get attention and coerce better caregiving. Both Antisocial and Histrionic Personality Disorders are characterized by manipulativeness, although antisocial subjects manipulate others in the pursuit of material gratification and histrionics manipulate to gain nurture. In evaluating the prospects of the persistence hypothesis, I pay closer attention to antisocial behavior, since that is better studied, more widely, discussed, and of greater philosophical interest.

8.5.2   The Explanation

Being able to manipulate others so that they further one's selfish ends is quite a useful trait to have, moral qualms aside. And, of course, psychopaths do not have moral qualms. But their ignorance or disregard of moral norms also makes it easier for them to commit acts that cause their arrest or attract the attention of clinicians. However, there could be many people who are equally unsavory and manipulative but who do not get into trouble or suffer adverse consequences. It is estimated, for instance, that fewer than 25 percent of those who commit nonviolent crimes are apprehended (McGuire et al. 1994). Such folk may cheat, deceive, and manipulate but be good enough at reading social cues and understanding the structure of reciprocal exchange that they can exploit the social system successfully. (Psychopaths appear to have no theory-of-mind deficit [Blair et al. 1996].) The persistence hypothesis claims that such individuals are adapted to exploit others via their sensitivity to, but disregard for, the system of social exchange, and that this is still adaptive in the current social environment.

The adaptive function of a psychological unit is the effect it has in virtue of which it is copied in successive generations (Godfrey-Smith 1994; Millikan 1989; Neander 1991). If it is indeed true that some personality disorders are adaptive strategies, we can give precisely this explanation of the mechanisms that generate the antisocial behavior of psychopaths or histrionics. Antisocial behavior is the adaptive function of these mechanisms, ensuring that enough psychopaths (or histrionics) make a good enough living to ensure that the mechanisms will be copied in subsequent generations. So people with these personality disorders are, on this view, functioning as they should; they do not have malfunctioning mental mechanisms. Nor is there reason to believe that the environment has changed in relevant ways since the time when the system was selected. The relevant environment in this case is social, and the current social environment, like the ancestral one, offers many opportunities to cheat and exploit one's fellows.

For this to be plausible it must turn out that, other things being equal, psychopathy is an effective way of passing along one's genes to the next generation. Indeed, a population with a minority of sociopaths may be in an evolutionarily stable state. The idea of an evolutionarily stable state comes from the application of game theory to evolutionary biology: it is a state of the population that cannot be invaded by a new strategy. That is, any mutant playing a strategy not currently played will get an average payoff per encounter that is lower than the prior average of the population.

It is possible for a strategy that seems very deviant to persist at a low level in a population that is in an evolutionarily stable state. Skyrms has shown how this is mathematically possible for apparently bizarre strategies such as "Mad Dog," which rejects a fair division of resources but accepts a grossly unfair one (1996, 29–31); that is, Mad Dogs punish those who play fair. It is not difficult to imagine the survival of more complex strategies that unfairly manipulate others.

These strategies will be useful provided two conditions apply. First, the subjects often must be able to disguise their cheating and deception, perhaps by exploiting and mimicking the signals that others use to convey cooperativeness (Frank 1988). Second, the antisocial behaviors must be maintained at a comparatively low level in the population. If too many people refuse to cooperate and deal fairly with others, then refusing to cooperate will accomplish nothing. We can expect an arms-race as cheaters evolve to be even better at exploiting others and the others evolve to become better at detecting cheaters and avoiding them.

8.5.3 Evaluating the Persistence Hypothesis

In the previous section I note that antisocial or histrionic strategies could be evolutionarily stable if certain conditions apply. However, this is so far a mathematical possibility only. Perhaps the lesson of Mad Dog is that the persistence of any behavior, no matter how bizarre, can be successfully simulated in a game-theoretic model. Some strategies, like Mad Dog, tend to be stable only with respect to a very finely tuned initial distribution of traits in the population: they thus depend heavily on a restricted range of starting points (Batterman et al. 1998). Other strategies are much more robust and will tend to evolve from a much wider range of initial conditions (Skyrms 2000).

So proponents of the persistence hypothesis should show that the strategy represented by their pet disorder is robust enough to evolve across a wide variety of game-theoretic situations, since confidence that it evolved may be shaken if the strategy requires very finely tuned initial conditions. And some effort must be made to show not just that the strategy is evolvable in a range of environments but that some of them are close to plausible human environments. And to do that requires a precise characterization of the strategy that the individuals are supposed to be playing. This has not been done yet— all we have been given is the speculation that they are psychopaths who do not get caught. Some researchers (Cleckley 1976; Hare 1999) have indeed argued that successful psychopaths are quite common, and the superficial charm and plausibility of psychopaths is widely admitted, as I noted. But these asides do not make for a clear description of the behaviors in question.

It may be that this points to a general difficulty with the whole game-theoretic approach, namely translating simple strategies into a plausible behavioral profile that might occur in the real world. However, if clear predictions about what a successful psychopath would do in a given situation could be made, it does not seem impossible to concoct an experimental study in which that behavior could be exhibited by a stool-pigeon placed among experimental subjects.

However, any behavioral study, based on adaptive thinking or not, would not show the existence of anything beyond what Mealey calls secondary sociopathy. This is supposed to be an expression not of adaptive antisociality but of a flexible ability to behave self-servingly in tough environments. Secondary sociopathy is an expression of amoral instrumental rationality, which recognizes that social norms exist but chooses to disregard them. It does not express a distinctive cognitive adaptation. If the persistence hypothesis is to work, it needs to connect with a theory of the specific cognitive mechanisms.

Studies have discovered significant cognitive differences between psychopaths and other people, notably a lack of response to frightening situations and a tendency to impulsiveness, as I review in chapter 5. But chapter 5 also shows that psychopathy may involve information-processing deficits. Psychopaths lack not just the moral emotions but also various planning abilities. If claims about cognitive deficits in psychopathy are correct, the persistence hypothesis is in trouble, because the cognitive mechanisms underlying its putatively adaptive strategy are ones that in fact appear to be far from functional. Again, the problem is that the form of the condition does not seem to mesh with its alleged adaptive role: the adaptive value of constantly failing to think ahead is hard to see. Yet psychopaths constantly fail to think ahead. The persistence strategy requires that there be a population with all the manipulativeness and amorality of psychopaths but without their cognitive deficits, and perhaps with a better-than-normal capacity to disguise their manipulative tendencies; the existence of such a population remains speculative. Grossman (1995, 180–185) argues that natural soldiers, who can kill without compunction and are not affected by the stress of combat (unlike 98 percent of combatants), are successful psychopaths. He claims that they share some of the same views of the world as psychopaths and tend to be very aggressive, but are able to fit into social structures with much greater ease and fight only when aroused in defense of ideals or people, rather than for purely selfish ends. (Being selfish and unmoved by causes, a psychopath would desert the night before a battle.) They are also levelheaded and capable of sticking to a plan. What is needed

304 | Chapter 8

if Mealey, McGuire, and Troisi are right is a population with the desire to prey on the rest of us that psychopaths normally exhibit, together with the coolness and rational powers of Grossman's natural soldiers. There might be, then, a population with the affective but not the cognitive profile of common psychopaths.

Another possibility is that the conjectured population does exist, obeying norms in a purely instrumental fashion out of cost-benefit reasoning, and breaking the rules when it suits them or when their guard has been lowered by alcohol. So it may be that successful psychopaths are able to inhibit some of the impulsiveness characteristic of their kind by exploiting other neural systems. They could have benefited, for example, from a middle-class upbringing that has stressed the gains to be had from delayed gratification. There is, above all, impressive agreement among psychopathy researchers that successful psychopaths exist, even though it is hard to reach such people for experimental purposes. So, it would be satisfying to have an explanation of how a psychopath could thrive. At the moment, though, the evolutionary view is no better off than the view that psychopathy is a disorder that some well-socialized individuals can learn to mask by exploiting other psychological capacities. It is hard, too, to think of a distinct prediction that the evolutionary game-theoretic view of psychopathy makes: Lalumiere, Harris, and Rice (2001) argue that the persistence view of psychopathy predicts their finding that psychopaths have normal births and not perinatal trauma, but although difficult births are implicated in some mental disorders, it does not seem in general true that if one believes a condition is pathological one should regard normal birth as counterevidence.

The last option is an indirect persistence strategy. I know of no attempt to explain any disorder as the by-product of a persistent adaptive trait, with the possible exception of some speculations that manic depression is somehow related to creativity. In any case, I think we have enough evidence now to evaluate these proposals.

8.6 Conclusion

In general, the evolutionary explanations examined here fail to stand up. The problem typically is that they fit only some symptoms of a disorder—psychopaths seem too irrational to be successful long-term manipulators, depressed people seem too unmotivated to be plotting a new strategy or too miserable to be reconciled with their lot, and schizotypal patients seem short on leadership skills. The failure is one of form not matching alleged function. In the case of some types of anxiety, though, the match is there, and these may be plausible candidates for a direct mismatch explanation, especially if fleshed out with assumptions about the distribution of anxiety in a population.

I have not given a general argument against either mismatch or persistence theories, but we can see what they need to succeed. First, function must match form. Arguments about evolutionary scenarios in psychology often focus on biological plausibility, but the psychological predictions the scenario makes are just as important, and usually easier to test, so why not start by evaluating them? Second, there should be some underlying neuropsychology that distinguishes possessors of a putative adaptive trait from the general population, or some underlying neuropsychology that shows how an adaptive response might be enabled. Again, the phobias do have a plausible neuropsychological basis, reflecting what we know about prepared forming of associations.

In general, the significance of evolution for psychopathology is likely to be at one remove. We are animals with an evolutionary past, and understanding our normal psychology likely will involve seeing how that past has shaped it. Hence, understanding breakdowns in that psychology ultimately will be, at least in part, based on our understanding of how evolved minds are organized.

The last three chapters have explored particular forms of those breakdowns. The remaining chapters show how the explanatory resources they call upon can be put together in a system of classification.

These diseases of the mind, forasmuch as they have their chief seat and organs in the head, are commonly repeated amongst the diseases of the head, which are divers, and vary much according to their site.

—Robert Burton, *The Anatomy of Melancholy* (1620)

9.1 Introduction

Previous chapters criticized the conceptual basis of the DSM's approach to mental illness and outlined the empirical and conceptual assumptions I prefer and the explanatory strategies that follow from them. This chapter turns from explanation to classification. In section 9.2 I elaborate on the DSM approach to classification more fully, with special attention to its neglect of causal information. This fails to cohere with the basic assumptions of the medical model, introduces heterogeneity into the taxonomy, and cuts psychiatry off from the other sciences of the mind/brain and from biomedicine more generally. Before I elaborate on these criticisms, I defend in section 9.3 the idea of causal taxonomy as a goal for psychiatry against two objections. The first is that the existing system works well enough and will improve according to its own lights, so that concerns about its limitations are beside the point. The second objection I rebut is that causal taxonomy is impractical.

I make three more explicit criticisms in section 9.4. First, the DSM-IV-TR is *incoherent*, in a specific sense. The DSM-IV-TR has a view of mental illness that relies on discriminating between disorders causally and describing their symptoms theoretically, and it tries to combine this with an official position that regards causal information and theoretically informed observation as impermissible. Second, the DSM-IV-TR is *heterogeneous*. The tensions that I point to by calling it incoherent have created a scheme that is liable to lump together dissimilar conditions and separate similar ones. Third, these problems are exacerbated by the *provincial* nature of the DSM-IV-TR, which focuses on conditions that belong to psychiatry as a result of the contingent processes I complain about in chapters 2 and 3. The DSM-IV-TR ignores conditions that do not have a home in this tradition, even when they are closely related. This isolation of psychiatry from the other sciences of the mind bolsters the existing skepticism about etiology in classification because it cuts psychiatry off from sources of causal information.

Many critics who are uneasy with the DSM argue that classification, in psychiatry as elsewhere, should be concerned with natural kinds. Section 9.5 addresses the issue of what natural kinds are and what their nosological significance might be. There are several theories of natural kinds on the market. Whether or not mental disorders are natural kinds depends on the adoption of a particular theory. If natural kinds have essences that stand in lawlike relations to their manifestations, then mental illnesses are not natural kinds. But if a natural kind is just any projectible category, then mental illnesses may be natural kinds. But the projectibility constraint is too easy to satisfy. A preference for one brand of natural kinds over another, however, is in many ways a matter of intellectual taste. It does not matter what conception of natural kinds we choose as long as a realist theory of classification can recognize mental illnesses. Classification, in psychiatry as elsewhere in science, should reflect the causal structure of the relevant domain. In chapter 10 I apply this lesson to the taxonomy of psychiatric illness.

9.2 The DSM Approach to Classification

9.2.1 Some Background

It is a familiar point that the transformation that led to the DSM-III was a reaction to the dominance of psychoanalysis in American psychiatry. The desire for better classification was a key part of the reaction. Dynamic psychiatry's somewhat cavalier attitude to nosology was unsatisfying. The rebirth of the medical model, and the attendant stress on classification and validation via the Feighner criteria and related schemes, led to a new, categorical system of classification in place of the first two editions of the DSM, which had been based very heavily on psychoanalytic theory (as Americanized by Adolph Meyer). Psychoanalysis regarded the symptoms of patients as different manifestations of a largely undifferentiated psychic crisis, and it saw the difference between pathological and normal individuals as a matter of degree rather than kind, one that could be collapsed by stress (Fenichel 1945). In light of this, psychoanalysts considered classification to be conventional, when they considered it at all. The slighting of taxonomy in the psychoanalytic tradition affected psychiatry in several ways: taxonomic research was moribund, over half of patients went undiagnosed in one pilot scheme using DSM categories (Cooper 2005), and the great historian of psychiatry Henri Ellenberger airily suggested that systems of classification said more about the unconscious of the classifier than the nature of mental illness (Ellenberger 1963).

The situation is much different now. Developed by working groups totaling over 1,000 specialists, the DSM-IV appeared in 1994. Smaller working groups subsequently convened to assess it in the light of later findings and literature reviews, and the DSM-IV-TR was the upshot of that renewed assessment. The result is a system distinguishing more than 300 disorders: it includes types of depression, personality disorders, addictions, sleep disorders, psychoses, sexual dysfunctions, and much else. It is very influential in other countries; for example, a standard British textbook of psychiatry (Gelder, Gath, and Mayou 1989) uses DSM categories on the grounds that the DSM is the established nosology.[1]

1. There is a competing franchise. The main alternative nosology is enshrined in the psychiatric section of the International Classification of Diseases (ICD), now in its tenth edition. As well as the ICD-10 (World Health Organization 1992b) itself, the ICD family includes other manuals that are standard modifications of the basic manual itself. In America, for example, the "clinical modification" of the ICD-9 (World Health Organization 1978) has for some years been the basis of patient records, insurance, and other administrative tasks (Bowker and Star 1999 includes much interesting history amid some obscure theorizing). The ICD system is explicitly practical. The ICD is intended as a source for the "systematic recording, analysis, interpretation and comparison of mortality and morbidity data collected in different countries or areas and at different times" (ICD-10, vol. 2, 2). Each chapter of the ICD-10 can be seen as a compendium of medical lore about a particular type of ailment, with the details of the lore contained in other volumes in the ICD "family." Psychiatry is served by the DCR-10 (World Health Organization 1992a) and the CDDG-10 (World Health Organization 1993), the wisdom in both being extracted and placed in ICD-10's chapter 5. Many of the complaints I raise about the DSM approach are also appropriately leveled at ICD-10's classification of mental disorders (though not at its other chapters). It is avowedly atheoretical in the same way: the CDDG-10 announces that its "descriptions and guidelines carry no theoretical implications" but are "simply a set of symptoms and comments that have been agreed, by a large number of advisors and consultants in many different countries, to be a reasonable basis for defining the limits of categories in the classification of mental disorders (2)."

9.2.2 How the DSM Works

The 300 mental disorders recognized by the DSM-IV are all supposed to meet the four criteria that I mentioned in chapter 6. The criteria together define a valid syndrome, which is not the same thing as a disease/disorder but a collection of symptoms that cluster together. The problem is to explain why the symptoms cluster together. And despite the criteria, very few of the DSM diagnoses are validated in the sense of construct validity, which tells us that some measurable property of an individual is highly correlated with diagnoses that assume its presence and not with those that assume its absence. All DSM diagnoses are supposed to be valid, but not many have actually been validated. In practice "mental disorder" refers to any condition recognized by the DSM-IV (and some that are not), whether or not the underlying cause is known, and whether or not the syndrome has been validated.

As I say in several places in this book, the DSM officially analyzes mental disorders as dysfunctions that we do not like, but it does not classify mental disorders based on a picture of normal function that specifies types of breakdown. The third and subsequent editions of the DSM have been conceived of as theory-neutral with respect to the causes of disorders. The current classification (DSM-IV-TR) aims to be operationalized, atheoretical, and purely descriptive. Diagnostic criteria are typically specified by a list of sufficient conditions (often disjunctive and with an occasional necessary condition thrown in) that describes the symptoms in the language of "clinical phenomenology." Clinical phenomenology draws on folk-psychological concepts, clinical concepts, and some that are a bit of both, such as self-esteem, delusion, anxiety, and depressed mood.

As the psychoanalytic influence in psychiatry decreased, it was expected that new bodies of theory would arise to put taxonomy on a sounder footing (Hempel 1963). That never happened, and the absence of a theory on which to base taxonomy became celebrated as a virtue rather than deplored as a handicap. DSM revisions have not embraced a clear theoretical basis in

Since the ICD-10 covers all of medicine, however, it is more explicit about the phony distinction between psychiatry and clinical neuroscience that is a recurring complaint in this book. Some disorders, such as Huntington's Chorea, that are included in the DSM-IV-TR are excluded from the ICD-10's chapter on mental and behavioral disorders and included in the subsequent chapter, on diseases of the central nervous system. The difference between the lack of a theoretical basis for classification in the mental disorders section of the ICD-10 and the causal mini-theories underlying the diseases in other chapters is very marked.

biological psychiatry, even though the individuals who have enacted the revision process usually have come from that camp. And, of course, there has been no systematic attempt to learn from the cognitive neurosciences. The proponents of DSM reform were wedded to biological psychiatry, but the diagnoses themselves are not tied to any theories about underlying causes. They are explicitly advertised as the product of "a descriptive approach that attempted to be neutral with respect to etiology" (DSM-IV-TR, xxvi). As well as their lingering attachment to broadly empiricist philosophical views about meaning and observation, the psychiatrists who masterminded the DSM-III had more practical grounds for embracing a theoretical neutrality perspective on etiology. The DSM-III architects hoped for a taxonomy that would enable practitioners from different backgrounds to talk to each other. They assumed this meant that the most they could hope for was agreement on symptoms and a discreet silence about causation. It was true too that the causes of most mental illnesses were unknown. In many cases, this is still true.

There are, however, obvious drawbacks to a causally neutral taxonomy of malfunctions that does not take into account information about normal functioning. Imagine for a moment trying to construct a classification system for malfunctions in some complex and well-engineered artifact built from numerous functional components—a television set, perhaps, or a computer network. The taxonomic task is to classify problems based entirely on clusters of user-salient symptoms, with no inkling of how the mechanism was designed to operate and without any theory of its component parts and their functions. This would almost certainly produce a set of categories that are *massively heterogeneous* from the point of view of someone who understands the system. A taxonomy based on observed symptoms, such as a darkened screen or a failure to print, constructed without any independent theory of normal function, would, on grounds of common manifestation, classify together problems caused by totally different underlying processes. A taxonomy based on observed symptoms also would fail to classify together problems with the same underlying cause if the problems manifest themselves in different ways under different conditions. This, near enough, is just what we should expect in a DSM-style classification of mental disorders. For surely the mind/brain, too, is a complex system in which many components interact. Thus, as Poland, von Eckhardt and Spaulding (1994, 254) conclude:

There is simply no reason to suppose that the features of clinical phenomenology that catch our attention and are the source of great human distress are also features upon which a science of psychopathology should directly focus when searching for regularities and natural kinds. Human interests and saliencies tend to carve out an unnatural domain from the point of view of nomological structure. Hence the relations between the scientific understanding of psychopathology and clinical responsiveness to it may be less direct than is commonly supposed. In insisting that classification be exclusively focused on clinical phenomenology, DSM not only undermines productive research but also undermines the development of effective relations between clinical practice and scientific understanding.

There is no reason to suppose that clinical phenomenology alone can discover the real distinctions among psychopathological states. Imagine how the state of health care would be if somatic medicine limited itself to symptoms in this way: we would classify together everyone who coughs as sufferers from "cough disorder" and thereby miss the fact that someone who coughs may be doing so for a number of very different reasons. She may just have something lodged in her throat, or be irritated by the smoky atmosphere in Rudy's Bar and Grill. Or she may be suffering from a variety of pathological conditions, including TB, bronchitis, or pulmonary edema. It is, to say the least, undesirable to fail to distinguish all these different causes of coughing, yet that is the result of relying exclusively on clinical phenomenology. Notice that "cough disorder" could be a very reliable diagnosis in the technical sense of conducive to agreement among observers. We could expect widespread agreement that somebody was coughing, but that would not indicate that any genuine condition had been found, nor would a diagnosis of cough disorder reduce uncertainty in anything but a psychological sense. This is a general point about classification in all spheres. Without some theory about what is being classified, there are no grounds for preferring one set of distinctions over another.

Let me soften these gibes a little. Of course, some theory must have guided the architects of DSM revisions, even if they are coy about it. And some syndromes may well correspond to homogenous categories in nature. Furthermore, although symptoms typically are thought of as defined by their clinical presentation, the relationships that exist among these symptoms can be quantified precisely using statistical tools. And yet, the great Paul Meehl was quite right to insist that only "dogmatism on the one hand, or confusion on the other, is produced by pretending to give operational definitions in which the disease entity is literally identified with the list of signs and symptoms. Such an operational definition is a fake" (1986, 222). I say more about all this in section 9.4 and in chapter 10. For now, I raise some of the practical problems it presents.

9.2.3 On Being Insane in Empiricist Places

In 1973 Robert Rosenhan published a notorious paper called "On Being Sane in Insane Places." It recounted how he and seven confederates—carefully briefed, deliberately unkempt—approached the staff at several mental hospitals. Each complained that they could hear a voice saying "empty," "hollow," and "thud" over and over again (Rosenhan 1973, 251). Rosenhan's team used pseudonyms, but otherwise answered all questions truthfully. The members of the team received diagnoses of schizophrenia, were admitted, and promptly stopped their complaints and acted normally. Nonetheless, they remained incarcerated for between one and seven weeks. Then they were discharged, still regarded as schizophrenics but in remission. One hospital, skeptical about the experiment, challenged Rosenhan to fool them too, and subsequently announced that they had detected 41 pseudopatients. In fact Rosenhan had sent none at all, and the hospital had declined to diagnose 41 (out of 193) people whom it usually would have admitted.

Rosenhan's paper had three themes. One was the neglect and lack of dignity that the patients endured in the asylums. Another was the claim that having once been labeled as a schizophrenic, one is subsequently treated as one, and one's behavior interpreted in that light. But his most controversial claim, which provoked a furious reaction from psychiatrists, was the claim that diagnoses do not reflect facts about the individuals who receive them, but "are in the minds of the observers" (Rosenhan 1973, 251). What mattered for diagnosis, Rosenhan claimed, was the attitude of the psychiatrist and the context of the interaction with the patient, not the patient's intrinsic properties. The strongest reaction came from some of the architects of the revived medical model and its attendant hopes for validation of diagnoses (Spitzer 1975). They wanted a system that went beyond impressionistic data and grouped symptoms into natural classes. The DSM-III, which was published at the time of the experiment, was the result.

Strictly speaking, Rosenhan cast no doubt on the medical model on conceptual grounds, even if he demonstrated the fallibility of diagnosis. Whether somebody harbors an abnormal pathological process is a different matter from whether we can discern it. But it misses the point somewhat to say in reply, as many do, that one might induce a doctor to give a prescription for, say, chronic back pain, just by claiming to have the problem. The point is that persistent reporting of back pain would trigger an investigation into what was causing it, one that could connect with our existing knowledge of the pathogenesis of chronic pain. We cannot yet use diagnosis as an input to causal investigation in psychiatry, and the current taxonomy is committed to a theoretical outlook that ensures that, within the letter of the law, we will be unable to do so in the future. Yet if a diagnosis is misplaced, it cannot be corrected without knowledge of normal and abnormal psychological activity.

Now, it is true that we know very little about the causes of mental disorders. But it is also true that, as Robert Kendell says, without defining diagnoses "at some more fundamental level than their clinical syndromes, we cannot maintain that one concept is right and another wrong" (Kendell 1975, 85). Kendell was writing at a time when substantial disagreement existed between psychiatrists over the correct taxonomy. The disagreement has largely abated since the reformation of DSM in the 1970s, at least in the sense that now everyone is using the same manual. But Kendell's question remains: what basis do we have for favoring DSM diagnoses over some other scheme, since it does not define conditions at a fundamental level?

One common response to this question stresses the improvements in validity and reliability that accrue with each edition of the DSM (Nathan and Langenbucher 1999). Yet most diagnoses remain unvalidated, and reliability (the independent agreement on a diagnosis across different tests), although it enables different people to agree, is not a satisfactory answer to Kendell's question. To see why, we need to turn to another issue, the extent to which a taxonomy accurately represents the underlying reality. I claim that to do the job required of it, a psychiatric taxonomy must reflect a theory about how the human mind works, but I am not saying that operationalism is never defensible. As Hempel noted, it seems perfectly reasonable to employ operational criteria when we are first building a theory and live in comparative ignorance of causes. But Hempel assumed that we would move to a causal taxonomy at a later stage; the mistake that the architects of the DSM have made is to have become locked into a system of classification by syndromes that is isolated from theories and discoveries that could produce a causal taxonomy. Classification on the basis of causal considerations should be our goal.

Facts do not thrust themselves upon us unadorned, and some expectations about what facts are important are bound to guide any scientific enterprise. So the complaint is not that clinicians are using some theory where they should be using none. Some theory is inevitable. This battle has been refought several times. In biological classification, pheneticists argued for taxonomies based on pure or direct observation with all features weighted equally. David Hull long ago pointed out in response that any assertion that two organisms exhibit "the same" trait must be based on some criteria about what counts as an interesting trait (1970). The pheneticist claimed that pure observation could serve as the basis for classifying species.[2] But pure observation is a myth, and even if we can observe without preconceptions, some theory is required to sort out the relevant observations from the irrelevant ones. Any two objects are similar or dissimilar in indefinitely many ways. So classification must proceed on the basis of some criteria that the classifier sees as salient.

A psychiatrist's decision that something about a patient is salient is likewise made on a theoretical basis. Many concepts used in DSM categories are ones we are not born knowing, and the average layman could not spot them. They are part of the mental equipment of clinicians, who are trained to use them. In that sense, they are the product of whatever theory one is brought up in. As I noted above, the interpretation of behavior as the expression of, for example, anxiety involves a theory-based understanding. As Meehl puts it, "What we *observe* in the psychoanalytic hour is speech (words, rate, pitch, timbre, volume, pauses), posture, gesture, facial expression. Everything else is theoretical inference" (1993, 710). Clinicians use interviews to assess patients, but the information they acquire in this way is sorted out using a mix of clinical lore and explicit theory.

Without a theory to justify distinctions among them, the members of a class can be as similar or as different as you like. It does not follow that the theory used in constructing a taxonomy must be causal, since we can distinguish things, relative to our interests, in many ways. Nor does it follow that, for some given aspect of reality, there is only one way to classify. Dupré (1993) is quite happy with the biological taxa that are recognized in ordinary language and in sundry nonscientific projects, even if they violate scientific boundaries, on the grounds that "the vocabularies of the timber merchant, the furrier or the herbalist may involve subtle distinctions among kinds of organisms; there is no obligation that these coincide with those of the taxonomist" (1993, 35).

An herbalist, a botanist, and a chef will classify plants very differently. However, I do think it likely that the following claim, though it will annoy some people, is true: *relative to a particular set of interests*, there is usually

2. For the history of fights over classification, see Hull 1988. The pheneticists lost out to the cladists, who argued that classification should reflect the history of life. Some cladists eventually argued that the methods used to reconstruct lineages could be divorced from their evolutionary rationale, thus opening themselves up to the same charges of arbitrary classification that had been leveled at the pheneticists (Ridley 1986).

only one way to classify some aspect of reality. An herbalist and a chef may classify plants differently, but there are facts about what plants do in cooking when combined with other plants, and there are facts about which plants dull pain and which ones induce death. Good cooks and good herbalists respect these facts. Relative to scientific interests, there is a right way to do things. Whether the scientific taxonomy has some ultimate metaphysical authority is not an issue I take a stand on.

So there are facts that determine what our classifications should look like once our interests are understood. In medicine, and therefore in psychiatry, the relevant facts have to do with the organization and functioning of the human animal in various environments, and the ways in which that functioning can depart from normal. A nosology should be causal, even if not every system of classification should be.

Although a nosology must be causally based if it is to distinguish valid conditions from each other, its fidelity to the causal structure of the domain of classification, is not, as I noted earlier, the only desideratum that a taxonomy must meet. It must be comprehensive, provide clinically significant information, and be usable. The problem is to find the correct balance between these desiderata. In the limit, a classification could consist of an indefinitely large number of taxa, each comprising one exhaustive description per patient. That would be very accurate, but quite useless. I argue in chapter 10 that a fruitful way to approach this problem is to sever the description of the reality of mental illness from its clinical implementation. We need to classify by exemplars, but we also may need to let clinicians give diagnoses that draw from several different exemplars at once, in order to fit the desirable simplicity of the scheme to the chaotic individuality of patients.

I say more about the details later on, after I discuss two objections to dropping the existing system in favor of a causal nosology. The first objection is that the current system, for all its faults, works well enough in practice, and the second objection is that a causally warranted taxonomy is an impractical goal. If they are sound, there is little point in bothering with an alternative approach.

9.3   Two Objections to the Idea of a Causal Taxonomy

My argument against the taxonomic basis of DSM continues with my further detailing its shortcomings. The arguments are intended to motivate a rival approach to nosology that reflects the causal structure of departures from normal psychological function. But before presenting the arguments, I defend the idea of causal taxonomy against two objections that are designed to preempt it. Section 9.3.1 defends a causal approach to taxonomy against the objection that my philosophical qualms are irrelevant since, despite its drawbacks, the present nosology works in practice. I suggest that in fact the current system licenses a number of questionable therapeutic strategies. Section 9.3.2 rebuts the objection that we do not have enough information about the etiology of mental illness to make nosology sensitive to causes, so the proposal that taxonomy should be based on causal structure is impractical. I argue that although there is no prospect of a complete causal inventory at any time we can foresee, nonetheless there are ways in which the promise of increased causal sophistication can reorient the foundations of psychiatric classification.

9.3.1 Is the DSM Acceptable (and Philosophical Conscience Irrelevant in Practice)?

Does the DSM work well enough in practice to make philosophical qualms pointless? After all, it is widely accepted that the chief advantage of classifying patients is that it allows for communication (Kendell 1975) and the current system, whatever philosophers say about its foundations, often does enable scientists and clinicians to agree on what they are dealing with when they both look at the same patient (that is, the diagnoses are often reliable, although few are very reliable). So the first objection to plans for taxonomic reform in the direction of greater realism is that the DSM-IV-TR is reliable, and has increasing validity, and that is a valuable achievement that we should not eliminate.

Second, some claim that practicing clinicians are aware of the drawbacks of DSM diagnoses and know how to finesse or ignore them in particular cases. They may, for example, simply treat some diagnoses as convenient labels for insurance purposes, but deal with patients using their own understandings of the problems. They can introduce causal hypotheses as a source of treatment ideas if they want, but they are not forced to do so. So, it might be suggested, we do not have to worry excessively about the philosophical basis of nosology as long as the system works out in practice, and we can trust in the pragmatism of clinicians.

I am unconvinced. It is hardly a defense of a system of classification to suggest, as the second objection does, that the system is acceptable because we know how to disregard it when we need to. We want a taxonomy that we can use, not one that we have to try to evade. And the freedom of clinicians to utilize their own theories in treatment is less attractive when one considers that it may mean the mother of an autistic child being told that her son's problems are her fault because the clinician she has consulted still takes his theory of autism from Bettelheim's claim that it is caused by "refrigerator mothers." As Sydney Morgenbesser is alleged to have said, pragmatism is great in theory, but it does not work in practice. This is a real problem in clinical psychology, which delivers most of the mental health care in America. It is routine for serious experimentalists and theorists to lament the gap between science and clinical practice (Dawes 1994; Hagen 1997; Loftus and Ketcham 1996; Lynn et al. 2002). The gap has permitted all sorts of theories and therapies that are empirically unsupported, or even known to be false, to proliferate in American psychotherapy. Ignoring the real causes of disorder in favor of a system that lets people make things up cannot help this situation. These objections tell most strongly against various forms of psychotherapy, but a number of drugs are prescribed without sufficient warrant, or with side effects that should be avoided. We need a taxonomy based on the best knowledge available.

And although the first objection is correct that many diagnoses are reliable, it is still true that diagnoses can be very reliable indeed but still of very limited value, especially if everyone is in agreement but they are all wrong. High reliability is desirable not as an end in itself but instrumentally, as a means to securing high validity, or as Kendell put it "the accuracy of the prognostic and therapeutic inferences derived from a diagnosis can never be higher than the accuracy with which, in a given situation, that diagnosis can itself be made" (1975, 27). Validity is the crucial issue.

In chapter 6 I explain that construct validity, which tests the occurrence of symptoms together in the presence of a diagnosis, is the most widely-used concept of validity, and I pointed out that there are situations in which the neglect of causes in construct validity can be perfectly reasonable, as in our hypothetical scheme to boost self-esteem. But the importance that reflective psychiatrists attach to validity often takes the form of hoping that it will uncover the nature of disease processes. What is wanted is a way of accurately representing the natural categories into which mental disorder falls. The reason we should be worried about classification is that this goal looks unattainable with the present system.

And notice too that the advantages of the present system may be retained. After all, I am not proposing that we burn every copy of the DSM-IV-TR. The DSM revision process has accumulated much information about the nature and interrelationships of the symptoms of mental illness. All this information will remain useful, and we can retain any true generalizations it supports. The point is, though, that we want to validate syndromes in terms of their causal basis, so generalizations about the relationship and course of symptoms, if valid, will turn out to be valid in virtue of the fact that they are effects of underlying destructive processes.

9.3.2 Is a Causal Approach to Nosology Impractical?

It is true that we lack confirmed, or even confident, hypotheses about the causes of many mental disorders. But in fact this is not a reason for keeping our best current theoretical accounts of how the mind works out of the DSM, since an understanding of normal function can serve as a basis for classification of illnesses as departures from normal function even where specific etiologies are unknown. Indeed, the lack of causal frameworks is partly because we lack any comprehensive theoretical edifice within which we can think about the normal mind and its breakdowns. That situation will not be remedied by refusing to embrace a theoretical framework until we know all about etiology. If a taxonomy is adopted without a view to its incorporating causal claims, then future research will tend to congeal around the atheoretical taxa and links to research into the causal structure of the domain will be harder to construct. So in fact it would be a good heuristic to base a nosology on departures from normal function despite the paucity of causal explanations of particular pathologies. Even a partial construction of a causal taxonomy of the right type will focus research in the right direction.

Put more precisely, we can distinguish between two situations, causal discrimination and causal understanding. Causal discrimination is possible when we have a syndrome that has a cause we know to be distinct from the cause of a different syndrome even if we are unsure about the nature of the causes in each case. Or, we might have good explanations of the two causal relationships; that is, causal understanding. The claim I am making is that classification can reflect causal discrimination in the absence of causal understanding. And it can use causal discrimination as source of hypotheses. If we have good reason to believe that two syndromes depend on different pathologies, then we can orient research around finding out what those differences are.

I do not mean to set the bar too low: obviously there always will be some reason to believe that autistics are different from manic depressives, and that something makes the difference. Causal discrimination, as I see it, occurs when enough information exists to generate a reasonable set of expectations about the causal topography, even if the details are still hidden. We might be able to classify disorders in gross classes even without clear details of the causal story in each case. We can classify Tourette's and the other movement disorders together as movement disorders and sketch in what we know about underlying processes even if we do not know in detail how they cause the disorders. Likewise, if autism is a theory-of-mind disorder, we can classify it as such even if we are ignorant of the development and functional architecture of normal theory-of-mind. It is noteworthy that our understanding of how theory-of-mind develops enables us to make predictions about the course of normal development and how that course is different for at least some autistic disorders. This is a step in the direction of what clinicians call predictive validity and it suggests possible interventions. Neither of these advantages follow from the present nosology. Calling autism a developmental theory-of-mind disorder is not the last word on the condition, but it does give us a degree of structure that can serve two purposes. It can provide a context in which future work on autism can proceed, filling in the details and, if necessary, revising out initial characterization of it. Classification based on causal discrimination also can reveal affinities among conditions in terms of broad similarities in their causal nature, even where the precise details are hidden.

9.3.3   Causal Discrimination and the DSM-IV

Causal discrimination is a genuine advance over clinical phenomenology as a basis for classification. After all, we already group conditions into classes based on some fact about them—mood disorders, for example. So is causal discrimination already built into the system?

To a certain extent it is. Rhetoric notwithstanding, it is clear that some existing diagnostic entities have some causal hypotheses smuggled in: "substance abuse," for example, obviously is a name for a process as well as a diagnostic term. The presence of such officially illicit causal terminology shows the extent to which the workings of the DSM defy generalization, even when the generalization, in this case about avoiding causal hypotheses, is the DSM's own official policy.

As things stand, then, conditions are in some cases grouped together in terms of coarse-grained causal descriptions that can guide research in outline, with the details to be filled in. They may survive the shift to a more comprehensively causal nosology, but the basis for their being grouped together will change, to reflect knowledge of actual causal processes. Mood disorders are so called because of a shared clinical phenomenology that makes disruptions to mood salient: the question is whether this shared salience, which certainly looks causal, reflects a shared process. So, yes, the existing system does sometimes group conditions into classes based on their sharing a clinical picture that makes a common cause look implicated (as in mood disorders), or cites as a cause some normal human activity that is disrupted or extreme (paraphilias, substance abuse). However, the classification currently assumes that we can learn about causes through reflection on clinical phenomenology, and that assumption may obscure underlying differences as often as it finds them. There are two differences between this rough causal grouping and the causal discrimination I refer to above, reflecting a different approach to finding causes and a different approach to the semantics of diagnostic terms.

Take autism again. Calling it a theory-of-mind disorder goes beyond a description of symptoms and adverts to a normal human capacity that is disrupted in the clinical case. Our ignorance of the details means that the underlying process gets picked out by a term that refers to a manifestation of it. "Theory-of-mind deficit" refers to that underlying pathology. The facts in virtue of which reference is made to that underlying pathology might be given by performance on standard tests, such as the false-belief test plus some others, like tests of eye movement. This is different from the current system, in which diagnostic labels refer to surface phenomena at the level of clinical phenomenology. Those labels are understood via facts about how patients appear but they also refer to phenomena at the same, surface, level. Causal discrimination looks for tests that pick out underlying processes and groups syndromes together depending on their common participation in those processes: it does not assume that causal significance can be determined by what looks like shared salience at the level of clinical phenomenology. And, as I note in the discussion of Hacking in chapter 7, the idea that diagnostic labels refer to inner pathologies by way of picking out manifest symptoms is an attractive one, provided we remember the causal interaction of process and symptom that makes the label usable.

So causal discrimination reflects a belief that, even in the absence of mature causal theories, we nonetheless can pull conditions apart by developing tests that discriminate between them and suggest different causal stories, even where the details of those causal stories remains obscure. A classic procedure for doing this is the neuropsychologist's employment of double dissociation. In this procedure, two complementary conditions are identified, each displaying a deficit missing in the other. A classic, though controversial case, is the absence of theory-of-mind in autism and its preservation in Williams's Syndrome, a rare genetic disease associated with physical abnormalities and severe cognitive impairments alongside emotional sensitivity, social skills, and linguistic competence[3] (Bellugi et al. 1994; Karmiloff-Smith et al. 1995) The conclusion typically drawn is that in one of these cases we see the absence or malfunction of a component of the normal mind that is preserved in other patients, or vice versa. The use of lesions as a way to make

---

3. The linguistic competence has not turned out to be as marked as was initially supposed by commentators (Pinker 1994) who saw Williams Syndrome as evidence for a "language module" (Cowie 1999). Nonetheless, theory of mind seems preserved in Williams Syndrome, along with a number of other aspects of social cognition.

predictions about the course of a condition rather than its cause is another example in neuropsychology of causal discrimination: attending to lesions allows us to argue that two similar conditions have different causes even when the causal stories are unknown.

Within psychiatry, the dominant approach that offers hope of causal discrimination is pharmacological dissection. Though rarely as clean as we would like (and indeed that is true of lesion and dissociation methods), it is another way to isolate what appears to be a discriminable causal process, in this case through its selective response to a drug. A classic example is the discovery in 1959 that imipramine, when administered to patients with anxiety neurosis, prevented panic attacks even though the other symptoms remained (McNally 1994). The result was recognition that panic disorder was a discriminable entity (even though it now appears most profitable to think of panic as cognitive, best treated through cognitive-behavior therapy).

Placing two syndromes in a taxonomy may require causal discrimination but it does not have to require causal understanding. Even if we are without any detailed causal information for a particular condition, a taxonomic enterprise designed around tests that let us make causal discriminations enables us to make some useful moves: we can, for example, be in a position to incorporate causal information about the condition as we gather it. This is not possible if we start out determined to keep causal information away from our taxonomy. A taxonomy that is causal in spirit also allows us to make effective differential diagnosis by pointing to causal information elsewhere in the taxonomy that applies to clinically similar conditions.

We can think of the stages through which a theory goes as reflected in changes in the models it employs, in terms of both content and use. A fairly sketchy model can discriminate among causes even if the details are not understood, and we can use it for predictive purposes. A more capacious model includes more causal information and secures greater understanding of the real nature of the phenomenon. Models become more capacious as more is revealed about the causal structures of interest, and the sketchy models guide the process of revelation and are revised by it.

Complete causal understanding will never be achieved about anything in the universe, but we can expect more and more elaboration of causal stories as causally discriminated conditions are better studied. The chief reason why we should aim for a causal dimension in taxonomy is that classification according to causes puts us in a much better position to move from causal discrimination to causal understanding. If taxonomy is to facilitate research, which everyone agrees is another of its key purposes, then it must be based on theories that provide causal understanding. It is increasingly apparent, as I point out in earlier chapters, that developments in several sciences are providing causal information that nosological revision can exploit.

I conclude that accepting the need for a causal taxonomy does not commit us to an impractical project, as long as we bear in mind that initially we will be looking to make causal discriminations among conditions. The proposal is not that we suddenly can furnish complete causal theories, just indications of where those theories are to be found. Causal discrimination can group conditions together even in the absence of sophisticated causal stories, and serve as the first stage in the complementary processes of theory building and taxonomy construction. The ultimate aim is to avoid the problems that beset the existing system.

9.4 Three Problems with the DSM Taxonomy

The claim I made in the last section was that classification can draw on causal discrimination in the absence of causal understanding. And it can use causal discrimination as source of hypotheses. If we have good reason to believe that two syndromes depend on different pathologies, then we can orient research around finding out what they are. I now turn to the detailed criticisms of the existing, noncausal classification system. The system of classification in the DSM is incoherent, heterogeneous, and provincial. It is *incoherent* in that it rests on a theory about the taxa of interest that requires symptoms to be expressions of underlying causes whilst at the same time it prohibits the mention of these underlying causes in the taxonomy. It is *heterogeneous* in that does not classify like with like at appropriate levels of explanation. And it is *provincial* in that it is cut off from much relevant inquiry. These complaints, though directed at the DSM in the ensuing discussion, reflect worries about the current state of biological psychiatry as a whole, since DSM-IV-TR is its flagship.

9.4.1 Incoherence

At various points I have suggested that the official self-image of the DSM-III and its successors is at variance with some aspects of their content. The DSM is a body of work informed by biological psychiatry, yet it deliberately tries to exclude that theoretical orientation from its own text. As I have noted, some diagnoses clearly bear the imprint of causal speculation, but even there the theorizing is folk-psychological rather than scientific. DSM-style psychiatry departs from what biological psychiatry, in its deference to medicine, seems to require.

Taken seriously, the assumption that disorders can be classified by symptomatology alone requires us to assume that a significant difference can exist between two individuals at the level of surface symptoms that does not reflect an underlying causal difference. And as Kendell pointed out a generation ago, "it is almost impossible to visualize how a discontinuity could exist at a symptomatic level without being a reflection of a more fundamental discontinuity elsewhere" (1975, 69). And in fact, the assumption is not taken seriously, since the syndromes included in the DSM-IV-TR are widely assumed to reflect underlying differences between individuals. But the classification is not supposed to depend on these underlying differences, which detaches the taxonomy from the model of disease that gives it its point. This reflects a mistaken assumption about the importance of operationalism, which is not even carried out properly, as I suggest in chapter 6.

None of this makes sense as an adoption of the methods of medicine, where diagnosis is causal. The basic assumption in medicine is that diagnosis is a matter of uncovering the causal antecedents of visible pathology, with conditions defined in terms of the underlying process as well as the surface features. If the medical model were the animating spirit of psychiatry, this realist, causal diagnostic picture would be the avowed aim. Paul Meehl says it better than anyone else (Meehl 1995, 267):

> Most critics don't even get it right. On the one hand they fault the DSM committees for using the medical model, which is dogmatically assumed to be inapplicable. But they ignore the fact that entities in the advanced specialties of medicine are not constructed like the DSM categories. The advance-science medical model does not identify disease taxa with the operationally defined syndrome: the syndrome is taken as evidentiary, not as definitory. The explicit definition of a disease entity in nonpsychiatric medicine is a conjunction of pathology and etiology and therefore applies to patients who are asymptomatic (which is why e.g., one can have a silent brain tumor or a staghorn kidney that never causes trouble during life and is only found postmortem). Perhaps we cannot blame psychologists ignorant of medicine for making this mistake, when some psychiatrists who are passionate defenders of the DSM don't understand how far it deviates from the optimal medical model. Accepting operationism (an erroneous philosophy of science) and the pseudomedical model (definition by syndrome only) engenders a wrongheaded research approach, unlikely to payoff in the long run.

The DSM understands mental disorders according to a disease model that sees symptoms as indicating hidden destructive processes within the organism. This core assumption has been inherited from general medicine even in the absence of knowledge about etiology and pathology in underlying neurological systems. As Guze puts it, "manifestations of illness are not random" but "often occur together and reflect disturbances involving different organs or even different parts of organs" (1992, 10). This understanding of disorder is what motivates the stress psychiatrists place on the validation of diagnoses. Diseases, on this model, are thought of as disruptions to underlying functions, with a characteristic and predictable course, outcome, and response to treatment. In addition, diseases tend to aggregate in families. This way of thinking about mental disorders is enshrined in DSM-style psychiatry and accounts for much of the attention paid to classification during the last quarter-century, on the grounds that "classification in medicine is called 'diagnosis'" (Goodwin and Guze 1996, xi). Yet, at the same time, psychiatrists affirm in the DSM-IV-TR that classification must not take into account what we know about the underlying processes and their failures to function. The model is not being applied in a consistent fashion.

I have said that in principle a psychiatric classification based on causal explanations of breakdowns of normal mental processes is desirable and attainable, especially if in the first instance we aim at casual discrimination before causal understanding. Medicine in general employs causal classifications. But in the DSM the medical model of disease is cut loose from its theoretical foundations. The result is that mental illnesses are construed as clusters of symptoms. The symptoms are assumed to be the effects of underlying causes, but those underlying causes are not permitted to enter into the classification. So although recent editions of the DSM assume that mental disorders are distinct and identifiable entities that should be validated, it leaves itself without the resources to say what kinds of distinct entities they are. The notion that disorders must be understood and classified as breakdowns of normal function is made by neuropsychology and by somatic medicine, but it is excluded from psychiatry. So it seems that the model of disorders that is being taken for granted in the DSM is one that has its conceptual home in a certain philosophical conception of taxonomy, but is being officially presented as something else. The idea that disorders are understood in terms of disruption to normal functioning, axiomatic in medicine and neuropsychology, ought to be axiomatic in psychiatry. It should be axiomatic not just because of its abstract desirability but because it reigns in the very somatic medicine that is the intellectual inspiration and role model of modern psychiatry, and it has an impressive track record in the wider medical sphere.

In chapter 4 I endorse the claim that psychiatry is a branch of medicine, suitably qualified as a claim about disease rather than methodology. As such, it should embrace the causal basis of medical diagnosis. Diagnosis is classification. Not all classifications are causal—we can distinguish things form another for many reasons, and with many rationales. But in medicine, as in science generally, we need causal taxonomies.

9.4.2 Heterogeneity

I begin this section by borrowing from Poland et al. a trenchant criticism of the antietiological model of classification that the DSM-III has bequeathed. However, although their criticism is correct, we can dissent from them in some respects. Poland et al. argue that the DSM-IV conceives of mental disorders as natural kinds, or, as they call them, "syndromes with unity." They criticize the syndromes with unity idea. I, on the other hand, think the idea has merit, provided it is better implemented, affiliated with explicit causal theorizing, and disentangled from the complications of the requirement that a taxonomy be composed of natural kinds. In particular, the requirement that symptoms be causal manifestations of underlying processes needs to be distinguished from the requirement that they stand in a lawlike relation to those underlying processes.

To begin with, then, Poland, von Eckhardt, and Spaulding claim that the DSM approach "is deeply flawed and not doing the work it should be doing" and that, as a result, the current situation regarding the classification of mental disorders "involves a crisis" (1994, 255). Moreover, as Poland and his colleagues go on to note, "DSM categories play pivotal roles in financing mental health care, maintaining hospital and clinical records, administering research funds, and organizing educational materials . . . concerned with psychopathology" (235). So there is more at stake than simply how to organize psychiatric knowledge: the practical import of classification in psychiatry is enormous.

The conception of mental disorders as syndromes with unity faces several problems. To begin with, it has a poor track record. Even sympathetic commentators admit that the DSM-IV was full of unvalidated diagnoses (Nathan and Langenbucher 1999). The more substantive worry about the syndromes-with-unity approach is that it seems to contradict what we know about how the component parts of human psychology cause distinct behaviors and other outward psychological phenomena. In other words, the sorts of things that clinical phenomenology detects just do not stand in the right relation to underlying processes for the syndromes-with-unity idea to work.

According to Poland et al., a classification scheme for psychopathology has two primary purposes. It should enhance the effectiveness of clinical activity, and it should facilitate scientific research concerned with psychopathology and its treatment. The DSM approach, they argue, does neither. This section is devoted to expounding and amplifying their critique. Though Poland and his colleagues offer a number of reasons for their skepticism, their main objection is that the DSM approach to classification is not guided by any theory about the structure and functioning of normal minds.[4]

Current taxonomic practice assumes that mental disorders are, first, clusters of associated symptoms forming syndromes and, second, expressions of malfunctions (and, third, harmful in some way, but I put that aside for now). Poland et al. (1994) call disorders that work like this "syndromes with unity," but they claim that there are in fact very few syndromes with unity in psychopathology. It is no surprise that I think that Poland et al. are correct in reading the DSM-III and its successors this way, but many of their objections still hold even if there is no such conception of mental illness informing the work—that is, even if the avowedly atheoretical underpinnings of recent editions of the DSM are in fact applied consistently. However, their criticism of the syndromes-with-unity idea needs qualification.

Poland et al. argue that there are few syndromes with unity because mental illnesses, at least as represented in DSM-IV-TR, lack "nomological unity" (1994, 245). They claim that syndromes with unity only exist if syndromes are the nomic expression of underlying processes. This is as a very strong requirement, namely that whenever a set of symptoms occurs, they do so because of some biological or psychological impairment that causes the symptoms as a matter of natural law. They argue, correctly, that the clinical manifestation of a pathology or malfunction is heavily context dependent: taking different forms depending on other aspects of the patient's psychology (252). Since they worry about the undermining of syndromes with unity by context dependence, it appears that they think that syndromes with unity must exhibit context-independent lawlike causal relations. Poland et al. call the lawful relations that syndromes with unity should display "tight" (253). In other words, a given cause must have the same effect almost regardless of circumstance. This amounts to a requirement that laws in psychiatry be

4. The great exception is the treatment of sleep disorders: several of the diagnoses included in the sleep disorders section are based around a clear underlying picture of normal sleep introduced, albeit very briefly and schematically, at the beginning of the section. Even so, some diagnoses in the sleep disorders section—like Nightmare Disorder, and maybe Jet Lag—are not just unproblematic scientific derivations. Jet Lag was included in the DSM-IV as a subtype of Circadian Rhythm Sleep Disorder but removed in the DSM-IV-TR, which still recognizes Circadian Rhythm Sleep Disorder but does not subtype it. Jet Lag is a diagnosis in the International Classification of Sleep Disorders (American Sleep Disorders Association 1990).

strict, or at least fundamental in the sense of chapter 4. As I say there, this condition is rarely met in the biomedical or cognitive sciences.

If the DSM-IV supposes that syndromes with unity exist, then Poland et al. are right to be skeptical, but they seem to think that no categorical classification of mental disease types is possible in the light of the nonexistence of syndromes with unity. This appears to rule out the possibility that patients with the same dysfunction can differ with respect to symptomatology but still belong to the same category, because the difference depends on the psychological context and hence violates the requirement that syndromes with unity exhibit tight laws. But the nonexistence of syndromes with unity does not show that a causal classification of mental illnesses is not possible. The DSM-IV-TR constraints, though, are too weak. They often overlook the possibility that patients who share an attribution made by clinical phenomenology can differ enough with respect to what causes it to require that they be put in different categories.[5] That is, Poland et al. fault the DSM-IV for assuming that it has uncovered syndromes with unity and then argue that, properly understood, syndromes with unity are not psychiatrically plausible. I agree with the first charge, but think that syndromes with unity need not be defined as strictly as Poland et al. define them, and that a similar idea can be defended. The best defense of the idea is to see psychiatric taxa as exemplars, which combine a stereotypical depiction of syndrome and course with a causal model that has an adjustable relation of fit with the real world. This gains us a middle ground between syndromes with unity, which group patients into law-governed categories that do not actually exist, and DSM categories, which are too anarchic.

DSM diagnoses are notoriously heterogeneous. Its categories group together very different symptom profiles as manifestations of the same behavioral or emotional disorder. Indeed, in many cases the DSM-IV-TR allows for two patients to be diagnosed with the same disorder despite having very few symptoms in common from the overall set. In some cases,

5. I say "often" because some symptoms, especially cognitive ones, may be parceled out across different categories in the DSM-IV. The DSM editions are the joint products of numerous committees, not all of which share an outlook. Hence, any generalization about the DSM-IV is likely to be met with numerous counterexamples. Nonetheless, like many other theorists, I think it is possible to discern an official line, which is present implicitly in the DSM-IV and less implicitly in the publications of its chief architects. The official line is, I think, kept to most exactingly with respect to the major cognitive and mood disorders that psychiatry is most closely associated with.

there can even be no overlap in symptoms across patients. By one reckoning there are 56 different ways to satisfy the criteria for Borderline Personality Disorder (Clarkin et al. 1983)! As one might expect, these heterogeneous categories are poor predictors of the patients' future course or response to treatment. The lack of rigor in clinical phenomenology only exacerbates this problem. Two patients could both be classified based on the same symptom, such as sadness or irritability, but vary greatly in the intensity with which it is present (Poland et al., 249). Or they could have the same symptom but nonetheless seem very different, since there are many different ways to manifest, for example, a loss of interest—one could stay in bed all day, one could flit restlessly from one thing to another. Stressing construct validity may obscure this point, since aspects of psychopathology may correlate with a diagnosis but nonetheless have different causes. What we really want is predictive validity (Kendell 1975, 40–42). Predictive validity measures the extent to which predictions made on the basis of a diagnosis are subsequently found to be correct. If two patients share a syndrome but do not share a common underlying abnormality, then a prediction applied to them both may only be true of one.

These problems reflect the heterogeneity of DSM syndromes. It seems unlikely that the same underlying causes explain an irritable adolescent who sleeps late, diets frantically, and lies around the house all day threatening to commit suicide on the one hand, and a sad middle-aged man who can not settle down to any of his normal hobbies, hardly sleeps, eats more and more, can not make love to his wife, and feels worthless. But both people satisfy the DSM criteria for Major Depression.

One problem with the DSM-IV-TR, then, is that the same symptoms (from the point of view of current diagnosis) may be produced by different causes. Indeed, as in biological medicine, even tokens of the same symptom type may be produced by different causes if the symptoms are defined phenomenologically, as pains or coughs, for example. Second, the same cause can fail to produce the same symptoms. The actual symptom it produces depends on other factors.

The case of anxiety illustrates the second type of complexity in the relation of symptom and cause. I note in chapter 4 that stress may be a general cause, in the sense of showing up in many explanations, without playing a fundamental explanatory role in any condition. Even intense emotional distress may not turn into pathological anxiety. It depends not just on the nature of the emotional reaction itself but on the network of support one has and one's own beliefs about how anxiety should be dealt with (McHugh and Slavney 1998). So a situation may trigger both an emotional response and a cognitive mechanism that stops the response from becoming pathological. The net effect of the interaction of the mechanisms is unpredictable. McHugh and Slavney's example illustrates what Elster calls "Type B mechanisms," in which different causal chains are triggered that "affect an independent variable in opposite directions, leaving the net effect indeterminate" (1999a, 3). In these cases even if we can predict the causal chains that will be triggered, we cannot predict the outcome.

Or it may be that the same system, in different situations, can initiate several responses, and it is indeterminate which of several causal chains may be triggered. Elster calls these cases "Type A" mechanisms (1999a, ch. 1). His example is that fear-elicited behavior in animals can take the form of either freezing or, on the other hand, increased activity, which may be either flight or fight. "But although we can identify the conditions that trigger freeze versus either flight or fight, we do not know which will trigger fight versus flight" (3). Whether flight or fight is triggered depends, it appears, on information processed by a unified flight/fright system about the "total stimulus context." But, as Elster points out, to say that what triggers one behavior or the other is the total stimulus context is just to say that the whole situation is such as to produce one behavior or the other. Classifying flight, fight, and freezing together as fear obscures the underlying differences, but it is permissible if we can specify the underlying differences and subtype according to known causal processes. Most DSM categories that exhibit similar heterogeneity do not depend on knowledge of the relevant causal pathways but on an intuitive sense that certain behaviors belong to the same class.

We can back away from the desire for nomic relations in favor of something less demanding. Suppose we have (as we often will) the following situation: cause C causes one of a set of symptoms, S1, S2, or S3, depending on whether some extra causes, E1, E2, or E3 are present. The tight laws Poland et al. require are absent in this situation. But, if C is necessary for S1, S2, or S3, then the presence of one of those symptoms does allow us to infer that C is present, along with one of the underlying extra causes. The claim that classification should reflect causal relationships is fully compatible with a state of affairs such as this, even though the claim that classification should reflect nomic connections is not. But that claim, as I have said, is much too strong.

The same behavior can have not just different causes, but such very different causes as to suggest that we are dealing with different conditions. Hare distinguished three behaviorally very similar groups on the basis of "personality structure, life history, response to treatment and prognosis" (1970, 9): these being psychopaths, acting-out aggressive neurotics, and delinquents raised in a subculture that praises antisocial behavior. But the current DSM diagnosis of antisocial personality disorder would cover all these individuals and a good part of the prison population besides.

The combination of heterogeneity with the conceptual confusion we have already looked at is what accounts for the flagrantly ad hoc quality of so many DSM diagnoses. As we saw in the discussion of Wakefield's criticisms, the diagnoses often are hedged with qualifications that make some concession to common sense, and they also may reflect a variety of theoretical assumptions required for the overall theory to be more plausible. The proliferation of small-scale, local hypotheses within the wider work is a clear sign that the overall theoretical system is in trouble. And the isolation of psychiatric classification from the wider enterprise of the sciences of the mind only makes matters worse.

To summarize this section: DSM-IV syndromes obscure the importance of underlying causes by lumping together patients who seem similar on the surface. This is an intellectual and therapeutic disaster. And although we cannot expect to see syndromes with unity, we have no reason yet to rule out the supposition that exemplars are explained by causal models that represent real, though not tightly lawful, causal relations.

9.4.3 Provinciality

The DSM-III exacerbated the problems discussed in the previous section by its deliberate isolation from most scientific research on the mind/brain. Strictly speaking, heterogeneity and provincialism are separate problems, since an attempt to lessen heterogeneity could proceed by looking for a theory of normal functioning employing only psychiatry's concepts and assumptions, and thus perpetuating its isolation. Psychiatrists do of course undertake research using the concepts and categories of their field, but those concepts and categories too often are isolated from wider bodies of research. This has two effects. First, concepts that exist only within psychiatry are not tested for consistency with the wider theoretical framework, and, second, psychiatric concepts of mental phenomena that are shared across the sciences of the mind do not get refined.

The concept of dissociation provides an example of the former problem. Dissociative disorders include Dissociative Identity Disorder, the successor to Multiple Personality Disorder. The change reflects the current wisdom that patients do not have more than one personality, but less than one personality; that is, their normal personality has not formed or is prone to fragmentation. The core concept, dissociation, is not defined in the DSM IV-TR, but the Dissociative Disorders are said to share an "essential feature," to wit, "a disruption in the usually integrated functions of consciousness, memory, identity, or perception" (DSM IV-TR, 519). Now, this concept of a cognitive disruption, which psychiatrists have inherited from the nineteenth century, certainly looks as though it could benefit from being compared with cognitive and other models of integrated functioning. The danger is that psychiatrists will treat dissociation as a well-understood phenomenon, because it plays a role in their own thinking, without seeing whether it is consistent with models of normal functioning assumed by the sciences that study memory, reasoning, planning, or other aspects of the normal, integrated psychological functioning.

The second problem is more pressing because few psychiatric terms are in fact technical coinings with no outside use. They tend to be taken from the wider culture (like "self-esteem," which many people think is a crossover from therapy but was in fact coined by Milton) or drawn from other sciences. But even when they are drawn from other sciences, technical terms in psychiatry tend to take on a life of their own. Attention is a focus of much research in cognitive psychology, but, as Poland et al. note, the DSM concept of attention is "unrelated to any sophisticated understanding of selectivity, vigilance, attentional capacity, and types of attentional systems" studied in cognitive psychology (1994, 246–247).

So provincialism has the consequence that DSM concepts and categories are not checked for consistency with our best theories of the mind. The result is a failure to learn from our understanding of psychological phenomena, thus reinforcing the ignorance of etiology that is used to defend the exclusion of theory from the DSM. Another problem arises because theories cannot be consistently applied if we respect disciplinary borders. Schizophrenia shares many features with neuropsychological and neurological disorders, such as bizarre delusions and motor problems. Depersonalization Disorder (DSM IV-TR, 530) is diagnosed if a patient persistently feels detached or estranged from the self, and a similar claim has been made (e.g., by Gerrans 2000) about the experiences of people with Cotard's Delusion, who claim that they are dead. The difference is that people with Depersonalization Disorder maintain intact reality testing: they know that they are not really automata, whereas it is claimed that Cotard's Delusion characteristically arises when individuals try to explain away their strange feelings.

It would be good to strive for more general theories that can take all these phenomena within their scope. If we are to do that we must align psychiatry more firmly with the other sciences of the mind. As I note in chapter 2 when discussing the models of abnormal motor control developed by Frith and colleagues, we need to ignore disciplinary boundaries in the interests of theoretical power and fruitfulness.

The best objection to the provincial isolation of the DSM-IV-TR is the simplest, and it is prefigured in most of the preceding discussion. Psychiatric conditions involve derangements of capacities that are studied by the sciences of the mind more generally. Cutting psychiatry off from these bodies of research is absurd. Aligning psychiatry with research into the sciences of the mind is bound to pay off.

9.4.4   What Nosology Ought to Be

The problems of incoherence, heterogeneity, and provincialism are probably not as bad in practice as I presented them as being. The ideology of the DSM is not always adhered to, and the good sense of many clinicians and researchers works to make things better than they might be. Science is bound to make its presence felt in the end, and of course the world is full of people researching into the various causes of psychiatric and neurological disorders. Nonetheless, the existing taxonomic system is not a happy one, and the flaws I complain about are potential obstructions to fruitful research. And it is worthwhile thinking about how the field might look once the synthesis of psychiatry and neuroscience I look forward to is in hand. I have talk about explanation in earlier chapters. Now it is time to wonder how classification might make use of the opportunities that synthetic neuropsychiatry will present.

The synthesis promises a classification system based on an account of the workings of the mind/brain. We must classify failures of normal function. In chapter 2 I suggest a two-stage picture, in which science uncovers the problems of the mind that then can be thought of as disorders or not based on our values. I take this chapter to provide further support for that picture by showing that contemporary psychiatry as it stands is ill-fitted to contribute to that project. I argued earlier that DSM-style psychiatry lacks a coherent picture of the mental just as common sense does, and its categories have developed haphazardly. The present objection is that the DSM approach to psychiatric classification is flawed and confused because it overlooks the necessity of basing psychiatric taxonomy on a causal theory. The DSM-IV-TR tries to provide an "operationalized" or "theory-neutral" taxonomy, but there is good reason to expect this approach to be fruitless: one of the few indisputable lessons of twentieth-century philosophy of science is that taxonomy must derive from a theory of the domain of inquiry that is being classified. In the case of psychiatry, we need to base classification of abnormal states on our best current theories of how the normal mind works. This difficulty besets not just the DSM-III and its successors but also any attempts at theory-neutral classification elsewhere in psychiatry (such as ICD-10) and, quite possibly, all forms of scientific inquiry. Just as in medicine generally, causal classification of psychopathology attends to a condition's etiology as well as its manifest symptoms.

So a theory of the mind should be the basis for a scientific classification that can serve as part of the first step in the two-step approach that I recommend. For purposes of understanding breakdowns in mental functioning, we do want to classify according to scientific and not lay theories. Where nonscientific interests enter my picture is in deciding which psychological breakdowns count as legitimate objects of attention in various enterprises. I imagine scientific classifications serving as input into nonscientific purposes. Those purposes might group together, for judicial reasons, say, a set of conditions that look odd from a scientific standpoint. But it is the scientific taxonomy that serves as the basis on which decisions are made.

What should that scientific taxonomy look like? Many discussions of this topic place importance on deciding whether psychiatric conditions are natural kinds. I have endorsed the view of Poland et al. that very few syndromes with unity exist in psychiatry. That does not mean that there are no determinate disease categories, only that if they do exist, they are not syndromes with unity. Whether the notion of natural kind is of any use in psychiatry is my next topic. Poland et al. want to make nosology properly scientific. Their skepticism about syndromes with unity is not informed by the belief that psychiatric categories are arbitrary, but by a conviction that natural kinds work in a particular way, and that psychopathology does contain not many kinds that work like that. Other writers, however, have argued that to deny that mental disorders are natural kinds casts doubt upon their objectivity or reality (D'Amico 1995; Reznek 1995). But that is only true if there are no scientific categories apart from natural kinds. Others have said that natural kinds must have essences and mental illnesses do not (Zachar 2000). The position I defend is that there are both natural kinds and other scientifically fruitful divisions in nature. Mental illnesses are kinds, even if they are not natural kinds. And whether they are seen as natural kinds is largely a matter of choice.

9.5 Kinds of Kinds

The many specific proposals about what makes a group of objects a natural kind, fall into four broad camps; I call these the *simple essentialist, nomological, refined essentialist,* and *ecumenical* views. The basic idea is that natural kinds comprise members that share the properties that define the kind. However, this leads us to an apparent problem immediately. Gold objects form a paradigmatic natural kind in virtue of being instances of gold, because all instances of gold share the property of having atomic number 79. So, having atomic number 79 is good enough to admit a wedding band into natural kindhood qua instance of gold, whereas if that same wedding band applied for membership in the natural kind club because it was made in Brooklyn, it would not get past reception. Products made in Brooklyn form a kind, but they do not form a natural kind. What separates the four views I distinguish is the answer they give to the question, what makes a kind a natural kind.[6]

9.5.1 Simple Essentialism

Simple essentialism is the view that what makes a collection of entities into a natural kind is some microstructural essence that accounts for their other properties. What makes gold a natural kind, for example, is the atomic structure of its instances. That shared atomic structure explains why gold objects share the same properties. All and only pieces of gold share the property of having atomic number 79, and that property they all share explains the presence of all the other salient properties of gold—being shiny, being malleable, and so on. Putnam, for example, famously argued that it was microstructure that distinguished water from its phenomenological twin XYZ (Putnam 1975). Kripke too seems to have taken the idea of microstructure very seriously. Each biological species, according to Kripke, has an essential microstructure that makes it the species that it is. An animal that looks and behaves just like a tiger nonetheless is not a tiger if it has the wrong physical microstructure, if it "differs from them internally enough that we should say that it is not the same kind of thing" (Kripke 1980, 121).

---

6. Boyd (1991) traces the history of natural kind—mongering back to Locke's real essences. The idea of an underlying microstructure explaining the salient properties of an entity certainly seems Lockean. However, Hacking (1991a) insists that the point of a natural kind is its connection with induction, and that was not something people worried about before Hume. Hacking thinks the notion of a natural kind surfaces in Mill and Venn in the mid-nineteenth century, and that is indisputably where the term "natural kind" originates. Hacking's and Boyd's rival histories rest on a dispute that is central to this section: is what makes a kind a natural kind its possession of an essence, or its utility in induction? Kinds with essences, I argue, are a proper subset of inductively useful kinds, and the dispute between Hacking and Boyd mirrors the dispute between refined essentialists and ecumenists over which kind of kind is natural.

Simple essentialism is the view that members of a natural kind share a microstructural essence. It is genetic rather than atomic structure that counts for species. Tony is a tiger because he has a tiger genome, says Kripke, since that genome is a tiger's essence. Note how simple essentialism is reductive: the essence explains the surface properties. Molecular structure causes gold to be shiny and water to be liquid and the molecular structure of the tiger genes make Tony's stripes.

Simple essentialism was enormously influential, as was the semantic theory Kripke and Putnam fashioned to accompany it. Yet it did not take long for problems to become apparent, and these problems led many to move to a different sort of essentialism. I turn to these criticisms in a moment, after dealing with a criticism of a different sort.

9.5.2 Nomological Natural Kinds

David Hull (1989) argues that natural kinds as Kripke and Putnam conceive them fail to meet the constraint that kinds be spatio-temporally unrestricted and immutable. Hull contends that laws of nature seek the deepest, immutable regularities that there are, and that natural kinds must be similarly immutable to play their role as the fundamental law-governed elements in the universe. On Hull's view chemical elements are natural kinds, in virtue of remaining the same throughout the universe. Chemical elements, says Hull, do not come into and go out of existence, even though their instances do. Nor can the kind lead be transmuted into gold, just as triangularity cannot evolve into circularity, although individual pieces of lead can be changed chemically. Natural kinds, says Hull, are eternal, not in the sense that they are always exemplified, but in the sense that the lawful structure of the universe means that "there is always a slot for them" (Hull, quoted in Callebaut 1993, 279).

Species, though, come into and go out of existence and evolve into other species. And all the species we know about are restricted to one particular corner of the universe. Therefore, concludes Hull, species cannot be natural kinds. Obviously, this account rules out the existence of natural kinds in the human sciences. No spatio-temporally unrestricted, exceptionless laws apply to humans qua humans, and a fortiori there are none in psychiatry. So if Hull is right, mental disorders cannot be natural kinds. Yet Hull's account of natural kinds misses the point that we often do not care whether natural kinds fit into spatio-temporally unrestricted laws. What counts is that they explain some of the causal structure of the world, even if only a local part of it. To people who think like this, the underlying structure of a kind is what's important, since it explains the clustering of observed properties.

Simple essentialism identifies this structure with microstructure, and says that the explanation must be reductive. I shall now argue that this is an untenable idea, and then introduce refined essentialism in 9.5.4. as the view that tries to keep the basics of simple essentialism while conceding that its account of essences needs to be rejected.

9.5.3 Difficulties with Simple Essentialism

Hull's nomological account of natural kinds was supposed to make trouble for simple essentialism, but the nomological view had troubles of its own. The objection to it was that natural kinds are marked out by an underlying causal structure that accounts for observed properties, and not their participation in exceptionless laws. A rival suggestion currently on the table is that the underlying structure is microstructure, as suggested by simple essentialism.

But simple essentialism faces a grave difficulty. The observable properties of a kind may not rely on a common underlying microstructure. For example, it is just false that there is one genetic structure which all members of a species share. There is a huge amount of genetic variation in nature, and the search for a unique genome common to all members of a species is a wild goose chase.

If a simple essentialist wants to stick to the suggestion that a natural kind must contain members who share a common reductive microstructure, then there is nothing for it but to deny that species are natural kinds. Wilkerson (1993) argues that the only genuine natural kinds in biology are those that share a genetic essence, and if there are no genetic essences then there are no natural kinds in biology.

Wilkerson insists that inductive success depends on natural kinds, yet if there are no natural kinds in biology his view renders incomprehensible the inductive success of biology. Members of species do not all have the same genome, but they do exhibit distinctive clusters of morphological, behavioral, and other properties. The difficulties for Wilkerson's view show what is unattractive about sticking to simple essentialism at all costs. The idea that there are no joints in the biological world except for individual organisms seems to deny that most biologists know what they are doing.

One response adopts an account of natural kindhood that rests on a different view of essences. This is what many philosophers have done. They have adopted a refined view of essence. Refined essentialism (Boyd 1989) keeps the idea that kinds have an essence but relaxes the restrictions on what the essence might be. The idea is that essences are the mechanisms that keep other properties in place. But it still does not follow that all kinds recognized by the sciences have essences. Hence, not all kinds are natural kinds.

But one can deny the distinction between kinds and natural kinds and say that all the kinds of any science are natural kinds. This is the ecumenist position that any relevant theoretical division in nature is a natural kind. Refined essentialism and ecumenism often are run together, which is a mistake.

9.5.4 Refined Essentialism

We owe refined essentialism mainly to Richard Boyd (1989, 1991). Boyd claims that a natural kind is a homeostatic cluster of properties: "either the presence of some of the properties in [the cluster] tends (under appropriate conditions) to favor the presence of the others, or there are underlying mechanisms which tend to maintain the presence of the properties in [the cluster], or both" (1989, 15). Boyd has singled out something very important, viz. the way in which clusters of properties may stick together homeostatically despite perturbations in the environment: "a natural kind is a cluster of properties, which, when realized together in the same substance, work to maintain and reinforce each other, even in the face of changes in the environment" (Kornblith 1993, 35).

Boyd's idea was that entities such as organisms and cells (Kornblith's examples) have systems that maintain their integrity in the face of environmental change. These systems participate in processes like the maintenance of body temperature or internal pressure, yet these causal processes are not necessarily explained via reduction to a more basic science. If we want to know how jackrabbits maintain their body temperature, we do not ask a chemist, we ask a physiologist, and she will mention long floppy ears in her answer, not atoms.

As Hacking notes (1991b), the homeostatic-property cluster account of natural kinds preserves and explains many of the characteristic concerns of the traditional literature on kinds. It serves to underwrite induction, for instance: if properties are held together homeostatically we can conclude on the basis of one property that others will typically occur with it. It also explains why there are divisions in nature: not all clusters of properties are stable, so some epistemically possible combinations will not be physically possible. The refined essentialist view subsumes the simple essentialist view, and preserves its insights: microstructure is one way to keep properties together. For all these reasons, we should conclude that refined essentialism is a better account of natural kinds than simple essentialism. Boyd's refined essentialism has, however, been overgeneralized. Natural kinds have been detected where one might not expect to find them. The assumption that almost any theoretically significant division is a natural kind is the position I call "ecumenism."

Kornblith (1993) reads Boyd as claiming that internal causal mechanisms maintain homeostatic property clusters. This might be quite rare, and nothing in Boyd's account rules out external forces as the maintainers of homeostasis. On the other hand, merely citing some mutually reinforcing properties is far too loose a condition unless some restrictions are spelled out. It would make Trotskyists a natural kind, for instance, since *being a Trot* tends to favor and reinforce the presence of some other properties (believing that socialism cannot be built in just one country, for instance). As Meehl (1992) points out, we can predict someone's attitudes and behavior with respect to political matters with perfect accuracy once we have established that they believe certain things about Trotsky. So Trotskyists are a social kind, in the language of chapter 7. If restrictions are put in to rule out Trotskyists as a natural kind, we end up with an account in terms of shared mechanisms. An ecumenical view lets natural kinds be maintained just by the sorts of historical contingencies that turn people into Trots, or keep groups like political parties united over time.

9.5.5 Ecumenism

A good statement of the ecumenist position is Griffiths's (1997). He argues that Boyd's view of natural kinds should be generalized for two reasons. The first is the psychology of concepts: people tend to see essences everywhere, and this "psychological essentialism" is an important part of normal concept formation (Keil 1989). But even if we see essences everywhere, that does not mean that all these essences actually exist.[7]

As well as arguing for the importance of the causal homeostatic approach to understanding concepts, Griffith also urges us to adopt it as a general account of what kinds are.

Categories from any special science that enter into the generalizations of that special science are now commonly regarded as natural kinds. Inflation, schizophrenia and

---

7. Kornblith (1993) lauds the tendency of humans to see (refined) essences everywhere because he thinks there really are essences everywhere, and our capacity to spot them is what makes us epistemically reliable. Yet there is evidence that many of our folk concepts do not determine natural kinds (see Atran 1990 on "folk biology"); we may simply be disposed to make mistakes about the natural world. In any case, it is clearly true that psychological essentialism leads people to look for essences where there are none. John Thorne, for instance, reports an acrimonious debate among scholars of Italian food about whether *risi e bisi* is *really* a risotto or a soup (Thorne 1996, xxiii–xxvii).

allopatric speciation events take their place alongside electrons and stars. In this more liberal perspective, natural kinds are not the most fundamental categories of nature, to be contrasted with categories that are useful but superficial. Instead, they are nonarbitrary categories, to be contrasted with arbitrary, nominalistic schemes of classification. (Griffiths 1997, 5–6)

But Griffiths's proposal that natural kinds are just nonarbitrary categories is in tension with the homeostatic cluster theory. Although it may be useful to distinguish allopatric speciation from other kinds of speciation, and inflation from other kinds of monetary phenomena, what reason do we have to assume that these categories have the homeostatic mechanisms that Boyd's account requires? Griffiths represents natural kinds as having causal homeostatic clusters and as inductively useful categories. This leads him to attenuate "essence": he argues, for example, that species are both individuals and natural kinds, and that their essences are historical (1997, 212; 2000). But even if species are picked out by a shared history, it is unclear how something's history can be its essence, a notion normally confined to Balkan politics.

Griffiths thinks that this liberalization of the notion of essence is justified by its resonance with Boyd's ideas and by its continuity with earlier theories of natural kinds (1997, 189). The idea that the essence of something can be external to it was never part of the tradition of natural kinds, but it is not ruled out by the idea of causal homeostatic mechanisms to which Boyd appeals. But if extrinsic forces are permitted to hold a kind together, then maybe chartered accountants are a natural kind, since they share properties in virtue of a historical fact—they passed the relevant exams. But now we start to wonder whether anything can be a natural kind. Properties cluster together in inductively and explanatorily useful ways all over the place, both inside and outside science, and it starts to look as though all kinds are natural, on the ecumenical account.

Why should we care, though? Why not break with tradition, do away with essences altogether, and embrace ecumenism? The great insight of ecumenism is the recognition of the many reasons for properties to cluster together, ranging from atomic structure to government declaration. So both refined essentialism and ecumenism are good concepts of kindhood. My taste runs to saying that the former picks out natural kinds and the latter does not, although it does pick out useful kinds. But I do not have a knockdown argument against the other positions. It is hard to see how to resolve the dispute over whether all kinds are natural, or only ones with simple essences are but that there are also other kinds of kinds.

Many criticisms of natural-kind concepts appeal to either the traditional understanding of natural kinds or to the role played in modern science by inductively useful categories of very different sorts. The basic idea is that there are many distinctions in nature and theories use them differently. Everyone, I think, can agree. Whether or not we call some or all of these natural kinds does not seem terribly consequential. What is the significance of this for psychiatry? The debate about whether mental disorders are natural kinds is similarly inconclusive, and it has obscured a more important issue: the availability of categorical classifications.

9.5.6 Does It Matter If Mental Disorders Are Not Natural Kinds?

If we take a refined essentialist view of mental disorders, then there may be some causal mechanisms that work to maintain symptoms in clusters that are largely unaffected by other circumstances. The fundamental explanations I discuss in chapter 4 are instances, and indeed there may be a close affinity between the processes tracked by fundamental explanations and the homeostatic processes envisaged by refined essentialism. Some disorders may take a form in which the final state is homeostatically buffered against normal environmental interference. However, many other mental illnesses will not come out as natural kinds on this view. In general, their forms and courses may be so varied that the exemplar will have to specify each component within a wide range of values, and be only partially applicable to different patient populations. The crucial question is not whether mental illnesses are natural kinds but whether they are, in Meehl's language, "taxonic" (Meehl 1992, 1995, 1999). The issue is whether mental illnesses can be usefully classified, and there are many ways to do that without being interestingly causal, or contributing to causal explanation.

Some mental illnesses may have simple reductive essences (think again of Huntington's), others may have essences if we take a more ecumenical view, and others if we are maximally ecumenical. Indeed, the last is most likely, given that most mental illnesses appear to be influenced by both environmental and intrinsic sources. What matters is whether we can come up with a basis for classification. Exemplars are idealizations, and it is those idealizations that we seek to explain in terms of the causal relations among elements in the world. I look at explanation in earlier chapters, and introduce the philosophical basis of classification in this chapter. Chapter 10 applies these lessons to psychiatric classification and asks what a causal taxonomy of mental illnesses might look like.

Studying archetypes is a fundamental task in nosography. But once the archetype is established the second nosographic operation begins: dissect the archetype and analyze its parts. One must, in other words, learn how to recognize the imperfect cases, the *formes frustes*.
—Jean-Martin Charcot, *Tuesday Lessons* (March 20, 1888)

10.1 Introduction

My discussion of the philosophical debate about taxonomy in the last chapter generated some desiderata that psychiatric classification does not meet. Specifically, contemporary nosology is not causal. In this chapter I bring the morals from these various critiques together and consider a positive agenda rooted in the philosophical idea that scientific classification depends on a theory. Disorders must be classified in clinical/etiological terms, defining diseases in terms of the systems that sustain the symptoms and the causes of the systems' being in that state.

This chapter sketches out one way to classify mental disorders. It does not give a comprehensive classification or reorganization of the contents of DSM-IV-TR or ICD-10, of course. Rather, I argue for a particular approach to nosology, give some examples of how it might work, and suggest how clinicians can exploit it in practice. The chapter also presents further rationales for a causal/explanatory nosology of disease types and a comparison of a nosology built on those lines to other proposals. The account is just a sketch, and years of empirical work are needed to fill in the details; quite possibly in a way that will overturn every specific proposal that I offer.

We can think of a system of classification as having inputs and outputs. The outputs are the generalizations it licenses, and the input is the theorizing about the domain that is being classified. (And, of course, we should imagine a loop back from output to input as inquiry directed by the classification system enables us to learn more about the domain.) Not every classification needs causal/explanatory inputs in order to be satisfactory, but a nosology does. In psychiatry we want to know what is actually wrong with people, and we want to predict what will happen to them and how we can intervene to prevent or mitigate regative symptoms. To do this we need to know the structure of the human mind and what can go wrong with it.

The cognitive sciences, social sciences, and life sciences are finding out what normal functioning is. Without ending its self-imposed exclusion from the wider project, psychiatry cannot carry out its own project of validating the disorders it recognizes, and it cannot carry out the project of establishing a taxonomy based on understanding breakdowns in normal functioning.

Here is one way to look at the outputs of classification (as I do in section 10.5): suppose we are designing a training manual for psychiatric residents, or a reaccreditation manual for practicing psychiatrists. Besides a true view, what would the exemplar approach offer in practice to these people that they do not have already in tacit and harmless theories that are simply laid over the observations and do not play a diagnostic role. (Assuming, that is, that clinicians' tacit theories are not positively harmful, as they might be in some cases. I am asking what the kind of approach that I outline can add to clinical lore, since it is aimed at solving basic research questions.)

In section 10.2 I distinguish my approach from a prototype-based approach. The difference is basically a matter of causal information, and I argue in section 10.3 that the logic of the medical model suggests that a nosology should be a set of causally defined categories. There are also independent reasons for thinking this is desirable in psychiatry. Section 10.4 looks at the question of whether taxonomy should be categorical or dimensional and provide some reasons for favoring the former, but this is an empirical issue that cannot be decided in advance of detailed information. Section 10.5 then discusses the clinical implications of my outlook, arguing that a set of taxa based around exemplars is not itself a clinical taxonomy, but it provides a wealth of information about possible interventions that the clinical perspective can exploit.

10.2 Exemplars, Dimensions, Prototypes

10.2.1 Introduction
In chapter 4 I argue that if psychiatry is to align itself more closely with the sciences, it should not privilege just one sort of explanation. It should adopt the full range of explanatory resources envisaged by the biopsychosocial

model and the more ecumenical versions of the medical model. I suggest that recent developments in the neurosciences were already tending in that direction. The medical model can survive the transformation because its core is a conception of disease as a destructive process, and this perspective on disease can be exploited by inquiries citing a wide variety of explanatory factors. The medical model of disease should be generalized to include etiologies explicable in terms drawn from the cognitive and social sciences. A classification system with those properties is categorical: that is, it treats disorders as discrete phenomena, qualitatively different from normal states in virtue of pathological causal histories. A dimensional system of classification represents disorders as falling between points on an axis, or as a location in multidimensional space, and not as discontinuous categories.

The existing systems, in the DSM-IV-TR and the ICD-10, are also categorical, so in that sense the approach I favor is not terribly revolutionary. As with many other discussions in this book, I take my approach to draw out the implications of the medical model in a more thoroughgoing way than much recent psychiatry seems inclined to do. Instead of construct validity, we should search for causal explanation, and those explanations will serve as the basis of taxonomy. I argue for classification by exemplar, in which a disorder is classified according to both its symptoms and the explanation of those symptoms at a variety of levels of explanation. Anyone who has not believed the discussion so far probably will not be convinced. But I do offer some arguments that especially pertain to classification. I begin by distinguishing an exemplar approach from a prototype approach, then move on to defend the exemplar approach before turning to dimensional classification, which may need to be recognized as part of the overall structure but can be accommodated along categories in an exemplar-driven system.

10.2.2 Exemplars and Prototypes

Exemplars are representations of symptoms and course, so strictly speaking it is not exemplars alone but explained exemplars that should make up a nosology. That is, the core of a nosology is a simple model of a disease process: most of the information belongs in the monograph, rather than in the classification, which involves a conjunction of pathology and etiology. I say more about this in section 10.3.1. My approach classifies representations, and sorts people into classes based on their resemblance to the exemplars. A particular patient may not correspond in every detail to an exemplar, and we may need to treat the exemplar as a theoretical guide to a clinical reality that varies in several respects (this is a modern version of Charcot's

[1987] distinction between *archetypes* and *formes frustes*). All exemplars share a property of simple neurological exemplars mentioned in section 6.2.2. The prototypical neurological exemplar, which I difine as one symptom plus one lesion, can be conceptualized as the profile of an archetypal patient with that symptom. But only in that one respect: the archetypal patient will suffer from several problems, but the patient's other psychopathology is ignored for purposes of classification, since we classify not people, but conditions. In some cases the fit between taxon and patient may be exact. There will be cases in which all patients share the same fine-grained disorder and have no other symptoms. That typically means that they share a specific deficit and a common causal pathway that accounts for it.

This is a clinical problem, and I discuss it more fully at the end of the chapter when I relate my approach to clinical concerns. Here, I note that we can approach this issue by bringing out the differences between a nosology organized around explained exemplars and one that is organized around prototypes. A patient can receive a diagnosis based on exhibiting only some features of an exemplar, even if the other features are absent. And a patient might receive several diagnoses, because he or she resembles several exemplars. It is common for people to show more than one set of symptoms, especially since being mentally ill often causes depression, so mood disorders are a common accessory to other mental illnesses. Causal information should remove a good deal of the uncertainty and heterogeneity surrounding the relations of people to disorders in the current system, but some people are bound to fit poorly, or to fit more than one exemplar, but none very well. We should expect some people to be hard to sort out.

This poor fit between taxon and patient is a recurrent complaint about the current system. This has often been dealt with by the introduction of intermediate categories. Schizoaffective Disorder (the concurrent presence of depressive or manic symptoms and symptoms of schizophrenia) is a clear example. Schizoaffective Disorder shows us that as well as apparent continua between normal and abnormal psychologies, there are also psychopathologies that shade into one another. Kraepelin's distinction between *dementia praecox* (schizophrenia) and manic-depression has been acknowledged by every taxonomy since. But as I note in chapter 6, when echoing Tuke's complaint (in Radden 1996), it is hard to say whether some people are melancholic maniacs or maniacal melancholics. As Radden notes, the existence of schizophrenia, affective disorders, and schizoaffective disorders is not a problem if we assume that cognition and affect can be diseased together as well as separately. But the greater problem comes if we believe that the schizophrenic spectrum shades into the spectrum of the mood disorders.

Many theorists argue that categorical classifications, which seek to group patients into distinct populations in virtue of some shared set of properties that demarcate them from other populations, tend to impose arbitrary distinctions along a spectrum where boundaries do not exist (Nathan and Langenbucher 1999; Widiger and Sankis 2000). Radden's point can be accommodated by looking for different causal stories that separate melancholic maniacs and maniacal melancholics. Such stories do not have to rely on a sharp separation between the reasoning faculty and the emotions. They just require some way to distinguish populations based on different causal histories.

A dimensional classification would not separate them but see them as occupying different spaces along various axes. Height and intelligence are the familiar examples. Dimensional classification, then, is advertised as superior to categorical classification because some (or all) sorts of psychopathology shade into each other, or lie along dimensions that span abnormality and normality. The personality disorders, for example, are usually thought of as extreme forms of normal traits.

I take up the issue of dimensional and categorical approaches in section 10.4. I argue that the arguments used to support dimensional classification as the default strategy are defective, and hence that although the issue is an empirical one, there remain good reasons for expecting the basic model of disease in medicine to apply to psychiatry, for the reasons I outline in section 10.3. That is a categorical model. At the moment, I want to argue that the case for dimensional classification across the board may rest on difficulties that reflect the current system, not ones that are intrinsic to categorical schemes. The current system is a categorical one, which defines conditions as syndromes based on clinical presentation rather than causal histories. It has many of the traits of a prototype-based scheme, although it is not worked out in statistical detail. A classification based on exemplar has some of the same properties, but supplements prototypical properties with causal information. The addition of this causal information makes all the difference.

The existing system, enshrined in the DSM-IV-TR, has many of the features of a prototype system, although it is never quite presented as such (Cantor and Genero 1986). Many DSM categories have fuzzy boundaries, being large sets of correlated properties. Prototype-based classifications appear to align well with the way human brains actually classify natural phenomena.[1] The basic idea is a simple one: we define a syndrome in terms of

1. Cantor and Genero 1986 is the most sophisticated defense in the clinical literature of a prototype-based nosology. They appeal to the extensive literature on concepts and classification within psychology.

core features, and what unites the patients who share a diagnosis is that they share enough features of the prototype. Apples and pears are prototypical fruits, for example, and dogs are prototypical animals, but tomatoes may not be a good match for fruit, and grasshoppers are not a good fit with animalhood. People will take longer to recognize the outliers as members of the category depending on their distance from the prototype. This threatens to make classification arbitrary and subjective unless it is supplemented by proposals about weighting features of the prototype and measuring degrees of divergence from them. The DSM-IV-TR does this informally by distinguishing essential features and nonessential ones for most conditions, and weighing all members of the second set equally. Or, as in the case of Major Depression, it presents a set of symptoms with the more prototypical ones listed first, even though the diagnosis can be made using only features that are further down the list.

A procedure like this does not enable us to see where to draw the line separating normal from clinical populations for a given trait unless we know what values of the relevant variable mark the difference. For example, we may know that complete muteness in a depressive is diagnostic of a psychotic depression, but not know where to draw the line separating depressives who are unresponsive yet not fully mute from nondepressed subjects or other kinds of depressives, who just do not feel like talking.

Now, detailed information that answers questions of this sort does not need to be built in to a classification system. Rather, it is part of the theory that explains how the system is to be handled. Or, as Hull (1970) puts it, the information is in the monograph, not the system. Most symptoms mentioned in the DSM-IV-TR can be detected by a standardized test, although very often clinicians employ an unstructured interview instead.

The worry, though, is justifying the basis on which these decisions are made. What rationale supports picking a measure, say a score above twenty on the psychopathy check list, as the gold standard for detecting the presence of psychopathology? The rationale is construct validity, which, as I indicate in chapter 6, may obscure differences in underlying pathology within a syndrome, but is not at all unreasonable for certain purposes, such as when the causes of a syndrome are unknown or not relevant. However, I also say in chapter 6 that ideally what we want is to take syndromes defined psychometrically and validate them by making their causal histories perspicuous.

Many of the failures of fit between conditions and patients, as well as much disagreement over diagnosis, are the result not of failure to meet a core symptom, but of a lack of clarity about the boundaries of the syndrome beyond that core symptom. My hope is that a combination of causal and statistical reasoning can go some way toward settling where the boundaries are.

10.2.3 The Trouble with Prototypes

A prototype of a mental disorder includes a number of characteristics of a good example of the category and adds to these core features a penumbra of less typical features. "Typical" here means that it satisfies the expectations of psychiatrists. The trouble is that this may not conform to how the world is, and to the extent that psychiatrists' expectations do match the world, it is because they map on to the causal structure of psychopathology. For example, it may be that sleep disorders and loss of appetite seem to psychiatrists to be at the depressive core, since they are more likely than the peripheral features to be produced by the same processes that cause lowered mood to begin with. There is some evidence that a prototype approach captures the way psychiatrists reason (Cantor and Genero 1986, 244). This is not surprising if our natural classificatory tendencies are guided by mentally represented prototypes; that is how people do it, and psychiatrists are people too.

The analogy with natural categorization (which may well be a matter of matching to prototypes) is suggestive, and it might make diagnosis easier for clinicians if nosology were reformed in an explicitly prototypical manner. But it is nonetheless the case that our natural categorization processes do not map onto the underlying structure of nature very well at all, and that what we really want is to get things right, not to make life easy. There is no reason to suppose that a prototypical nosology will work unless the construction of the prototype is guided not by the untutored judgments of the classifiers (which is how cognitive psychologists determine how we represent prototypes in our head) but by what the world is like. And if we are going to classify by what the world is like we might as well go the whole hog and forget about prototypes.

Exemplars work by including all the symptoms that can arise from a particular set of causal processes. Like prototypes, they are idealized representations, but unlike prototypes any methodological stress on some symptoms is a matter of causal priority. Similarly for diagnosis: including all the symptoms allows us to match patients to exemplars by censoring the model so as to capture only the limited set of its constituent parts that apply. By articulating causal explanations as variously as we need to, the hope is that we will eventually be able discriminate more finely among conditions that are currently lumped together by DSM's fuzzy categories, by distinguishing syndromes on the basis of etiology and pathology.

10.3 Exemplar Nosology

Categories like exemplars that organize symptoms causally are a better bet than prototypical categories, but how should the nosology be structured? I begin with exemplars and then discuss a complication, which is that some conditions should perhaps be classified dimensionally, rather than categorically. I set aside some issues that the prospect of classifying by models of exemplars raises. For example, there are issues about how to represent taxonomic information in a way that secures the best mix of perspicuousness and ease of use. Essen-Möller argued in a series of papers (e.g., 1963, 1971) that a psychiatric taxonomy needs to list both the clinical presentations and causal histories of disorders as independent axes, because we usually want to generalize across both sets of phenomena in psychiatry. Unless the system had two independent dimensions, one listing pathologies and one etiologies, he argues, the retrieval of the relevant information would become too complex. I will put these issues aside to focus on more basic principles, I starting with a simple analogy between physical and mental diseases designed to show that the logic of the medical model suggests that classification should be done by a mixture of etiology and pathology.

10.3.1 Classifying Diseases

My picture of psychiatry insists that mental illnesses are diseases, and so the simplest way to get a handle on classifying mental illnesses is by analogy with physical disease. And the simplest way to understand the taxa of medicine is as conjunctions of etiology and pathology (Meehl 1973, 285–289). The picture is one in which an etiology causes a pathology that causes the clinical syndrome (although "cause" here will have to be understood in terms of realization in some cases). A disease entity is distinct if it varies from another with respect to either pathology or etiology, although in fact the convergence of two different etiological processes on exactly the same pathology seem unheard of in general medicine, since a pathology is state of a physical system and for two such states to be literally the same they would have to be indistinguishable at the molecular level. But different etiologies will have different physical effects at some level—such as the molecular—even if they cannot be distinguished easily at a systems level. A different pathology, remember, is not a different set of symptoms, since the symptoms are manifestations of the pathology. Tokens of the same symptom may well be caused by different pathologies. The case of one etiological agent having different pathological consequences, depending on the overall state of the system, is not at all uncommon. The same infectious agent, for instance, may affect many different organs. Ultimately the key commitment of modern medicine is that all disease has a physical site.

The model of disease as pathology and etiology translates to psychiatry quite straightforwardly. The main difference is that the presence of intentional phenomena in the mind produces extra levels of explanation. Still, there seems no difference in principle between thinking of pathology not just in terms of properties of tissue, like lesions or disrupted synaptic connections, but also in terms of properties of cognitive systems. So "site" may have to be interpreted quite loosely as a component of cognitive architecture. The hope, as I have said, is to understand the computational properties of cognitive parts, which are brain areas with a particular job to do, although these brain areas may cross-classify standard taxonomies of brain regions. The physical realizations of the "what" and "where" streams, for instance, count as physical parts, although they are each distributed across several anatomically distinct regions.

As with pathology, nosology needs a conception of etiology that covers phenomena at a number of levels of explanation. As well as interactions between genes and biological agents like germs, for example, we may want to recognize a wide variety of factors, including disruptions to the normal functioning of cognitive systems, psychodynamic factors, or even marital difficulties. A huge number of variables have been shown to influence depression, for example.

When discussing schizophrenia in chapter 6 I sketch a simple explanatory structure in which a combination of causes, none of which are individually necessary or sufficient, combine in various ways to produce an end state in which neural disconnection affects a number of cognitive capacities. The causal interactions in that story represent the various particular ways in which the developmental story can be told, and it is the developmental story that gives the etiology of schizophrenia, and the end state represents the pathology, which can be reached in various ways by different combinations. Schizophrenia, then, is a neurodevelopmental disorder of cognition, whereas Alzheimer's is a degenerative disorder, since even though it may have similar cognitive properties to schizophrenia, it represents a degeneration from normal function, whereas in Schizophrenia normal function is never reached. So the two conditions illustrate distinct etiologies for dementia. Korsakoff's Syndrome illustrates a third etiology, in which a vitamin deficiency (usually developed by alcoholics who neglect nutrition) causes lesions of the mamillary bodies and the medial dorsal nucleus, two components of the emotional circuit postulated by Papez in the 1930s. The result is severe memory impairment.

The distinctions between etiology and pathology in general medicine can be a bit strained, and in psychiatry they might be seem more so. Some physical processes or agents, such as toxins, are natural fits for etiology rather than pathology in biological medicine, whereas in psychiatry the discovery that some variable is involved in a disorder may leave us unilluminated about the sort of role it plays without further scrutiny. Indeed, for a cognitive variable such as a negative, self-accusatory attributional style, it is not clear prima facie whether it should be thought of as pathology, etiology, or symptom.

In this situation the only recourse is to look. We discover, for instance (Wilson et al. 1996, 214) that the "depressive attributional style" often does not show up in individuals before they become depressed, varies over time, leaves when the depression lifts, and does not predict relapse, all of which suggest that it is a symptom, rather than a cause, of depression.

Perhaps an example will be helpful at this point. One of the disorders featured at the beginning of the book, OCD, shows how etiology can separate disorders.

10.3.2 Obsessive-Compulsive Disorder

Infection is one of the causes of mental retardation, and the paradigmatic nineteenth-century mental disorder, Neurosyphilis, was an infectious disease. More recent research has suggested that childhood Obsessive-Compulsive Disorder (OCD) also may be the result of infection. For my purposes, the details, despite their intrinsic interest, are of less importance than the consequences of this discovery, which has led to diagnostic reform along just the lines one would expect to follow from a treatment of classification as a conjunction of pathology and etiology.

The background is that resemblances were noted between Obsessive-Compulsive Disorder and Sydenham's Chorea, which is known to be an unusual complication of an untreated streptococcal infection.[2] In Sydenham's Chorea, strep antibodies attack the basal ganglia, with a number of results including the jerky flapping of the arms that once earned Sydenham's Chorea the name "St. Vitus's dance." Of particular interest to OCD researchers,

---

2. Sydenham's Chorea is not a DSM-IV-TR diagnosis, but it does appear among the neurological disorders in ICD-10.

however, was the finding that obsessive thoughts and compulsive behavior also feature in Sydenham's Chorea. Scrutiny of children with OCD led to the discovery that they too were suffering from symptoms caused by strep infections.

The result was that a new disorder was recognized: Pediatric Autoimmune Neuropsychiatric Disorder Associated with Streptococcus, or PANDAS (Swedo et al. 1998).[3] This reform was not made on the basis of any differences in symptoms between children with OCD and adults. But OCD is remarkably similar across all ages (and cultures, too). The only difference traditionally recognized between adult and childhood OCD is the degree of insight the patient has into his or her own symptoms: adults usually are aware that their rituals and obsessions are bizarre, whereas children may not be. This is readily explained by children's lesser socialization, which means that the nature of OCD is more or less a constant across ages and cultures. It is this feature of the disease that makes it one of the best candidates for a natural kind among the DSM diagnoses. The cognitive basis of OCD is poorly understood but widely believed to be very similar across all patients; it is possible, however, that PANDAS has a cognitive basis which differs from that of other OCD cases. Nevertheless, despite the robust co-occurrence of symptoms, the particular etiology of PANDAS has led to its classification as a separate disorder. This is evidence of the appeal, on theoretical and clinical grounds, of the two-dimensional picture that conjoins pathology and etiology. Strep infection is the etiological agent in both cases, but Sydenham's Chorea and PANDAS differ with respect to some key aspects of the pathology.

I do not want to explore any more details about physical etiologies, except to note that they are very diverse. The brain may disturbed in many ways, and these can cause psychological dysfunction in adults as well as in children. Psychological disturbances may result from damage to the brain caused by strokes or injuries or by other kinds of physical trauma. Apperceptive agnosia, for instance, often has been noted in subjects who have experienced carbon-monoxide poisoning (Farah 1990). Various physical disorders also can cause fully developed minds to malfunction. Metabolic disorders that interfere with the synthesis of neurotransmitters are an obvious example. Autoimmune responses of the sort believed to be responsible for Multiple Sclerosis can lead to the demyelination of nerve tissue, which slows the transmission of impulses. Late-onset genetic disorders like Huntington's are

---

3. More colloquially, Pediatric Autoimmune Disorder after Strep.

another example. Rather than running through all these possibilities at length, I simply note that a two-dimensional picture may need to advert to any of them—and indeed, more than one, since many disorders, likely including schizophrenia, stem from multiple causes that exercise a common influence over the sustaining pathology.

Corresponding to the three levels of symptom, pathology, and etiology are distinct sorts of inquiries that feed into one another. The result is the construction of a taxon.

10.3.3 Building a Taxon

So far I mostly have looked at conditions that have been around for a while, but PANDAS, as a subtype of OCD, is new. The production of a new taxon involvers certain epistemic moves. The opening stages often are fortuitous. The initial attention paid to strep as a possible causal factor in childhood OCD was motivated by similarities with Sydenham's Chorea, but the demarcation of Panic Disorder from Anxiety Neurosis was triggered by an entirely unexpected effect of imipramine (McNally 1994). There are more principled ways to look for what Meehl (1992, 1999) calls "latent taxonicity," or a hidden causal structure below a set of symptoms. Statistical methods can be used to go beyond merely identifying a cluster of symptoms, to detect an underlying causal structure based on relationships between variables. Meehl's preferred method was "cut kinetics," which is based on the idea that if subjects were accurately classified in groups by means of one variable, then there would be another variable that also would distinguish them (1999, 169–170). He assumes that if several such tests are mutually supporting they show the existence of a genuine causal structure. Recently, there has been an explosion of interest in Bayes nets, which can discern causal information from probabilistic relations (Glymour 2001, Pearl 2000, Woodward 2002). The initial choice of symptoms to subject to statistical treatment is guided by methodological considerations that may come from anywhere, but these assumptions can be tested statistically.

Yet, despite Meehl's view, classification in psychopathology is not *just* "a problem in applied mathematics" (1995, 266). It is also a biological problem. Once the causal relations among symptoms, or among symptoms and other variables of interest, have been revealed, it is necessary to explain the actual realizations of those relations. This involves discerning the physical basis of the symptoms and understanding their causal connections in physical detail. Only when that is done do we have a model of the system that can be used for classification.

10.3.4 Some Results

I do not presume to know which disorders eventually will be included in a causal-pathological taxonomy and which will not. But even if the existing set of disorders survive, we will need some revision, in the sense that disorders may be grouped together in unexpected ways. As things stand, we classify by syndrome rather than pathology, and reinterpreting conditions so as to strictly reflect the conjunction of pathology and etiology might change the picture quite dramatically. For example, if we classify in terms of pathology and etiology, instead of phobias and panic disorders we might talk in terms of different disorders of the evolved-threat response system. Another rearrangement, if we bought into a different explanation, could be to group Panic Disorder together with whatever delusions could be explained via a bias theory (at the level of intentional rather than mechanistic explanations). I state my skepticism, in chapter 5, about whether bias theories could work, but the point I want to draw attention to now is that panic, on Clark's cognitive model, shares crucial features with bias models of delusions, in that both posit an interaction between an experience and a cognitive response to that experience. Another way of approaching classification would be to put panic in with Schizophrenia if Schizophrenia receives the explanation that Frith envisages, since both conditions share a failure of self-monitoring: thought and action are inadequately policed in Schizophrenia, while in Panic Disorder it is the monitoring of the autonomic nervous system that is at fault.

So, there are the basics of the picture of taxonomy I recommend, and which I think captures important aspects of the way the medical model is developing. Despite the strictures of the DSM, a huge volume of work is being done on the causal bases of mental illness, and if that is joined together with the cognitive neurosciences and developed according to the medical conception of disease that informs the research, we can expect something like this picture.

However, many psychiatrists argue that the categorical basis I have assumed will win out is in fact wrongheaded. Some of the arguments in favor of the rival, dimensional approach to classification are not satisfactory, but the issue is an empirical one, and there are precedents from medicine for disorders being conceived of dimensionally.

10.4 Dimensional and Categorical Classification

Dimensional conceptions have been defended for many psychiatric diagnoses. Widiger and Sankis (2000) note that personality disorders often are viewed as extreme forms of normal personality types and argue that

356 | Chapter 10

dimensional classifications may be widely applicable. They defend their claim by stressing the likelihood that most mental disorders result from complicated genetic factors and are not simple Mendelian conditions like Huntington's. We have seen that this is true, and in fact their argument would be strengthened if they noted that most mental disorders are the product of both environmental and genetic factors, not just interacting genes. A combination of environmental and genetic factors can indeed result in continuous quantities, and I do not want to impose a general constraint on classification in advance of empirical inquiry. Forcing continuous phenomena into a categorical framework can cause all sorts of problems, and nobody should try to lay down the law given the present state of psychiatry. Science must discover the phenomena and respond to the facts by building the most appropriate classification/representation scheme.

However, an exemplar-based scheme of classification and explanation can accommodate dimensional conditions. And some arguments in favor of dimensional approaches appear to assume that categories are an appropriate basis for classification only where a disorder depends on a specific etiology (which, in my terms, is akin to insisting that categories depend for their existence on fundamental causal factors). This is unmotivated and leaves the burden of proof where I have placed it.

10.4.1 Exemplars Can Accommodate Categories and Dimensions

In fact, any nosology could in principle incorporate both categories and dimensions (Kendell 1975, 130), which is just as well since there may be both categories and dimensions in psychopathology. But whether the right taxa are categorical or dimensional, it is still necessary to represent the symptoms of mental illnesses and model the processes that give rise to them. An exemplar represents the symptoms and course. We then have to attach a model to the exemplar, but the model can vary. Many aspects of neurotic illnesses, for example, long have been regarded as extreme forms of symptoms that are exhibited by most of us at one time or another, and the personality disorders can be viewed similarly. Many theorists have seen concepts of neurotic disorder as representing idealized personality types that are, at bottom, extreme forms of some quality that is present or absent to a greater degree in the normal population (Kendell 1975, 134; Widiger and Sankis 2000). I claim that exemplars are idealizations, and if the correct explanation of the idealization situates at one end of continuum of normal function, that is not a problem for the overall exemplar view.

Indeed, it may be that the question of categorical versus dimensional classification is a matter of convenience, since one may be able to define a taxon in terms of measurements along various dimensions or in categorical terms.

However, a persistent tendency in psychiatry (and personality psychology, too) has been to assume that categorical classification only makes sense if we can find specific etiologies. The reasoning appears to be that since a category is either present or absent, there must be some causal factor to account for it that is either present or absent. This reasoning often leads scholars to tie the prospects for a categorical taxonomy to our prospects of uncovering one gene, or a small set of genes, that explain the presence of a disorder. If, as seems likely for most diagnoses, we can find no such gene, theorists who reason along these lines conclude that the categorical approach to classification is unlikely to work. Widiger and Sankis (2000) argue quite explicitly that dimensional classification is more appropriate because complex psychopathologies are unlikely to have specific etiologies, and Kendell (1975, 123–124) insists that the abandonment of the search for a unique cause for each disease means the end of the concept of diseases as distinct entities and its replacement by a loose sense of a disease as a convenient cluster of observable effects at different levels.

But the conclusion that giving up on specific etiology means abandoning the idea of diseases as specific afflictions of physical systems is doubly unwarranted: first, it may be a problem only if we retain the current syndromes as taxa. Existing syndromes may well be the product of several different etiologies. Second, and more important, the conclusion just does not follow. Even where there are no specific etiologies (and I agree that usually there will not be), it does not follow that categorical classification is indefensible. Nor does it follow that where it is appropriate it is not an attempt to capture part of the causal structure of nature but merely a way to mark out some related observations of interest. The idea that a disease is a distinct destructive process afflicting a system is entirely consistent with its having been caused by numerous interacting factors.

A specific etiology, remember, is something like a major gene, a stroke, or some other precipitating event that permits the construction of a fundamental explanation. It ensures the presence of a syndrome if it is present. But it is not part of the categorical view that breakdowns in mental structures, if they exist, are under the control of single genes. From the absence of a specific etiology it simply does not follow that we can tell whether a mental disorder is a joint in nature or an extreme variant of normal function.

10.4.2 Categories and Causes

I claim that a classification system built around explained exemplars can accommodate dimensional taxa if we want to recognize them, and that whether we do is not decidable before we have the facts. But the expectation that dimensional taxa will be the norm does not follow from the unavailability of specific etiologies or fundamental explanations in psychiatry. The justification of this claim starts by looking at genes, since the argument often is defended by appeal to a shortage of single-major-locus conditions.

A straightforward Mendelizing condition, or a condition close enough that we might as well think of it as Mendelizing, is one in which the presence of one mutation permits us to develop a fundamental explanation. The argument that categorical classification depends on specific etiology assumes that such an explanation must apply in every case in order to give rise to natural categories. However, suppose a condition is under the control of many genes. It still may exhibit a statistical character that entitles us to call it a genuine break in nature. Suppose schizophrenia depends on the interaction of many different genes, but the system "has a quasi-step function, where in a certain narrow region there is a rise in probability of illness from zero to some substantial probability" (Meehl 1999, 168). It is not merely a matter of taste to say that in this case schizophrenia is distinguished from normality by a genuine break in nature, especially since exactly the same reasoning is widespread in medicine. Elsewhere, Meehl (1992, 127) uses the example of scurvy, which has zero probability when levels of ascorbic acid are at or above a normal level, but always develops when the level dips below the crucial threshold of deficiency.

We have no trouble treating scurvy as a distinct category. And we point to an ascorbic acid deficiency as its cause, even though it is not simply present or absent in the way a mutation is. So there is in principle no problem with applying the medical model of disease when we have a surface discontinuity that is stepwise dependent on a causal factor, as in scurvy, nor in applying it when the presence or absence of symptoms depends on many underlying discontinuous properties, like whether or not a harmful form of a gene is present. Typically, in mental illness we should expect to see both these causal relations put together. There will be many causes, and a combination of some pushes the system into the state that we think of as pathological. And note that the test of whether the system is in that state is often (though not always) whether it exhibits, as a proximal cause of the symptoms, a detectable underlying pathology, which might be a state of anything from brain tissue to self-defeating thoughts.

The same arguments apply to any set of causal relations. A combination of genes, developmental accidents, and cognitive properties may mean that if Jules and Jim begin drinking heavily together after work every day, Jules will become dependent on alcohol and Jim will not. If so, there is some set of properties that Jules exhibits that demarcates him from Jim at the proximal level of underlying pathology even though there may be a period of time during which they are behaviorally indistinguishable. These may be psychological properties such as cravings, or some subtle changes in the brain.

Categorical classification is possible in the absence of specific etiologies, or even necessary ones. Although there may be a specific set of causes in Jules's case that explain why Jules becomes an alcoholic, these causes may not be the same for all alcoholics. If they are not, then the relevant etiological factor that determines the shape of the alcoholism exemplar will be a model that specifies various pathways, each reachable through the interactions of numerous different physical, psychological, and environmental variables, which converge on a proximal cause of alcohol dependence. That proximal cause itself may not be a generalization that specifies even a roughly lawlike relation between some neurological or psychological abnormality and the symptoms of alcoholism. It may be a set of generalizations about brain systems or cognitive properties of the alcoholic.

So, a diverse set of distal causal interactions could be different ways to get to upset proximal mechanisms that sustain the symptoms of a disorder. Even with this much chaos, there is no reason to hold that categorical classification is inapplicable if the set of proximal causes are states of the system that look abnormal (on a theory of what the system normally looks like) or that raise the probability of symptoms (where those symptoms are values of some variables that we have decided on theoretical grounds are the bad values for those variables to have).

In the situation just described, there are different stories about proximal mechanisms and different stories about distal histories, but no clear relations between them. I have not presented the story in a way that relates one possible history to one possible final state; instead, the relations are noisy in both directions. The project of subtyping a disorder assumes that there are stories to tell that take one way for the developmental or other distal facts (if there are some) to turn out, and one way for the proximal facts to turn out, and a distinctive relation between them that explains a distinctive form of the symptoms. If those conditions hold, we have a specific diagnosis within the more general one, with an associated model that represents one pathway through the overall model of the exemplar. It may be that we can do this for only one symptom, if the symptom (say, delusions of bodily control in schizophrenia) depends on a failure in a specific system. Or it may be that we can do if for a set of symptoms that together make up a recognizable form of the superordinate exemplar.

Furthermore, two conditions with similar causes might nonetheless differ with respect to some factor that occurs only in one and is therefore able to ground a distinction even if it explains nothing: neuroticism often is regarded as the quintessential dimensional trait, and it is implicated in both mood and anxiety disorders, which often occur together. Nonetheless, anxiety does and depression does not come with a distinctive set of autonomic symptoms like sweating and choking. The presence of these distinguishes the two conditions even though they are specific effects, rather than specific causes.

So far, I have defend the idea that categorical, rather than dimensional, classification, is possible in principle even in the absence of specific etiologies. I have claimed that a surface discontinuity might depend on a discontinuity in some underlying system, or it might not. But what if we have a continuity both on the surface and in values of the underlying variable? Suppose that changes in the value of some variable was manifested behaviorally not stepwise but smoothly, so that there was no obvious cut off either way. But suppose that we nonetheless felt that some values of the behavioral variable were sufficiently deviant to merit a diagnosis. Then we would have a dimensional disorder, since now the clinically significant values would shade into the normal ones without any break. Notice that the debate usually concerns whether there is a break between normal function and abnormal functioning (categorical), or whether normal and abnormal individuals exist along a shared dimension. In this sense, the disease model is presumptively categorical, but it admits of exceptions like hypertension. There is another way to address the topic by considering all of the cases mentioned in chapter 6, in which the same syndrome contained different symptoms. Do we know enough about these disorders to say that, if they were relevantly discontinuous, they are categorically distinct disorders or points along a dimension, even if the dimension as a whole is one of abnormality, with no shading into normality?

In the latter case, where we ask about a categorical abnormality that has a dimensional structure, the result will hinge on whether there are ways to subtype the disorder etiologically. I am not willing to say that we know enough about many disorders to be confident that, if they are relevantly discontinuous, there are distinct disorders.

10.4.3 Dimensions and Causes

The classic example of a dimensionally conceived medical condition is hypertension, which is the end of a continuous biological distribution caused by a large number of interacting genetic and environmental factors. No one keels over when blood pressure reaches the value at which high blood pressure is diagnosed, nor does the system suddenly shift states. It just creates greater statistical risk.

Mental retardation is a classic dimensional mental illness. But this dimensional classic can be subtyped more finely by the introduction of etiology. In fact, we can distinguish a dimensional diagnosis made on the basis of cuts along the relevant output variable (IQ) and a series of categorical distinctions among forms of mental retardation that are based on distinct etiologies and associated extra psychopathology, as well as low IQ.

Mental retardation is distinguished from dementia by age of onset: retardation is intellectual impairment that occurs before adulthood, whereas dementia is an affliction of maturity. In addition to the point about age of onset, DSM IV-TR (41) contains two criteria for mental retardation:

significantly subaverage general intellectual functioning (Criterion A) that is accompanied by significant limitations in adaptive functioning in at least two of the following skill areas: communication, self-care, home living, social/interpersonal skills, use of community resources, self-direction, functional academic skills, work, leisure, health, and safety (Criterion B).

It is hard to think of an area of waking life that does not fall somewhere under B, Criterion A is vague as stated. It is made more precise by statistical means. The DSM-IV followed the orthodoxy in distinguishing four grades of severity; mild (IQ 50–70), moderate (IQ 35–50), severe (IQ 20–35) and profound (IQ below 20). An IQ of 70 is conventionally chosen as the cutoff point for retardation because it is two standard deviations below the mean. The DSM-IV-TR (41) notes a number of biological factors that can predispose a child to retardation, and says that retardation "has many different etiologies and may be seen as a final common pathway of various pathological processes that affect the functioning of the central nervous system."

So why not exploit these etiological differences for a more precise diagnosis? For the sake of argument, assume that psychology advances to the stage at which "subaverage intellectual functioning" can be understood in terms of cognitive dysfunction. At present, the idea of g, or general intelligence, is ambiguous between a structural reading and a reading in terms of resources. A structural reading claims that there is some central system in the mind that controls cognitive performance and that g varies depending on the level of central functioning. A resource-based reading locates g in something like processing speed, which could be higher or lower across a whole array of different structures, and hence g might exist even in a purely modular mind. Testing for general intelligence using a battery of tests designed to tap into different capacities does not discriminate between these two readings, so we have to imagine some future psychology that can tell us whether a particular system is faulty or if central cognitive systems are globally impaired. Perhaps IQ declines as more and more systems develop faults, or progressively simpler cognitive tasks simply become too difficult with widespread processing problems. Whatever the details (and, of course, they remain entirely speculative), assume that we can understand the cognitive basis of retardation well enough to explain currently recognized symptoms of mental retardation in cognitive terms. The question is, would that suffice for an adequate taxonomy?

The current taxonomy that subtypes retardation in terms of the symptoms is a one-dimensional classification. On the DSM approach, everybody who suffers from mental retardation is classified together on the basis of cognitive dysfunction and its consequent problems in living, even though it is noted that the diagnostic features can arise in many different ways. This is a contrast to the picture presented when etiology is added.

Gelder, Gath, and Mayou (1989, 838–840) survey some of the social and environmental factors that can contribute to low IQ, and then go on to list twenty-one distinct syndromes according to a combination of symptoms and etiology. Some symptoms, such as epilepsy and "failure to thrive," crop up several times; others are distinctive features of a single diagnosis, such as the cat-like cry that characterizes sufferers from *cri du chat*, a form of retardation caused by a deletion on chromosome 5. The etiologies they survey fall into five distinct groups (plus a "miscellaneous" section): chromosomal abnormalities, inborn errors of metabolism, autosomal dominance or recession, infection, and cranial malformation. Each of these is itself causally various, but the classification provides more information than the DSM provides, and it admits of further etiological specification if necessary.

Instead of just opting for a one-dimensional picture and noting that central impairments have diverse physical bases, suppose that we can take a cognitive explanation of retardation in terms of systemic impairment and those five physical histories. That would give us a two-dimensional classification that cites both characteristic symptoms, more or less present, and physical and cognitive explanations for the presence or absence of particular symptoms. There is much more information included in the two-dimensional account, and it allows us to make a more precise diagnosis instead of the general one of "mental retardation," even among patients who *ex hypothesi* share a general cognitive dysfunction. We can classify patients according to what actually caused the cognitive deficit rather than by just the deficit and the resulting symptoms. This is particularly important in dealing with moderate retardation, which can have either of two very different histories.

IQs below 50 are very uncommon (about 0.3% of the population) but they are six times more frequent than would be expected on the basis of a perfectly normal distribution of intelligence, whereas moderate retardation occurs only slightly more than would be expected statistically. This asymmetry in the IQ curve indicates a bimodal distribution—the second, small bump is made of people with neurological defects. The moderately retarded, with IQs above 50, may be either at the upper end of this small mode, or at the lower end of the normal distribution. In the former case, retardation results from an affliction that one might find in Gelder, Gath, and Mayou's list of twenty-one. In the latter case, there is no distinct disease at all, just medically normal people of unusually low intelligence. If we continue to imagine that the cognitive deficits underlying retardation are understood, we can translate this distinction among patients as follows: we have, on the one hand, a group of people whose cognitive processes are at the tail end of the normal distribution. Perhaps their cognitive architecture does not work as well as that of their luckier peers because of accidents of education, nutrition, and so on. These people are like individuals who are unusually short because of contingencies of early development. In this case, we would have to acknowledge the existence of a dimensional disorder—one way of explaining disorder, then, must be via the citing of a normal distribution.

On the other hand, we have the people who fall into the second population, whose different conditions account for the greater-than-expected number of people with severe retardation. These unfortunates have something qualitatively abnormal wrong with them. Their cognitive architecture could not have developed to a point where it functioned normally. They are like people who are unusually short because they suffer from diseases ensuring that they would be short in almost any environment we can envisage. (Most of the actors who played the Munchkins in *The Wizard of Oz* had such a disease, a genetic problem called achondroplasia. They were not genetically normal people who happened to be short. They were short because of a defective allele.)

So there is a distinction between people who have low IQs because they have low-functioning but normal minds, and those whose retardation stems from physical problems that prevented normal development. This distinction seems like an important one, but we cannot make it using DSM criteria. Again, the heterogeneity of DSM categories stands in the way of what we need from classification. In this case, just differentiating what underlies the symptoms will not suffice—we may need to differentiate what underlies the cognitive problems, too. The DSM approach, does not prohibit searching for the causal structures that underpin differential diagnosis but classification in DSM-IV-TR, unlike that of Gelder, Gath, and Mayau conveys less information that it ought to.

An objection is that the difference may not be something we can ground with a psychological theory in the long run. I imagine discovering a difference between "low-functioning normal" and "developmentally abnormal" that could be captured in terms of a theory of cognitive function. Instead, the category of low-functioning normal might contain people with a disorder of many small causes rather than the one big cause that affects the populations that meet a specific diagnosis. But since everyone's IQ score is the product of myriad interactions between genetic and environmental factors, everybody fits into the category of those whose intelligence is affected by many small causes. So even without a grounding in psychological theory, we still are not in a position to sort out the low scores from the normal ones by assigning them to different categories. If we decide to count mental retardation as a disorder, which is very reasonable given its implications for all manner of activities, we will in some cases cite explanatory mechanisms similar to the PKU genotype or some other distinct cause that marks a categorical condition. However, the residual cases will need to be understood dimensionally, as the lower end of intellectual functioning.

This matters for the debate between categorical and dimensional approaches in the following ways. We have seen that the existence of an axis of functioning that shades into normality is not a barrier to making discriminations among subjects along that axis. On the picture I recommend, we recognize several exemplars of mental retardation. Some (such as *cri du chat*) would be identifiable by a peculiar symptomatology, others might have symptoms in common, but in every case we can imagine the exemplar as a model with room for specifications of symptoms and a variety of possible explanatory structures. One of the exemplars easily could be the exemplar of a dimensionally conceived disorder. In principle, an exemplary set of symptoms, defined as output variables, could depend on causal input variables in a way that does not permit us to distinguish input or output discontinuities, and force us to recognize a dimensional taxon. It is the smooth rather than stepwise relationship between the two variables, rather than the absence of specific etiologies, that force the decision. As I said, questions determining how often it will be forced upon us are for the science to decide.

Let me finish, then, by asking how the taxonomy whose basis I have sketched feeds into the demands of the clinic. If a nosology cannot be turned to the service of useful knowledge, after all, there is little reason to embrace it.

10.5   Accommodating the Clinicians

10.5.1   Scientific and Clinical Perspectives
The picture I have developed is built to facilitate research into the causes of mental illness. I end with a discussion of how this epistemic task feeds into more practical clinical demands. Unlike optimists such as Kendell (2002) and Schaffner (2002), I am skeptical that in the end we will see a single classification system for mental disorder that suits both researchers and clinicians. I do, though, think that clinical projects can exploit the resources offered by a research-driven classification that aims to reflect the causal structure of the human world. The clinical niceties and the demands of explanation and classification are not at odds, as we can see once the requirements of the clinician are understood.

In an important recent book, Spaulding, Sullivan, and Poland (2003) have worked backward from the demands of rehabilitation, as it were, to provide a philosophical justification for rethinking the conceptual foundations of psychiatry. They argue that a rehabilitation-based approach exposes the flaws in the medical model. However, although the perspective I adopt does abstract away from individual variation and is not constructed with therapy or rehabilitation in mind, it nonetheless represents the right balance between the demands of research and therapy. Although it is chiefly a way of formulating a research project, it does not commit us to denying that the clinical perspective is useful. Furthermore, the explanatory structure I envisage can be used as a set of resources for clinical and therapeutic projects. Spaulding, Sullivan, and Poland present a very clear and attractive way of thinking about the clinical enterprise, so I shall state that and then contrast it with the way I prefer.

Spaulding, Sullivan, and Poland (2003) argue that their synthetic approach has a different conceptualization of mental illness to that offered by the medical model. Rather than seeing mental disorders as diseases, their approach sees them as problems that have to be overcome in order to restore normal functioning. They seem to argue that psychiatry should be oriented around rehabilitation and that this requires rethinking the assumptions that currently gave the field its shape. Subject to the qualifications I have made in the last three chapters, the assumptions they criticize are ones I share. I defend the assumptions of the medical model on the following counts: the disease conception of mental illness is consistent with Spaulding, Sullivan, and Poland's approach; generalization is necessary; and assessment of patients and predictions about the effects of treatment must be based on statistical rather than idiographic methods. A clinician might look at a group of patients who are very similar in terms of the exemplars they instantiate, and yet still see people who need to be treated in different ways depending on the different opportunities for effective intervention in each case. Nothing in my interpretation of the medical model rules this out.

As they portray it, then, the medical model sorts patients into disease classes and proceeds from there, just as I have asserted. Spaulding, Sullivan, and Poland (2003) confront the medical model's idea that proper diagnosis is the key to treatment with a rival conception of what treatment depends on that does not rely on diagnosing a particular medical condition. In their view, treatment ought to be based on the identification of the patient's desires and aspirations. Once the identification of these desires and aspirations has been made, the treatment team can work back from them to identify those features of the patient's current lot that obstruct their realization.

These ideas may be very helpful and attractive in a clinical setting. They offer a means of working out the approach to take with an individual patient in the light of the difficulties of fitting a patient to a classification, which I have already noted. I do, however, dissent from the contentions they make regarding how assessment of patients and establishment of treatment programs should proceed. I also deny, and argue against, the idea that a rehabilitation-based account needs to reject the foundations of the medical model. But this is not because I reject their conception of the clinical project. Their clinical perspective can live happily alongside the research-driven explanatory projects of the type I envisage as providing the basis for explanation and classification in psychiatry. The relation that clinical and causal-explanatory projects bear to each other is more complicated than they believe it to be. I think this is generally correct. A research taxonomy and a clinical repertoire of concepts must be related in some way, but they need not be identical. The research projects, in my way of viewing the matter, can provide a set of resources that clinical projects like Spaulding, Sullivan, and Poland (2003) rehabilitation-driven "synthetic paradigm" may draw on. Doing things this way, I will argue, preserves what is valuable about the medical model but at the same time preserves the distance from it that Spaulding et al. wish to preserve.

10.5.2 Exemplars as a Clinical Resource

Spaulding, Sullivan, and Poland's (2003) clinical projects have a different nature than the explanatory project I have been discussing. Their fundamental contention is that the medical model and their own "integrated rehabilitation paradigm" are at odds on the nature of disease (2003, 17). Whereas the medical model sees mental illness as medical diseases to be explained in terms of biological abnormalities, their integrated paradigm views mental illness not as a medical disease in need of cure but as a disability that imposes barriers between the patient and the type of life the patient wants to lead. But there is no real conflict. We can distinguish between supplementing and replacing the medical model. A supplementation view simply says that the medical model conceives of mental illness in a way that may not always apply in therapeutic settings, where other ways of thinking about mental illness are required to supplement it. If that is their view, it fits with the two-stage picture and associated conceptual pluralism that I advocate in chapter 3. I am happy to agree with them in that case. But Spaulding, Sullivan, and Poland criticize the medical model in ways that sometimes suggest not just that they find it ill-fitted to their picture of rehabilitation (e.g., 2003; 258–259), but that they want to replace it with their own picture.

The supplementation thesis is attractive; the replacement thesis is not. Consider the conception of "problem" that plays an important role in Spaulding, Sullivan, and Poland's view. A "problem," for them, is anything that stands between the patient and the achievement of a long-term goal, which is some aspect of normal or desirable human functioning (2003, 46–47), and can be almost anything from a "recreational skills deficit" (313) to a biological impairment in the central nervous system. Problems that appear to involve biological and cognitive dysfunction are the clearest examples of the competing assertions of the medical model and the rehabilitation paradigm. What Spaulding, Sullivan, and Poland do, in effect, is to use "recognized psychiatric symptoms" (2003, 308) as resources that can be extracted and applied to individual patients. A particular "attribution problem," for example, might be a paranoid delusion such as is often seen in schizophrenia, or the Capgras Delusion, or the sort of mood-linked misinterpretation of others' behavior that occurs in major depressive episodes. Any of these attribution problems, in their picture, could be diagnosed in a patient together with myriad neurological problems or problems in living. The concept of an attribution problem serves them as a way of tracking very different causal histories. What the histories have in common is a shared property of obstructing the realization of some important aspect of the patient's goal, as determined by the patient in concert with treatment team.

The characterization is of individuals as (1) potentially presenting with symptoms from different DSM diagnoses jumbled together in one person and (2) with psychiatric symptoms relating to different problems in living depending on the general shape of patients lives. This is a clear statement of the clinician's lament that patients' symptoms and impairments tend to be smeared over several DSM diagnoses and fit none very well. Spaulding, Sullivan, and Poland's statement of the issue also points to a solution, which is to assess patients according to the extent to which they fit symptoms drawn from different diagnoses. Not all the problems they refer to are symptom types drawn from the DSM-IV-TR, but if psychiatry is synthesized more widely with cognitive and behavioral sciences, as I advocate, then a number of the cognitive and behavioral deficits they mention might be gathered under the psychiatric banner.

This exploitation of psychiatric diagnosis as a pool of symptoms that can be applied to particular patients in different ways is consistent with a number of concepts of what symptoms are. The approach I recommend, of seeing symptoms as the manifestation of underlying pathologies, derives from my construal of the medical model and is only one way to think of the resources made available for clinical use. However, it is consistent with the rehabilitative venture that Spaulding, Sullivan, and Poland are embarked on, and that is enough to show that the replacement thesis is not required by the clinical project.

So it is consistent with their view to see the medical model as a source of input to rehabilitation, affording the clinician a set of resources. Even if patients do not show all the features of an exemplar, they may nonetheless exhibit some of the features of some exemplars. Therefore, the clinician can exploit the theoretical knowledge associated with the exemplars. Exemplars are idealizations. Some patients may fit them precisely, others may fit them imprecisely, and still others may present with symptoms that correspond to aspects of several exemplars. This fits the comment that Spaulding, Sullivan, and Poland make about one of their problem types, "episodic neurophysiological dysregulation of the central nervous system" (2003, 305). Within this problem, they say, are various abnormalities "traditionally associated with schizophrenia, schizoaffective disorder, severe or psychotic depression, delusional disorder, and bipolar disorder" (2003, 305) This is a set of related psychotic categories that, on my account, it is the business of exemplar explanation to differentiate and supply with causal histories. These distinct causal histories then are available for the clinician as a guide for intervention. It is the clinician's job to intervene in order to counteract or prevent the normal outcome of the pathological processes that cause exemplars to have their clinical profiles and natural histories. Without such information about causal processes, how can intervention be planned or evaluated? The taxa that Spaulding, Sullivan, and Poland need cross-classify the taxonomy of exemplars, but they need the exemplars to make sense, since symptoms as explained by different causal histories are abstracted away from the remainder of the exemplar if they have properties that explain the role they play in preventing the patient's rehabilitation. That role is functionally defined, as a problem of a certain sort. But features of the explanation previously given to the condition are what explain the properties in virtue of which the symptom plays the functional role. What makes a group of symptoms tokens of the same problem are their causal powers. And to figure out what those are, we need to know the properties of a symptom as explained by the model. As I have said, exemplars are idealizations, and they are explained by models that have an adjustable relation to reality. One of the things we can do with the models is to focus on only a single causal pathway and follow it through to a particular symptom. In Spaulding, Sullivan, and Poland language, causal pathways from different models can be brought together as tokens of a problem.

When dealing with an individual patient, it is also possible to exploit the indeterminacy of the model in another way. A clinician can treat the causal pathways of interest as schema to be filled in with the specific details of interest in the particular case. Each path through the model can be realized by numerous different causal histories, so the clinician can use the pathway to a given sympton as a way to look for and organize the relevant details of the patient's life.

This appreciation of the causal powers, as I say in chapter 4, does not mean that intervention must be directed in any particular way. Even if we have established that a symptom is best explained in terms of one main causal factor, such as neurotransmitter abnormality, it does not follow that treatment must be directed at directly manipulating that causal factor. However, treatments usually are tested on populations, and success is much more common when treating someone as a instance of a category than when trying to craft unique or idiosyncratic responses to all individual patients (Dawes 1994; Meehl 1957).

The inevitability of individual variation does not mean that generalization is not also inevitable. Patients must be sorted into classes. The idea that individual clinical profiles should be the focus of psychiatry is unhelpful if we want to explain psychopathology. What is needed is an explanatory story that generalizes across particular manifestations to make their shared causal structure apparent. All individuals differ when we reach the fine details of the mental functioning and personal history that jointly conspire to cause psychopathology, so a maximally fine-grained classification might recognize as many disorders as there are patients. In that case, the wealth of detail swamps any ability we have to make generalizations. So as we ascend toward more helpful groupings of patients, we in fact lose information—we may recognize different classes of depression, but within each class there exist people who have become depressed differently, through the myriad and sometimes banal tragedies that we all encounter. Knowing the details in each case may help a clinician, but what is wanted in taxonomy is not enough detail to uniquely specify every individual case, but the right level of generality to group patients into useful classes.

The desire to be right about what we classify must be balanced against the reasons why we want to classify in the first place; in psychiatry, we want to group people by their shared conditions, but those conditions must be shared at the right grain—we shouldn't group together alcoholics and schizophrenics, but neither should we separate the gin drinkers from the bourgeois consumers of dry white wine just because they are addicted to different kinds of alcohol. We sort people into classes not because they are not individuals, but because we have the best chance of developing useful knowledge to apply to the individual case when we see what that case has in common with its relatives.

10.6 Summing Up

My discussion of the exemplar approach's clinical utility brings us back to the social and moral matters that I mentioned earlier but promised to avoid. So this is a good place to take stock. I hope I have shown that psychiatry is of philosophical interest even apart from these normative matters that I have scanted, and that the methods of recent philosophy of science can both illuminate the current state of biological psychiatry and be informed by it in turn. Psychiatry touches upon the philosophy of the cognitive and biological sciences in many ways, ranging across first-order questions like rationality, deviance, and experience as well as second-order questions such as explanation, theory-building, and natural kinds. I have only scratched the surface of many of these issues, including most notably the role of new technologies, such as brain imaging, in the accumulation and interrogation of the data out of which psychiatry will spin its new theories. My focus has been on broad issues, often seen from a long way off, but the message would have been unaffected had I dwelled further on the details.

The message, above all, is that psychiatry is deeply entangled with the sciences of the mind, and even the social sciences, and is better understood once we apply within it the philosophical tools that these other disciplines have helped to forge. The entanglement is becoming steadily more intimate, but the field's own self-image, to judge from the DSM and other conceptual statements, has not kept up. This estrangement does psychiatry no good at all, and impoverishes kindred disciplines like cognitive science. If psychiatry embraces its future as applied cognitive neuroscience then it will be better placed to take clinical and theoretical advantage of the inquiries that already are blurring its official boundaries.

Philosophers, too, can reflect on how so many different inquiries are jumbled together in the study of psychopathology, and use this reflection to consider further the complex and inexact methods of the human sciences.

Ultimately, of course, the point of psychiatry is not intellectual illumination but the relief of suffering. I have talked about the changing intellectual structure of psychiatry, but this structure, as the present chapter argues, will have to provide the basis for our clinical struggles. Nearly everything is left to be done, and I hope that philosophers will try to redress that.

Bibliography

Abbasi, K., and K. S. Khan, 2004. India versus Pakistan and the power of a six: An analysis of cricket results. *British Medical Journal* 328: 800.

Adolphs, R. 2003. Cognitive neuroscience of human social behavior. *Nature Reviews Neuroscience* 4: 165–178.

Allen, C., and M. Bekoff. 1997. *Species of mind: The philosophy and biology of cognitive ethology.* Cambridge, MA: The MIT Press.

Ainslie, G. 2001. *Breakdown of will.* Cambridge: Cambridge University Press.

American Psychiatric Association. 1952. *Diagnostic and statistical manual of mental disorders.* 1st ed. Washington, DC: American Psychiatric Association.

American Psychiatric Association. 1968. *Diagnostic and statistical manual of mental disorders.* 2nd ed. Washington, DC: American Psychiatric Association.

American Psychiatric Association. 1980. *Diagnostic and statistical manual of mental disorders.* 3rd ed. Washington, DC: American Psychiatric Association.

American Psychiatric Association. 1983. Statement on the insanity defense. *American Journal of Psychiatry* 140: 681–688.

American Psychiatric Association. 1987. *Diagnostic and statistical manual of mental disorders.* 3rd ed., rev. Washington, DC: American Psychiatric Association.

American Psychiatric Association. 1994. *Diagnostic and statistical manual of mental disorders.* 4th ed. Washington, DC: American Psychiatric Association.

American Psychiatric Association. 2000. *Diagnostic and statistical manual of mental disorders.* 4th ed., textual rev. Washington, DC: American Psychiatric Association.

American Sleep Disorders Association. 1990. *International classification of sleep disorders: Diagnostic and coding manual.* Diagnostic Classification Steering Committee, M. J. Thorpy, Chairman. Rochester, MN: American Sleep Disorders Association.

Andreasen, N. C. 1984. *The broken brain.* New York: Harper & Row.

Andreasen, N. C. 1997. Linking mind and brain in the study of mental illnesses: A project for a scientific psychopathology. *Science* 275: 1586–1593.

Andreasen, N. C., P. Nopoulos, D. S. O'Leary, D. D. Miller, T. Wassink, and M. Flaum 1999. Defining the phenotype of schizophrenia: Cognitive dysmetria and its neural mechanisms. *Biological psychiatry* 46: 908–920.

Andreasen, R. O. 1998. A new perspective on the race debate. *British Journal for the Philosophy of Science* 49: 188–225.

Ankeny, R. A. 2000. Fashioning descriptive models in biology: Of worms and wiring diagrams. *Philosophy of science* 67 (suppl.): S260–S272.

Ankeny, R. A. 2002. Reduction reconceptualized: Cystic fibrosis as a Paradigm case for molecular medicine. In *Mutating concepts, evolving disciplines: Genetics, medicine, and society*, ed. L. S. Parker and R. A. Ankeny, 127–141. Dordrecht, the Netherlands: Kluwer.

Antony, M. M., and D. H. Barlow. 2002. Specific phobias. In *Anxiety and its disorders*, D. H. Barlow, 380–417. 2nd ed. New York: Guilford Press.

Appiah, K. A. 1996. Race, culture, identity: Misunderstood connections. In *Color conscious: The political morality of race*, ed. K. A. Appiah and A. Gutmann, 30–105. Princeton, NJ: Princeton University Press.

Ashizawa, T., L-J. Wong, C. S. Richards, C. T. Caskey, and J. Jankovic. 1994. The CAG repeat size and clinical presentation in Huntington's disease. *Neurology* 44: 1137–1143.

Atran, S. 1985. Causal constraints on categories and categorical constraints on biological reasoning across cultures. In *Causal cognition*, ed. D. Sperber, D. Premack, and A. J. Premack, 205–233. Oxford: Clarendon Press.

Atran, S. 1990. *Cognitive foundations of natural history*. Cambridge: Cambridge University Press.

Bandura, A. 1986. *Social foundations of thought and action: A social cognitive theory*. Upper Saddle River, NJ: Prentice-Hall.

Barlow, D. H. 2002. *Anxiety and its disorders*. 2nd ed. New York: Guilford Press.

Baron-Cohen, S. 1995. *Mindblindness*. Cambridge, MA: The MIT Press.

Baron-Cohen, S., A. Leslie, and U. Frith. 1985. Does the autistic child have a theory of mind? *Cognition* 21: 37–46.

Baron-Cohen, S., A. Leslie, and U. Frith. 1986. Mechanical, behavioral, and intentional understanding of picture stories in autistic children. *British Journal of Developmental Psychology* 4: 113–125.

Baron-Cohen, S., and J. Swettenham. 1996. The relationship between SAM and ToMM; Two hypotheses. In *Theories of theories of mind*, ed. P. Carruthers and P. Smith, 158–168. Cambridge: Cambridge University Press.

Barondes, S. 1998. *Mood genes*. New York: W. H. Freeman.

Bateson, P., and P. Martin. 1999. *Design for a life*. London: Jonathan Cape.

Batterman, R., J. D'Arms, and K. Górny. 1998. Game theoretic explanations and the evolution of justice. *Philosophy of Science* 65: 76–102.

Baxter, L. R., J. M. Schwartz, K. S. Bergman, M. P. Szuba, B. H. Guze, J. C. Mazziotta, A. Alazraki, C. E. Selin, H-K. Ferng, P. Munford, and M. E. Phelps. 1992. Caudate glucose metabolic rate changes with both drug and behavior therapy for obsessive-compulsive disorder. *Archives of General Psychiatry* 49: 681–689.

Beatty, J. 1982. Classes and cladists. *Systematic Zoology* 31: 25–34.

Bechtel, W. 2002. Decomposing the brain: A long term pursuit. *Brain and Mind* 3: 229–242.

Bechtel, W., and J. Mundale. 1999. Multiple realizability revisited: Linking cognitive and neural states. *Philosophy of Science* 66: 175–207.

Bechtel, W., and R. Richardson. 1993. *Discovering complexity*. Princeton, NJ: Princeton University Press.

Beck, A. T., A. J. Rush, B. F. Shaw, and G. Emery 1979. *Cognitive therapy of depression*. New York: Guilford Press.

Beck, J. 1996. *Cognitive therapy: Basics and beyond*. New York: Guilford Press.

Becker, G., and K. Murphy. 1988. A theory of rational addiction. *Journal of Political Economy* 96: 675–700.

Beckham, E. E., W. R. Leber, and L. K. Youll. 1995. The diagnostic classification of depression. In *Handbook of Depression*, ed. E. E. Beckham and W. R. Leber, 36–60. 2nd ed. New York: Guilford.

Bellugi, U., P. Wang, and T. L. Jernigan. 1994. Williams syndrome: An unusual neurophysiological profile. In *Atypical cognitive deficits in developmental disorders: Implications for brain function*, ed. S. Broman and J. Grafman, 23–56. Cambridge, MA: The MIT Press.

Bermudez, J. L. 1998. Philosophical psychopathology. *Mind and language* 13: 287–307.

Bermudez, J. L. 2001. Normativity and rationality in delusional psychiatric disorders. *Mind and Language* 16: 457–493

Bigelow, J., and R. Pargetter. 1987. Functions. *Journal of Philosophy* 84: 181–196.

Binmore, K. 1994. *Game theory and the social contract, vol. 1: Playing fair.* Cambridge MA: The MIT Press.

Binmore, K. 1998. *Game theory and the social contract, vol. 2: Just playing.* Cambridge MA: The MIT Press.

Bishop, M., and J. D. Trout. 2004. *Epistemology and the psychology of human judgment*. New York: Oxford University Press.

Blackwood, N. J., R. J. Howard, R. P. Bental, and R. M. Murray. 2001. Cognitive neuropsychiatric models of persecutory delusions. *American Journal of Psychiatry* 158: 527–539.

Blair, R. J. 1995. A cognitive developmental approach to morality: Investigating the psychopath. *Cognition* 57: 1–29.

Blair, R. J. 1997. Moral reasoning and the child with psychopathic tendencies. *Personality and Individual Differences* 26: 731–739.

Blair, R. J., L. Jones, F. Clark, and M. Smith. 1997. The psychopathic individual: A lack of responsiveness to distress cues? *Psychophysiology* 34: 192–198.

Blair, R. J., C. Sellars, I. Strickland, F. Clark, A. Williams, M. Smith, and L. Jones. 1996. Theory of mind in the psychopath. *Journal of Forensic Psychiatry* 7: 15–25.

Blakemore, S-J., D. M. Wolpert, and C. D. Frith. 2002. Abnormalities in the awareness of action. *Trends in Cognitive Sciences* 6: 237–242.

Blashfield, R. K. 1989. Alternative taxonomic models of psychiatric classification. In *The validity of psychiatric diagnosis*, ed. L. Robins and J. E. Barrett, 19–34. New York: Raven Press.

Bleuler, E. 1911. *Dementia praecox, or The group of schizophrenias*, trans. J. Zinkin (1950). New York: International Universities Press.

Bolton, D., and J. Hill. 1996. *Mind, meaning, and mental disorder: The nature of causal explanation in psychiatry*. New York: Oxford University Press.

Boorse, C. 1975. On the distinction between disease and illness. *Philosophy and Public Affairs* 5: 49–68.

Boorse, C. 1976a. What a theory of mental health should be. *Journal for the Theory of Social Behavior* 6: 61–84.

Boorse, C. 1976b. Wright on functions. *Philosophical Review* 85: 70–86.

Boorse, C. 1977. Health as a theoretical concept. *Philosophy of Science* 44: 542–573

Boorse, C. 1997. A rebuttal on health. In *What is disease?* ed. J. M. Humber and R. F. Almeder, 3–143. Totawa, NJ: Humana Press.

Borch-Jacobsen, M. 1999. What made Albert run. *London Review of Books* 21 (11): 9–10.

Bowker, G. and S. L. Star. 1999. *Sorting things out: Classification and its consequences*. Cambridge, MA: The MIT Press.

Boyd, R. 1980. Scientific realism and naturalistic epistemology. In *PSA 1980*, vol. 2, ed. P. D. Asquith and R. N. Giere, 613–662. East Lansing, MI: Philosophy of Science Association.

Boyd, R. 1989. What realism implies and what it does not. *Dialectica* 43: 5–29.

Boyd, R. 1991. Realism, anti-foundationalism and the enthusiasm for natural kinds. *Philosophical Studies* 61: 127–148.

Brion, S., and C. P. Jedynak. 1972. Troubles du transfert hemispherique. *Revue Neurologique Paris* 126: 257–266.

Bridgman, P. 1927. *The logic of modern physics.* New York: MacMillan. Reprinted 1962.

Brown, P. 1990. The name game: Toward a sociology of diagnosis. *Journal of Mind and Behavior* 11: 385–406.

Callebaut, W. 1993. *Taking the naturalistic turn.* Chicago: University of Chicago Press.

Cameron, N. 1944. Experimental analysis of schizophrenic thinking. In *Language and thought in schizophrenia,* ed. J. S. Kasanin, 50–64. Berkeley: University of California Press.

Campbell, J. 2001. Rationality, meaning, and the analysis of delusion. *Philosophy, Psychiatry, Psychology* 8: 89–100.

Campbell, W. G. 2003. Addiction: A disease of faulty volition caused by impaired memory. *Journal of Addictive Diseases* 22: 3A.

Cantor, N., and N. Genero. 1986. Psychiatric diagnosis and natural categorization: A close analogy. In *Contemporary directions in psychopathology: Towards the DSM-IV,* ed. T. Millon and G. L. Klerman, 233–256. New York: Guilford Press.

Carruthers, P. 2003. Moderately massive modularity. In *Minds and persons,* ed. A. O'Hear. *Royal Institute of Philosophy Supplement* 53: 67–89.

Cath, D. C., P. Spinhoven, T. C. A. M. Van Woerkom, B. J. M. Van de Wetering, C. A. L. Hoogduin, A. D. Landman, R. A. C. Roos, and H. G. M. Rooijmans. 2001. Gilles de la Tourette's syndrome with and without obsessive-compulsive disorder compared with obsessive-compulsive disorder without tics: Which symptoms discriminate? *Journal of Nervous and Mental Diseases* 189: 219–228.

Charcot, J-M. 1987. *Charcot, the clinician: The Tuesday lessons.* Trans. C. G. Goetz. Philadelphia: Lippincott, Williams & Wilkins.

Cherniak, C. 1986. *Minimal rationality.* Cambridge, MA: The MIT Press.

Cheung, M. S-P., P. Gilbert, and C. Irons. 2004. An exploration of shame, social rank and rumination in relation to depression. *Personality and Individual* Differences 36: 1143–1153.

Christodoulou, G. N. 1991. The delusional misidentification syndromes. *British Journal of Psychiatry* 159 (suppl. 14): 65–69.

Churchland, P. M. 1981. Eliminative materialism and the propositional attitudes. *Journal of Philosophy* 78: 67–90.

Clark, A. 1997. *Being there: Putting brain, body and world together again.* Cambridge, MA: The MIT Press.

Clark, D. M. 1986. A cognitive approach to panic. *Behavior Research and Therapy* 24: 461–470.

Clark, D. M. 1988. A cognitive model of panic attacks. In *Panic: Psychological perspectives*, ed. S. Rachman and J. Maser, 71–90. Hillsdale, NJ: Erlbaum.

Clark, D. M. 1997. Panic disorder and social phobia. In *The science and practice of cognitive behavior therapy*, ed. D. M. Clark and C. G. Fairburn, 121–153. Oxford: Oxford University Press.

Clark, D. M., and A. Ehlers. 1993. An overview of the cognitive theory and treatment of panic disorder. *Applied and Preventative Psychology* 2: 131–139.

Clarkin, J., T. Widiger, A. Francis, S. Hurt and M. Gilmore. 1983. Prototypic typology and the borderline personality disorder. *Journal of Abnormal Psychology* 92: 263–275.

Cleckley, H. 1988. *The mask of sanity*. 5th ed. Augusta, GA: E. S. Cleckley. Orig. pub. 1976.

Cohen, J. 1997. The arc of the moral universe. *Philosophy and Public Affairs* 26: 91–134.

Cohen, D. J., and J. F. Leckman. 1999. Introduction: The self under siege. In *Tourette's syndrome: Tics, obsessions, compulsions*, ed. J. F. Leckman and D. J. Cohen, 1–20. New York: John Wiley.

Collini, S. 1999. Grievance studies. In *English pasts: Essays in history and culture*, 252–268. Oxford: Oxford University Press.

Cooper, R. 2005. *Classifying madness: A philosophical examination of the diagnostic and statistical manual of mental disorders*. Dordrecht, the Netherlands: Springer.

Cosmides, L., and J. Tooby. 1997. The modular nature of human intelligence. In *The Origin and Evolution of Intelligence*, ed. A. B. Scheibel and J. W. Schopf, 71–101. Sudbury, MA: Jones & Bartlett.

Cottingham, J. 1998. Descartes' treatment of animals. In *Descartes*, ed. J. Cottingham, 225–233. New York: Oxford University Press.

Cowan, W. M., and E. R. Kandel. 2001. Prospects for neurology and psychiatry. *Journal of the American Medical Association* 285: 594–600.

Cowie F. 1999. *What's within? Nativism reconsidered*. New York: Oxford University Press.

Craske, M. G., and D. H. Barlow. 2001. Panic disorder and agoraphobia. In *Clinical handbook of psychological disorders*, ed. D. H. Barlow. 3rd ed. New York: Guilford Press.

Cronbach, L., and P. Meehl. 1955. Construct validity in psychological tests. *Psychological Bulletin* 52: 281–302.

Cronkite, R. C., and R. H. Moos. 1995. Life context, coping processes, and depression. In *Handbook of Depression*, ed. E. E. Beckham and W. R. Leber, 569–587. 2nd ed. New York: Guilford.

Culver, C. M., and B. Gert. 1982. *Philosophy in medicine.* New York: Oxford University Press.

Cummins, R. 1975. Functional analysis. *Journal of Philosophy* 72: 741–764.

Cutting, J., and M. Shepherd. 1987. Introduction. In *The Clinical Roots of the Schizophrenia Concept*, ed. J. Cutting and M. Shepherd. Cambridge: Cambridge University Press.

Damasio, A., and H. Damasio. 1989. *Lesion analysis in neuropsychology.* New York: Oxford University Press.

Damasio, A. 1994. *Descartes' error.* New York: G. P. Putnam's.

D'Amico, R. 1995. Is disease a natural kind? *Journal of Medicine and Philosophy* 20: 551–569.

D'Arms, J., R. Batterman, and K. Górny. 1998. Game theoretic explanations and the evolution of justice. *Philosophy of Science* 65: 76–102.

Davidson, A. 2001. *The emergence of sexuality: Historical epistemology and the formation of concepts.* Cambridge, MA: Harvard University Press.

Davidson, D. 1980. Mental events. In *Essays on actions and events*, 207–227. Oxford: Clarendon Press.

Davidson, D. 1993. Thinking causes. In *Mental causation*, ed. J. Heil and A. Mele, 3–19. Oxford: Clarendon Press.

Davies, M., and M. Coltheart. 2000. Introduction: pathologies of belief. In *Pathologies of belief*, ed. M. Coltheart and M. Davies, 1–46. Oxford: Basil Blackwell.

Davies, W., A. R. Isles, and L. S. Wilkinson. 2001. Imprinted genes and mental dysfunction. *Annals of Medicine* 33: 428–436.

Davis, K. L., D. G. Stewart, J. I. Friedman, M. Buchsbaum, P. D. Harvey, P. R. Hof, J. Buxbaum, and V. Haroutunian. 2003. White matter changes in schizophrenia: Evidence for myelin-related dysfunction. *Archives of General Psychiatry* 60: 443–456.

Dawes, R. 1994. *House of cards: Psychology and psychotherapy built on myth.* New York: Free Press.

De Renzi, E. 2000. Prosopagnosia. In *Patient-based approaches to cognitive neuroscience*, ed. M. J. Farah and T. E., Feinberg, 85–95. Cambridge, MA: The MIT Press.

Drummond, J. M., F. A. Issa, C-K. Song, J. Heberholz, S-R. Yeh, and D. H. Edwards. 2002. Neural mechanisms of dominance hierarchies in crayfish. In *The crustacean nervous system*, ed. K. Wiese, 124–135. Berlin: Springer.

Dugas, M. 1986. La maladie des tics: D'Itard aux neuroleptiques. *Revue de Neurologie* 142: 817–823.

Dupré, J. 1993. *The disorder of things.* Cambridge, MA: Harvard University Press.

Dupré, J. 2001. *Human nature and the limits of science.* New York: Oxford University Press.

Dupré, J. 2006. Fact and value. Forthcoming in *Value-free science: Ideal or illusion,* ed. H. Kincaid, J. Dupré, and A. Wylie. New York: Oxford University Press.

Dutton, D. 1996. What are editors for? *Philosophy and Literature* 20: 551–566.

Egan, F. 1995. Computation and content. *The Philosophical Review* 104: 181–203.

Ehlers, A., and P. Breuer. 1992. Increased cardiac awareness in panic disorder. *Journal of Abnormal Psychology* 101: 371–82.

Ellenberger, H. F. 1993. Fallacies of psychiatric classification. In *Beyond the unconscious: Essays of Henri F. Ellenberger in the history of psychiatry,* ed. M. Micale, 309–327. New Haven: Yale University Press. Orig. pub. 1963.

Ellenberger, H. F. 1970. *The discovery of the unconscious.* New York: Basic Books.

Elliott, C. 2003. *Better than well. American medicine meets the American dream.* New York: W.W. Norton.

Ellis, H. D., J. Whitley, and J. P. Luauté. 1994. Delusional misidentification: The three original papers on the Capgras, Frégoli, and intermetamorphosis delusions. *History of Psychiatry* 5: 117–146.

Ellis, H. D., A. W. Young, A. H. Quayle, and K. W. de Pauw. 1997. Reduced autonomic responses to faces in Capgras delusion. *Proceedings of the Royal Society: Biological Sciences* B264, 1085–1092.

Elman, J., E. Bates, M. Johnson, A. Karmiloff-Smith, D. Parisi, and K. Plunkett. 1996. *Rethinking Innateness.* Cambridge, MA: The MIT Press.

Elster, J. 1993. *Political psychology.* Cambridge: Cambridge University Press.

Elster J. 1999a. *Alchemies of the mind.* Cambridge: Cambridge University Press.

Elster J. 1999b. *Strong feelings: Emotion, Addiction, and human behavior.* Cambridge, MA: The MIT Press.

Engel, G. L. 1961. Is grief a disease—A challenge for medical research. *Psychosomatic Medicine* 23: 18–22.

Engel, G. L. 1977. The need for a new medical model: A challenge for biomedicine. *Science* 196: 129–136.

Engel, G. L. 1981. The clinical application of the biopsychosocial model. *Journal of Medicine and Philosophy* 6: 101–123.

Engel, G. L. 1992. How much longer must medicine's science be bound by a seventeenth century world wiew? *Psychotherapy and Psychosomatics* 57: 3–16.

Enoch, M., and W. Trethowan. 1979. *Uncommon psychiatric syndromes.* 2nd ed. Bristol: John Wright & Sons.

Essen-Möller, E. 1961. On classification of mental disorders. *Acta Psychiatrica Scandinavica* 37: 119–126.

Essen-Möller, E. 1971. Suggestions for further improvement of the international classification of mental disorders. *Psychological Medicine* 1: 308–311.

Farah, M. J. 1990. *Visual agnosia*. Cambridge, MA: The MIT Press.

Farah, M. J., and T. E. Feinberg. 2000. Visual object agnosia. In *Patient-based approaches to cognitive neuroscience*, ed. M. J. Farah and T. E. Feinberg, 79–84. Cambridge, MA: The MIT Press.

Feighner, J. P., E. Robins, S. B. Guze, R. A. Woodruff, G. Winokur, and R. Munoz. 1972. Diagnostic criteria for use in psychiatric research. *Archives of General Psychiatry* 26: 57–63.

Feinberg, T. E. and D. M. Roane. 2000. Misidentification syndromes. In *Patient-based approaches to cognitive neuroscience*, ed. M. J. Farah and T. E. Feinberg, 155–161. Cambridge, MA: The MIT Press.

Fenichel, O. 1945. *The psychoanalytic theory of neuroses*. New York: Norton.

Festinger, L. 1957. *A theory of cognitive dissonance*. Evanston, IL: Row, Peterson.

Fingarette, H. 1988. *Heavy drinking: The myth of alcoholism as a disease*. Berkeley: University of California Press.

Finger, S. 1994. *Origins of neuroscience*. New York: Oxford University.

Fish, F. 1962. *Fish's schizophrenia*. 3rd ed. M. Hamilton, ed. Bristol: John Wright 1984.

Fodor, J. 1983. *Modularity of mind*. Cambridge, MA: The MIT Press.

Fodor, J. 1987. *Psychosemantics*. Cambridge, MA: The MIT Press.

Fodor, J. 1995. A theory of the child's theory of mind. In *Mental simulation*, ed. M. Davies and T. Stone, 109–122. Oxford: Basil Blackwell.

Fodor, J. 2000. *The mind doesn't work that way*. Cambridge, MA: The MIT Press.

Frances, A., A. H. Mack, M. B. First, T. A. Widiger, R. Ross, L. Forman, W. Wakefield Davis. 1994. DSM-IV meets philosophy. *Journal of Medicine and Philosophy* 19: 207–218.

Frank, R. H. 1988. *Passions within reason*. New York: W.W. Norton.

Friedman, M. 1974. Explanation and scientific understanding. *Journal of Philosophy* 71: 5–19.

Friston, K. J. 1996. Theoretical neurobiology and schizophrenia. *British Medical Bulletin* 52: 644–655.

Frith, C. D. 1992. *The cognitive neurospychology of schizophrenia*. Hove, UK: Lawrence Erlbaum.

Frith, C. D, S-J. Blakemore, and D. Wolpert. 2000a. Abnormalities in the awareness and control of action. *Philosophical Transactions of the Royal Society, Series B*; 355: 1771–1788.

Frith, C. D, S-J. Blakemore, and D. Wolpert. 2000b. Explaining the symptoms of schizophrenia: Abnormalities in the awareness of action. *Brain Research Reviews* 31: 357–363.

Frith, C. D., and E. Johnstone. 2003. *Schizophrenia: A very short introduction*. New York: Oxford University Press.

Frith, U. 2003. *Autism: Explaining the enigma*, 2nd ed. London: Blackwell.

Fulford, K. W. M. 2002. Contentious and non-contentious evaluative language in psychiatric diagnosis. Dateline 2010. Report to the chair of the DSM-VI task force from the editors of *Philosophy, Psychiatry, and Psychology*. In *Descriptions and prescriptions: Values, mental disorders and the DSMs*, ed. J. Z. Sadler, 323–361. Baltimore, MD: Johns Hopkins University Press.

Garcia, J., and R. Koelling. 1966. Relation of cue to consequence in avoidance learning. *Psychonomic Science* 4: 123–124.

Garner, D. M., P. E. Garfinkel, D. Schwartz, and M. Thompson. 1980. Cultural expectations of thinness in women. *Psychological Reports* 47: 483–491.

Gelder, M., D. Gath, and R. Mayou. 1989. *Oxford textbook of psychiatry*, 2nd ed. Oxford: Oxford University Press.

Gerrans, P. 1999. Delusional misidentification as sub-personal disintegration. *Monist* 82: 590–608.

Gerrans, P. 2000. Refining the explanation of Cotard's delusion. In *Pathologies of belief*, ed. M. Coltheart and M. Davies, 111–122. Oxford: Basil Blackwell.

Geuss, R. 2001. *History and illusion in politics*. Cambridge: Cambridge University Press.

Ghiselin, M. 1974. A radical solution to the species problem. *Systematic Zoology* 23: 536–544.

Ghiselin, M. 1997. *Metaphysics and the origin of species*. Albany: SUNY Press.

Ghiselli, E., J. Campbell, and S. Zedeck. 1981. *Measurement theory for the behavioral sciences*. San Francisco, CA: W. H. Freeman.

Gibbard, A. 1990. *Wise choices, apt feelings*. Cambridge, MA: Harvard University Press.

Giedd, J. N., J. L. Rapoport, M. J. Kruesi, C. Parker, M. B. Schapiro, A. J. Allen, H. L. Leonard, D. Kaysen, D. P. Dickstein, and W. L. Marsh. 1995. Sydenham's chorea: Magnetic resonance imaging of the basal ganglia. *Neurology* 45: 2199–2202.

Giere, R. 1988. *Explaining science: A cognitive approach*. Chicago: University of Chicago Press.

Giere, R. 1999. *Science without laws*. Chicago: University of Chicago Press.

Gilbert, P. 2003. Evolution, social roles, and the differences in shame and guilt. *Social Research* 70: 1205–1230.

Gilovich, A. 1991. *How we know what isn't so*. New York: Free Press.

Gilles de la Tourette, G. 1885. Etude sur une Affection Nerveuse Caractérisée par de l'Incoordination Motrice Accompagnée d'Echolalie et de Coprolalie (Jumping, Latah, Myriachit). *Archives de Neurologie* 9: 19–42, 158–200.

Glymour, C. 1992. Freud's Androids. In *The Cambridge companion to Freud*, ed. J. Neu, 44–85. Cambridge, Cambridge University Press.

Glymour, C. 2001. *The mind's arrows*. Cambridge, MA: The MIT Press.

Godfrey-Smith, P. 1994. A modern history theory of functions. *Nous*, 28: 344–362.

Godfrey-Smith, P. 1996. *Complexity and the function of mind in nature*. Cambridge: Cambridge University Press.

Godfrey-Smith, P. 1998. Functions: Consensus without unity. In *The philosophy of biology*, ed. D. L. Hull and M. Ruse, 280–292. Oxford: Oxford University Press.

Goldberg, E. 2001. *The executive brain*. New York: Oxford University Press.

Goldstein, K. 1944. Methodological approach to the study of schizophrenic thought disorder. In *Language and thought in schizophrenia*, ed. J. S. Kasanin, 17–40. Berkeley: University of California Press.

Goldstein, A. 2001. *Addiction*. 2nd ed. New York: Oxford University Press.

Goodwin, D., and S. Guze. 1996. *Psychiatric Diagnosis*. 5th ed. New York: Oxford University Press.

Goodwin, F., and K. R. Jamison. 1990. *Manic-depressive illness*. New York: Oxford University Press.

Gopnik, A., L. Capps, and A. Meltzoff. 2000. Early theories of mind: What the theory theory can tell us about autism. In *Understanding other minds: Perspectives from developmental cognitive neuroscience*, ed. S. Baron-Cohen, H. Tager-Flusberg, and D. J. Cohen, 50–72. New York: Oxford University Press.

Gopnik, M., and M. Crago. 1991. Familial aggregation of a developmental language disorder. *Cognition* 39: 1–50.

Gopnik, M. 1990a. Dysphasia in an extended family. *Nature* 344: 715

Gopnik, M. 1990b. Feature blindness: A case study. *Language Acquisition* 1: 139–164.

Gotlib, I. 1992. Interpersonal and cognitive aspects of depression. *Current Directions in Psychological Science* 1: 149–154.

Gottesman, I. I. 2002. Defining Genetically Informed Phenotypes for the DSM-V. In J. Z. Sadler (ed) *Descriptions and Prescriptions: Values, Mental Disorders and the DSMs*. Baltimore, MD: Johns Hopkins University Press: 291–300.

Gottesman, I. I., and T. D. Gould. 2003. The endophenotype concept in psychiatry: Etymology and strategic intentions. *American Journal of Psychiatry* 160: 636–645.

Graham, G. 2002. Recent work in philosophical psychopathology. *American Philosophical Quarterly* 39: 109–134.

Graham, G., and L. Stephens, eds. *Philosophical psychopathology*. Cambridge, MA: The MIT Press.

Grandin, T., and M. Scariano. 1986. *Emergence: Labeled autistic*. Tunbridge Wells: Costello.

Grant, B. W. 1999. *The condition of madness*. Lanham, MD: University Press of America.

Gray, J. A. 1987. *The psychology of fear and stress*, 2nd ed. Cambridge: Cambridge University Press.

Gray, J. A., and N. McNaughton. 2000. *The neuropsychology of anxiety*, 2nd ed. New York: Oxford University Press.

Green, M. F. 2001. *Schizophrenia revealed*. New York: W.W. Norton.

Greenfield, J. G., and J. M. Wolfsohn. 1922. The pathology of Sydenham's chorea. *Lancet* 2: 603–606.

Greenwood, J. 2003. Social facts, social groups, and social explanation. *Nous* 37: 93–112.

Greenspan, P. 2001. Good evolutionary reasons: Evolutionary psychiatry and women's depression. *Philosophical Psychology* 14: 327–338.

Griffiths, P. 1997. *What emotions really are*. Chicago: University of Chicago Press.

Griffiths, P. 1999. Author's response. *Metascience* 8: 49–58.

Grossman, D. 1995. *On killing*. Boston: Back Bay Books.

Gutting, G. 1989. *Michel Foucault's archaeology of scientific reason*. Cambridge: Cambridge University Press.

Guze, S. B. 1992. *Why psychiatry is a branch of medicine*. New York: Oxford University Press.

Hacking, I. 1979. Michel Foucault's immature science. *Nous* 13: 39–51.

Hacking, I. 1986. Making up people. In *Reconstructing individualism: Autonomy, individuality and the self in Western thought*, ed. T. C. Heller, M. Sosna, and D. E. Wellbery, 222–236. Stanford: Stanford University Press.

Hacking, I. 1991a. A tradition of natural kinds. *Philosophical Studies* 61: 109–126.

Hacking, I. 1991b. On Boyd. *Philosophical Studies* 61: 149–154.

Hacking, I. 1992a. Making up people. In *Forms of desire: Sexual orientation and the social constructionist controversy*, ed. E. Stein, 69–88. New York: Routledge. Reprinted in I. Hacking, *Historical ontology*: Cambridge, MA: Harvard University Press, 2002.

Hacking, I. 1992b. World-making by kind-making: Child abuse for example. In *How classification works: Nelson Goodman among the social sciences*, ed. M. Douglas and D. Hull, 180–238. Edinburgh: Edinburgh University Press.

Hacking, I. 1993. Working in a new world: The taxonomic solution. In *World changes: Thomas Kuhn and the nature of science*, ed. P. Horwich, 275–310. Cambridge, MA: The MIT Press.

Hacking, I. 1995a. The looping effect of human kinds. In *Causal cognition*, ed. D. Sperber, D. Premack, and A. Premack. Oxford: Oxford University Press.

Hacking, I. 1995b. *Rewriting the soul*. Princeton, NJ: Princeton University Press.

Hacking, I. 1998. *Mad travelers*. Charlottesville: University of Virginia Press.

Hacking, I. 1999. *The social construction of what?* Cambridge, MA: Harvard University Press.

Hagen, E. H. 1999. The functions of postpartum depression. *Evolution and Human Behavior* 20: 325–359.

Hagen, M. 1997. *Whores of the court: The fraud of psychiatric testimony and the rape of American justice*. New York: Harper Collins.

Hale, A. S. and N. R. Pinninti. 1994. Exorcism-resistant ghost possession treated with clopenthixol. *British Journal of Psychiatry* 165: 386–388.

Happé, F. 2000. Parts and wholes, meaning and minds. In S. Baron-Cohen, H. Tager-Flusberg, and D. J. Cohen. 2000. *Understanding other minds: Perspectives from developmental cognitive neuroscience*, 203–221. New York: Oxford University Press.

Hare, R. D. 1965. Psychopathy, fear arousal, and anticipated pain. *Psychological Reports* 16: 499–502.

Hare, R. D. 1970. *Psychopathy: Theory and research*. New York: Wiley.

Hare, R. D. 1986. Twenty years of experience with the Cleckley psychopath. In *Unmasking the psychopath*, ed. W. H. Reid, D. Dorr, J. I. Walker, and J. W. Bonner III. New York: W.W. Norton.

Hare, R. D. 1991. *The Hare psychopathy checklist-revised*. Toronto: Multi-Health Systems.

Hare, R. D. 1999. *Without conscience: The disturbing world of the psychopaths among us*. New York: Guilford Press.

Harrison, P. J., and M. J. Owen. 2003. Genes for schizophrenia? Recent findings and their pathophysiological implications. *The Lancet* 361: 417–419.

Harman, G. 1986. *Change in view: Principles of reasoning.* Cambridge, MA: The MIT Press.

Harman, G. 1995. Rationality. In *Thinking: An invitation to cognitive science,* vol. 3, ed. E. E. Smith and D. Osherson. Cambridge, MA: The MIT Press.

Harpending, H. C., and J. Sobus. 1987. Sociopathy as an adaptation. *Ethology and Sociobiology* 8 (suppl.): 563–572.

Harvey, M. 1995. Translation of "Psychic paralysis of gaze, optic ataxia, and spatial disorder of attention" by Rudolf Balint. *Cognitive Neuropsychology* 12: 261–282.

Haslanger, S. 1995. Ontology and social construction. *Philosophical Topics* 23: 95–125.

Haslanger, S. Forthcoming. Social construction: The "debunking" project. In *Socializing metaphysics,* ed. F. Schmitt. Lanham, MD: Rowman and Littlefield.

Hastings, M. 1998. The Brain, circadian rhythms, and clock genes. *British Medical Journal* 317: 1704–1707.

Hatfield, G. 1999. Mental functions as constraints on neurophysiology: Biology and psychology of vision. In *Where biology meets psychology; Philosophical essays,* ed. V. Hardcastle, 252–271. Cambridge, MA: The MIT Press.

Hawkes, K., J. F. O'Connell, N. G. Blurton Jones, H. Alvarez, and E. L. Charnov. 1998. Grandmothering, menopause, and the evolution of human life histories. *Proceedings of the National Academy of Science* 95: 1336–1339.

Hempel, C. 1965. Fundamentals of taxonomy. In *Aspects of scientific explanation,* 137–154. New York: The Free Press.

Horwitz, A. V. 2002. *Creating mental illness.* Chicago: University of Chicago Press.

Hitchcock, C. 2003. Of Humean bondage. *British Journal for the Philosophy of Science* 54: 1–25.

Hobson, J. A., and J. Leonard. 2001. *Out of its mind: Psychiatry in crisis.* Perseus.

Horwitz, A. V. 2002. *Creating mental illness.* Chicago: University of Chicago Press.

Hull, D. 1970. Contemporary systematic philosophies. *Annual Review of Ecology and Systematics* 1: 19–54.

Hull, D. 1989. On human nature. In *The metaphysics of evolution,* 11–24. Albany: SUNY Press.

Hull, D. 1988. *Science as a process.* Chicago: University of Chicago Press.

Hull, D. 2001. *Science and selection.* Cambridge: Cambridge University Press.

Hume, D. 1966. *Enquiry concerning the principles of morals.* New York: Open Court. Orig. pub. 1777.

Hurley, S. L. 2000. *Consciousness in action.* Cambridge, MA: The MIT Press.

Hurley, S. L. 2001. Perception and action: Alternative views. *Synthese* 129: 3–40.

Hursthouse, R. 1999. *On virtue ethics.* Oxford: Oxford University Press.

Ingleby, D. 1980. Understanding "mental illness". In *Critical psychiatry: The politics of mental health*, 23–71. New York: Pantheon.

Jackson, F. 1998. *From metaphysics to ethics.* Oxford: Basil Blackwell.

Jamison, K. R. 1995. *An unquiet mind.* New York: Vintage.

Jaspers, K. 1959. *General psychopathology*, trans. J. Hoenig and M. W. Hamilton. 7th ed. Baltimore, MD: Johns Hopkins University Press 1997.

Jellinek, E. 1960. *The disease concept of alcoholism.* New Haven, CT: Hillhouse Press.

Johnston, M. 1993. Objectivity refigured: Pragmatism without verificationism. In *Reality, representation, and projection*, ed. J. Haldane and C. Wright, 85–130. New York: Oxford University Press.

Kagan, J. 1994. *Galen's prophecy: Temperament in human nature.* New York: Basic Books.

Kahneman, D., P. Slovic, and A. Tversky, eds. 1982. *Judgment under uncertainty: Heuristics and biases.* Cambridge: Cambridge University Press.

Kandel, E. R. 1998. A new intellectual framework for psychiatry. *American Journal of Psychiatry* 155: 457–469.

Kandel, E. R. 1999. Biology and the future of psychoanalysis: A new intellectual framework for psychiatry revisited. *American Journal of Psychiatry* 156: 505–524.

Kandel, E. R. 2000. The molecular biology of memory storage. In *Les Prix Nobel 2000*, ed. T. Frangsmyr, 283–373. Stockholm: The Nobel Foundation.

Kaplan, J. M. 2000. *The limits and lies of human genetic research.* New York: Routledge.

Kapur, S. 2003. Psychosis as a state of aberrant salience: A framework linking biology, phenomenology, and pharmacology in schizophrenia. *American Journal of Psychiatry* 160: 13–23.

Karmiloff-Smith, A. 1992. *Beyond modularity: A developmental perspective on cognitive science.* Cambridge, MA: The MIT Press.

Karmiloff-Smith, A., E. Klima, U. Bellugi, J. Grant, and S. Baron-Cohen. 1995. Is there a social module? Language, face-processing and theory of mind in individuals with Williams Syndrome. *Journal of Cognitive Neuroscience* 7: 196–208.

Keeley, B. 1999. Fixing content and function in neurobiological systems: The neuroethology of electroreception. *Biology and Philosophy* 14: 395–430.

Keil, F. 1989. *Concepts, kinds, and cognitive development.* Cambridge, MA: The MIT Press.

Keller, M. B., P. W. Lavori, T. I. Mueller, J. Endicott, W. Coryell, R. M. A. Hirschfeld, and T. Shea. 1992. Time to recovery, chronicity, and levels of psychopathology in major depression—A 5 year prospective follow up of 431 subjects. *Archives of General Psychiatry* 49: 809–816.

Kendell, R. E. 1975. *The role of diagnosis in psychiatry*. Oxford: Blackwell Scientific Publications.

Kendell, R. E. 2002. Five criteria for an improved taxonomy of mental disorders. In *Defining psychopathology in the 21st century; DSM-V and beyond*, ed. J. E. Helzer and J. J. Hudziak, 3–17. Washington, DC: American Psychiatric Publishing.

Kendler, K. S. 1990. Towards a scientific psychiatric nosology. *Archives of General Psychiatry* 47: 969–973.

Kendler, K. S. 2001. A psychiatric dialogue on the mind-body problem. *American Journal of Psychiatry* 158: 989–1000.

Kendler, K. S., R. C. Kessler, M. C. Neale, A. C. Heath, and L. Eaves. 1993. The prediction of major depression in women: Towards and integrated etiologic model. *American Journal of Psychiatry* 150: 1139–1148.

Kennedy, I. 1983. *The unmasking of medicine*. London: Allen and Unwin.

Kennedy, N., R. Abbott, and E. S. Paykel. 2003. Remission and recurrence of depression in the maintenance era: Long-term outcome in a Cambridge cohort. *Psychological Medicine* 33: 827–838.

Kennet, J. 2002. Autism, empathy, and moral agency. *Philosophical Quarterly* 52: 340–357.

Kenny, A. 1963. *Action, emotion, and will*. New York: Humanities Press.

Kessler, R., K. McGonagle, S. Zhao, C. Nelson, M. Hughes, W. Eshleman, and K. Kendler. 1994. Lifetime and 12-month prevalence of DSM-III-R psychiatric disorder in the United States. *Archives of General Psychiatry* 51, 8–19.

Kim, N., and W. Ahn. 2002. Clinical psychologists' theory-based representations of mental disorders predict their diagnostic reasoning and memory. *Journal of Experimental Psychology: General* 131: 451–476.

King, R. A., J. F. Leckman, L. Scahill, and D. J. Cohen. 1999. Obsessive-compulsive disorder, anxiety and depression. In *Tourette's syndrome: Tics, obsessions, compulsions*, ed. J. F. Leckman and D. J. Cohen, 43–62. New York: John Wiley.

Kitcher, P. 1984. 1953 and all that: A tale of two sciences. *Philosophical Review* 93: 335–373.

Kitcher, P. 1985. *Vaulting ambition: Sociobiology and the quest for human nature*. Cambridge, MA: The MIT Press.

Kitcher, P. 1997. *The lives to come: The genetic revolution and human possibilities*. Rev. ed. New York: Simon & Schuster.

Kitcher, P. 1998. Function and design. In *The philosophy of biology*, ed. D. L. Hull and M. Ruse, 258–279. Oxford: Oxford University Press.

Kitcher, P. 2001. *Science, truth, and democracy*. New York: Oxford University Press.

Kitcher, P. 2003a. Race, ethnicity, biology, culture. In *In Mendel's mirror: Philosophical reflections on biology*, 230–257. New York: Oxford University Press.

Kitcher, P. 2003b. Battling the undead: How (and how not) to resist genetic determinism. In *In Mendel's Mirror: Philosophical Reflections on Biology*, 283–300. New York: Oxford University Press.

Kornblith, H. 1993. *Inductive inference and its natural ground*. Cambridge, MA: The MIT Press.

Kovacs, J. 1998. The concept of health and disease. *Medicine, Health Care, and Philosophy* 1: 31–39.

Kraepelin, E. 1987. Dementia praecox. In *The clinical roots of the schizophrenia concept*, ed. J. Cutting and M. Shepherd, 13–24. Cambridge: Cambridge University Press. Orig. pub. 1896.

Kraepelin, E. 1990. *Psychiatry: A textbook for students and physicians*. 6th ed. Ed. J. Quen. Trans. H. Metoui and S. Ayed. Canton, MA: Science History Publications. Orig. Pub. 1899.

Kraepelin, E. 1919. *Dementia praecox and paraphrenia*. Ed. G. M Robertson. Trans. R. M. Barclay. Edinburgh: E & S. Livingstone.

Kravitz, E. A. 2000. Serotonin and aggression: Insights gained from a lobster model system and speculations on the role of amine neurons in a complex behavior. *Journal of Comparative Physiology A* 186: 221–238.

Kripke, S. 1980. *Naming and necessity*. Oxford: Basil Blackwell.

Kruger, J., and D. Dunning. 1999. Unskilled and unaware of it: How difficulties in recognizing one's own incompetence lead to inflated self-assessment. *Journal of Personality & Social Psychology* 77: 1121–1134.

Kurlan, R., C. Daragjati, P. G. Como, M. P. McDermott, K. S. Trinidad, S. Roddy, C. A. Brower, and M. M. Robertson. 1996. Non-obscene complex socially inappropriate behavior in Tourette's syndrome. *Journal of Neuropsychiatry and Clinical Neurosciences* 8: 311–317.

Kushner, H. I. 1999. *A cursing brain? The histories of Tourette's syndrome*. Cambridge, MA: Harvard University Press.

Laing, R. D. 1965. *The divided self*. Harmondsworth, UK: Penguin.

Lalumière, M., G. Harris, and M. Rice. 2001. Psychopathy and developmental instability. *Evolution and Human Behavior* 22: 75–92.

Langdon, R., and M. Coltheart. 1999. Mentalising, schizotypy, and schizophrenia. *Cognition* 71: 43–71.

Langdon, R., and M. Coltheart. 2000. The cognitive neurospychology of delusions. In *Pathologies of belief,* ed. M. Coltheart and M. Davies, 183–216. Oxford: Basil Blackwell.

Leckman, J. F., R. A. King, and D. J. Cohen. 1999. Tics and tic disorders. In *Tourette's syndrome: Tics, obsessions, compulsions,* ed. J. F. Leckman and D. J. Cohen, 23–42. New York: John Wiley.

Leshner, A. I. 1997. Addiction is a brain disease, and it matters. *Science* 278: 45–47.

Leslie, A. 1987. Pretence and representation: The origins of "theory of mind." *Psychological Review* 94: 412–426.

Leslie, A. 1992. Autism and the "theory of mind" module. *Current Directions in Psychological Science* 1: 18–21.

Leslie, A. 1994. ToMM, ToBY, and Agency: Core architecture and domain specificity. In *Mapping the mind,* L. Hirschfeld and S. Gelman. Cambridge: Cambridge University Press.

Leslie, A., and L. Thaiss. 1992. Domain specificity in conceptual development: neuropsychological evidence from autism. *Cognition* 43: 225–251.

Lewis, D. 1972. Psychophysical and theoretical identification. *Australasian Journal of Philosophy* 50: 249–258.

Lewis, D. A., and P. Levitt. 2002. Schizophrenia as a disorder of neurodevelopment. *Annual Review of Neuroscience* 25: 409–432.

Lewis, D. A., and J. A. Lieberman. 2000. Catching up on schizophrenia: Natural history and neurobiology. *Neuron* 28: 325–334.

Lewontin, R. 1974. The analysis of variance and the analysis of causes. *American Journal of Human Genetics* 26: 400–411.

Lhermitte, F. 1983. "Utilization behavior" and its relation to lesions of the frontal lobes. *Brain* 106: 237–255.

Lilienfeld, S. O., and L. Marino. 1995. Mental disorder as a Roschian concept: A critique of Wakefield's "harmful dysfunction" analysis. *Journal of Abnormal Psychology* 104: 411–420.

Lilla, M. 2001. *The reckless mind: Intellectuals in politics.* New York: New York Review Books.

Lloyd, E. 2005. *The case of the female orgasm: Bias in the science of evolution.* Cambridge, MA: Harvard University Press.

Loftus, E., and K. Ketcham. 1996. *The myth of repressed memory.* New York: St Martin's Press.

Longino, H. 2001. *The fate of knowledge.* Princeton, NJ: Princeton University Press.

Lycan, W. 1990. The continuity of levels of nature. In *Mind and cognition,* ed. W. Lycan, 77–96. Oxford: Basil Blackwell.

Lykken, D. 1995. *The antisocial personalities*. Hillsdale, NJ: Erlbaum.

Lynn, S. J., S. O. Lilienfeld, and J. M. Lohr, eds. 2002. *Science and pseudoscience in clinical psychology*. New York: Guilford Press.

Mace, C. 2002. Survival of the fittest? Conceptual selection in psychiatric nosology. In *Descriptions and Prescriptions: Values, Mental Disorders, and the DSMs*, ed. J. Z. Sadler, 56–75. Baltimore, MD: Johns Hopkins University Press.

MacIntyre, A. 1982. *After virtue: A study in moral theory*. London: Duckworth.

Macmillan, M. 2000. *An odd kind of fame: Stories of Phineas Gage*. Cambridge, MA: The MIT Press.

Maguire, E. A., D. G. Gadian, I. S. Johnsrude, C. D. Good, J. Ashburner, R. S. J. Frackowiak, and C. D. Frith. 2000. Navigation-related structural change in the hippocampi of taxi drivers. *Proceedings of the National Academy of Sciences of the United States of America* 97: 4398–4403.

Maher, B. A. 1988. Anomalous experience and delusional thinking: The logic of explanations. In *Delusional beliefs*, ed. T. F. Oltmanns and B. A. Maher, 15–53. New York: Plenum.

Mallon, R. Forthcoming. Social constructionism, social roles, and stability. To appear in *Socializing metaphysics*, ed. F. Schmitt. Lanham, MD: Rowman and Littlefield.

Mallon, R., and S. Stich. 2000. The odd couple: The compatibility of evolutionary psychology and social construction. *Philosophy of Science* 67: 133–154.

Marchetti, C, and S. Della Salla. 1998. Disentangling the alien and anarchic hand. *Cognitive Neuropsychiatry* 3: 191–208.

Margolis, J. 1980. The concept of mental illness. In *Mental illness: Law and public policy*, B. A. Brody and H. T. Engelhardt, 3–23. Dordrecht, the Netherlands: D. Reidel.

Marks, I. M. 1987. *Fears, phobias, and rituals*. Oxford: Oxford University Press.

Marks, I. M., and R. Nesse. 1994. Fear and fitness: An evolutionary analysis of anxiety disorders. *Ethology and Sociobiology* 15: 247–261.

Marr, D. 1982. *Vision*. San Francisco, CA: W. H. Freeman.

Marshall, J. C. and P. W. Halligan. 1996. Towards a cognitive neuropsychiatry. In *Method in madness: Case studies in cognitive neuropsychiatry*, ed. P. W. Halligan and J. C. Marshall, 3–11. Hove, UK: Psychology Press.

Marshall, P. J., Y. Bar-Haim, and N. A. Fox. 2004. The development of P50 suppression in the auditory event-related potential. *International Journal of Psychophysiology* 51: 135–141.

Martin, J. B. 2002. The integration of neurology, psychiatry, and neuroscience in the 21st century. *American Journal of Psychiatry* 159: 695–704.

Maudsley, H. 1867. *Physiology and pathology of the mind*. New York: Appleton. Reprinted 1977, Washington, DC: University Publications of America.

Maudsley, H. 1870. *Body and mind*. London: MacMillan.

Maynard Smith, J. 1982. *Evolution and the theory of games*. New York: Cambridge University Press.

McCarthy, L., and J. Gerring. 1994. Revising psychiatry's charter document: DSM-IV. *Written communication* 11: 147–192.

McGinn, C. 1989. Can we solve the mind-body problem? *Mind* 98: 349–366.

McGlashan, T. H., and R. E. Hoffman. Schizophrenia as a disorder of developmentally reduced synaptic connectivity. *Archives of General Psychiatry* 57: 637–648.

McGuire, M., F. Fawzy, J. Spar, R. Weigel, and A. Troisi. 1994. Altruism and mental disorders. *Ethology and sociobiology*, 15: 299–321.

McGuire, M., and A. Troisi. 1998. *Darwinian psychiatry*. New York: Oxford University Press.

McHugh, P., and P. Slavney. 1998. *The perspectives of psychiatry*. 2nd ed. Baltimore, MD: Johns Hopkins University Press.

McKay, P., P. J. McKenna, and K. Laws. 1996. Severe schizophrenia: What is it like? In *Method in madness*, ed. P. W. Halligan and J. C. Marshall. Hove, UK: Psychology Press.

McKenna, P. J. 1994. *Schizophrenia and related syndromes*. New York: Oxford University Press.

McNally, R. 1994. *Panic disorder*. New York: Guilford Press.

Mealey, L. 1995. The sociobiology of sociopathy: An integrated evolutionary model. *Behavioral and brain sciences* 18: 523–599.

Meehl, P. E. 1957. When shall we use our heads instead of the formula? *Journal of Counseling Psychology* 4: 268–273.

Meehl, P. E. 1973. Why I do not attend case conferences. In *Psychodiagnostics: Selected papers*, ed. P. E. Meehl, 225–303. Minneapolis: University of Minnesota Press.

Meehl, P. E. 1977. Specific etiology and other forms of strong influence: Some quantitative meanings. *Journal of Medicine and Philosophy* 2: 33–53.

Meehl, P. E. 1986. Diagnostic taxa as open concepts. In *Contemporary Directions in Psychopathology: Towards the DSM-IV*, ed. T. Millon and G. L. Klerman, 215–231. New York: Guilford Press.

Meehl, P. E. 1990. Toward an integrated theory of schizotaxy, schizotypy and schizophrenia. *Journal of Personality Disorders* 1–99.

Meehl, P. E. 1992. Factors and taxa, traits and types, differences of degree and differences in kind. *Journal of Personality* 60: 117–174.

Meehl, P. E. 1993. Philosophy of science: Help or hindrance? *Psychological Reports* 72: 707–733.

Meehl, P. E. 1995. Bootstraps taxometrics: Solving the classification problem in psychopathology. *American Psychologist* 50: 266–275.

Meehl, P. E. 1999. Clarifications about taxometric method. *Applied and Preventive Psychology* 8: 165–174.

Megone, C. 1998. Aristotle's function argument and the concept of mental illness. *Philosophy, Psychiatry & Psychology* 5: 187–201.

Mellor, C. S. 1970. First-rank symptoms of schizophrenia. *British Journal of Psychiatry* 117: 15–23.

Meloy, J. R. 1988. *The psychopathic mind*. Northvale, NJ: Jason Aronson.

Menzies, P. 1996. Probabilistic causation and the pre-emption problem. *Mind* 417: 85–118.

Meyer, J. H., S. McMain, S. H. Kennedy, L. Korman, G. M. Brown, J. N. DaSilva, A. A. Wilson, T. Blak, R. Eynan-Harvey, V. S. Goulding, S. Houle, and P. Links. 2003. Dysfunctional attitudes and 5-HT2 receptors during depression and self-harm. *American Journal of Psychiatry* 160: 90–99.

Micale, M. 1995. *Approaching hysteria*. Princeton, NJ: Princeton University Press.

Miller, G. A. 1996. How we think about cognition, emotion, and biology in psychopathology. *Psychophysiology* 33: 615–628.

Millikan, R. 1989. In defense of proper functions. *Philosophy of Science* 56: 188–202.

Milner, A. D. 1998. Neuropsychological studies of perception and visuomotor control. *Philosophical Transactions of the Royal Society of London, B* 353: 1375–1384.

Mineka, S., M. Davidson, M. Cook, and R. Keir. 1984. Observational conditioning of snake fear in rhesus monkeys. *Journal of Abnormal Psychology* 93: 355–372.

Mineka, S., R. Keir, and V. Price. 1980. Fear of snakes in wild and laboratory-reared rhesus monkeys. *Animal Learning and Behavior* 8: 653–663.

Mineka, S., and A. Tomarken. 1989. The role of cognitive biases in the origins and maintenance of fear and anxiety disorders. In *Aversion, avoidance, and anxiety: Perspectives on aversively motivated behavior*, ed. L. Nilsson and T. Archer. Hove, UK: Lawrence Erlbaum.

Mishara, A. 1994. A phenomenological critique of commonsensical assumptions in DSM-III-R: The avoidance of the patient's subjectivity. In *Philosophical perspectives on psychiatric diagnostic classification*, ed. J. Sadler, O. Wiggins, and M. Schwartz, 129–147. Baltimore, MD: Johns Hopkins University Press.

Mishkin, M., L. G. Ungerleider, and K. A. Macko. 1983. Object vision and spatial vision: Two cortical pathways. *Trends in Neurosciences* 6: 414–417.

Moore, M. S. 1980. Legal conceptions of mental illness. In *Mental illness: Law and public policy*, B. A. Brody and H. T. Engelhardt, 25–69. Dordrecht, the Netherlands: D. Reidel.

Moss, L. 2003. *What genes can't do*. Cambridge, MA: The MIT Press.

Mumford, D. B., A. M. Whitehouse, and I. Y. Choudry. 1992. Survey of eating disorders in English-medium schools in Lahore, Pakistan. *International Journal of Eating Disorders* 11: 173–184.

Mumford, D. B., A. M. Whitehouse, and M. Platts. 1991. Sociocultural correlates of eating disorders among Asian schoolgirls in Bradford. *British Journal of Psychiatry* 158: 222–228.

Murphy, D. 2001. Hacking's reconciliation: Putting the biological and sociological together in the explanation of mental illness. *Philosophy of the Social Sciences* 31 (2): 139–162.

Murphy, D. 2003. Adaptationism and psychological explanation. In *Evolutionary psychology: Alternative approaches*, ed. F. Rauscher and F. Scher, 161–184. Dordrecht, the Netherlands: Kluwer.

Murphy, D., and S. Stich. 1999. Eliminating emotions? *Metascience* 8: 13–25.

Murphy, D., and S. Stich. 2000. Darwin in the madhouse: Evolutionary psychology and the classification of mental disorders. In *Evolution and the human mind*, ed. P. Carruthers and A. Chamberlain, 62–92. Cambridge: Cambridge University Press.

Murphy, D., and R. L. Woolfolk. 2000a. The harmful dysfunction analysis of mental disorder. *Psychiatry, Philosophy, Psychology* 7: 241–252.

Murphy, D., and R. L. Woolfolk. 2000b. Conceptual analysis versus scientific understanding: An assessment of Wakefield's folk psychiatry. *Psychiatry, Philosophy, Psychology* 7: 271–293.

Murphy, J. G. 1972. Moral death: A Kantian essay on psychopathy. *Ethics* 82: 284–298.

Nagel, E. 1961. *The structure of science*. New York: Harcourt, Brace.

Nathan, P. E., and J. W. Langenbucher. 1999. Psychopathology: Description and classification. *Annual Review of Psychology* 50: 79–107.

Neander, K. 1991. Functions as selected effects: The conceptual analyst's defense. *Philosophy of Science* 58: 168–84.

Nesse, R. M. 1998. Emotional disorders in evolutionary perspective. *British Journal of Medical Psychology* 71: 397–415.

Nesse, R. M. 1999. Testing evolutionary hypotheses about mental disorders. In *Evolution in health and disease*, ed. S. C. Stearns, 261–266. New York: Oxford University Press.

Nessee, R. M. 2000. Is depression an adaptation? *Archives of General Psychiatry 57*: 14–20.

Nesse, R. M., and G. C. Williams. 1995. *Why we get sick*. New York: Times Books.

Neu, J. 1980. Minds on trial. In *Mental illness: Law and public policy*, ed. B. A. Brody and H. T. Engelhardt, 73–105. Dordrecht, the Netherlands: D. Reidel.

Newman, J. 1998. Psychopathic behavior: An information processing perspective. In *Psychopathy: Theory, research, and implications for society*, ed. D. J. Cooke, A. E. Forth, and R. D. Hare, 81–104. Dordrecht, the Netherlands: Kluwer.

Nichols, S. 2002a. How psychopaths threaten moral rationalism, or is it irrational to be amoral? *The Monist* 85: 285–304.

Nichols, S. 2002b. On the genealogy of norms: A case for the role of emotion in cultural evolution. *Philosophy of Science* 69: 234–255.

Nichols, S. 2002c. Norms with feeling: Towards a psychological account of moral judgment. *Cognition* 84: 221–236.

Nichols, S., and S. P. Stich. 2003. Mindreading: *An integrated account of pretence, self-awareness, and understanding other minds*. New York: Oxford University Press.

Nisbett, R., and L. Ross. 1980. *Human inference: Strategies and shortcomings of social judgment*. Englewood Cliffs, NJ: Prentice-Hall.

North, C. S., J-E. M. Ryan, D. A. Ricci, and R. D. Wetzel. 1993. *Multiple personalities, multiple disorders*. New York: Oxford University Press.

Nunn, J. 1996. *Ancient Egyptian medicine*. London: British Museum Press.

Offer, D., and M. Sabshin. 1980. Normality. In *Comprehensive textbook of psychiatry*, vol. 2, ed. H. I. Kaplan, A. M. Freedman, and B. J. Sadock, 608–613. New York: Williams & Wilkins.

Ofshe, R., and E. Watters. 1996. *Making monsters: False memories, psychotherapy, and sexual hysteria*. Berkeley, CA: University of California Press.

Ohman, A. 1979. Fear relevance, autonomic conditioning, and phobias: A laboratory model. In *Trends in behavior therapy*, ed. P. O. Sjoden, S. Bates, and W. W. Dockens, 107–133. New York: Academic Press.

Ohman, A. 1986. Face the beast and fear the face: Animal and social fears as prototypes for evolutionary analyses of emotion. *Psychophysiology* 23: 123–145.

Omi, M., and H. Winant. 1994. *Racial formation in the United States: From the 1960s to the 1990s*. New York: Routledge.

O'Reilly, R. C., and Y. Munakata. 2000. *Computational explorations in cognitive neuroscience*. Cambridge, MA: The MIT Press.

Orford, J. 1985. *Excessive appetites*. Chichester, UK: Wiley.

Orphanides, A., and D. Zervos. 1995. Rational addiction with learning and regret. *Journal of Political Economy* 103: 739–758.

Oyama, S. 2000. *The ontogeny of information*. 2nd ed. Durham, NC: Duke University Press.

Palmer, 1999. *Vision science: From photons to phenomenology*. Cambridge, MA: The MIT Press.

Papineau, D. 1994. Mental disorder, illness, and biological dysfunction. *Philosophy, Psychology and Psychiatry: Royal Institute of Philosophy Supplement* 37: 73–82.

Parkin, A. J. 1996. The alien hand. In *Method in madness: Case studies in cognitive neurospychiatry*, P. W. Halligan and J. C. Marshall, 173–183. Hove, UK: Psychology Press.

Parnas, J., F. Schulsinger, H. Schulsinger, S. A. Mednick, and T. T. Teasdle. 1983. Behavioral precursors of schizophrenia spectrum. *Archives of General Psychiatry* 39: 658–664.

Parry-Jones, W. L. L., and B. Parry-Jones. 1995. Eating disorders: Social section. In *A history of clinical psychiatry*, ed. G. Berrios and R. Porter, 602–611. New York: New York University Press.

Patterson, P. 2002. Maternal infection: A window on neuroimmune interactions in fetal brain development and mental illness. *Current Opinion in Neurobiology* 12: 115–118.

Peacocke, C. 1994. *A study of concepts*. Cambridge, MA: The MIT Press.

Pearl, J. 2000. *Causality*. Cambridge: Cambridge University Press.

Perenin, M. T., and A. Vighetto. 1988. Optic ataxia: A specific disruption in visuo-motor mechanisms. *Brain* 111: 643–674.

Phillips, J. 2000. Conceptual models for psychiatry. *Current Opinion in Psychiatry* 13: 683–688.

Phillips, W., S. Baron-Cohen, and M. Rutter. 1998. Understanding intention in normal development and in autism. *British Journal of Developmental Psychology* 16: 337–348.

Piateli-Palmarini, M. 1994. *Inevitable illusions*. New York: Wiley.

Pinker, S. 1984. *Language learnability and language development*. Cambridge, MA: Harvard University Press.

Pinker, S. 1994. *The language instinct*. New York: W. Morrow & Co.

Pinker, S. 1997. *How the mind works*. New York: W. W. Norton.

Pinker, S. 2005. So how *does* the mind work? *Mind and Language* 20: 1–24.

Plomin, R., and P. McGuffin. 2003. Psychopathology in the postgenomic era. *Annual Review of Psychology* 54: 205–228.

Poland, J., B. von Eckhardt, and W. Spaulding. 1994. Problems with the DSM approach to classifying psychopathology. In *Philosophical psychopathology*, ed. G. Graham and G. L. Stephens, 235–260. Cambridge, MA: The MIT Press.

Porter, R. 1987. *Mind-forg'd manacles: A history of madness in England from the Restoration to the Regency*. Cambridge, MA: Harvard University Press.

Price, B. H., R. D. Adams, and J. T. Coyle. 2000. Neurology and psychiatry: Closing the great divide. *Neurology* 54: 8–14.

Price, J., L. Sloman, R. Gardner, P. Gilbert, and P. Rohde. 1994. The social competition hypothesis of depression. *British Journal of Psychiatry* 164: 309–315.

Putnam, H. 1975. The meaning of "meaning." In *Mind, language, and reality*: 215–271. Cambridge: Cambridge University Press.

Pylyshyn, Z. 1984. *Computation and cognition*. Cambridge, MA: The MIT Press.

Pyszczynski, T., and J. Greenberg. 1987. Self-regulatory perseveration and the depressive self-focussing style: A self-awareness theory of reactive depression. *Psychological Bulletin* 102: 122–138.

Quartz, S., and T. Sejnowski. 2002. *Liars, lovers, and heroes: What the new brain science reveals about who we are*. New York: HarperCollins.

Radden, J. 1996. Lumps and bumps: Kantian faculty psychology, phrenology, and twentieth century psychiatric classification. *Philosophy, Psychiatry, Psychology* 3: 1–14.

Railton, P. 1986. Moral realism. *Philosophical Review* 95: 163–207.

Raine, A. 1993. *The psychopathology of crime: Criminal behavior as a clinical disorder*. San Diego: Academic Press.

Ramachandran, V. S., and W. Hirstein. 1998. The perception of phantom limbs—The D. O. Hebb lecture. *Brain* 121: 603–630.

Rawls, J. 1971. *A theory of justice*. Oxford: Oxford University Press.

Regard, M., and T. Landis. 1997. "Gourmand syndrome": Eating passion associated with right anterior lesions. *Neurology* 48: 1185–1190.

Reznek, L. 1995. Dis-ease about kinds. *Journal of Medicine and Philosophy* 20 (5): 571–584.

Ridley, M. 1986. *Evolution and classification: The reformation of cladism*. New York: Lonngman's.

Ringel, T. M., A. Heidrich, C. P. Jacob, and A. J. Fallgeter. 2004. Sensory gating deficit in a subtype of chronic schizophrenic patients. *Psychiatry Research* 125: 237–245.

Robert, J. S. 2000. Schizophrenia epigenesis? *Theoretical Medicine and Bioethics* 21: 191–215.

Robertson, M. M., and S. Baron-Cohen. 1998. *Tourette syndrome: The facts.* New York: Oxford University Press.

Robinson, D. N. 1998. *Wild beasts and idle humours: The insanity defense from antiquity to the present.* Cambridge, MA: Harvard University Press.

Robinson, T. E., and K. C. Berridge. 2003. Addiction. *Annual Review of Psychiatry* 54: 25–53.

Robins, E., and S. B. Guze. 1970. Establishment of diagnostic validity in psychiatric disorders: Its application to schizophrenia. *American Journal of Psychiatry* 126: 983–987.

Robins, L., and J. E. Barrett. 1989. Introduction. In *The validity of psychiatric diagnosis*, ed. L. Robins and J. E. Barrett, New York: Raven Press.

Rohde, P., P. M. Lewinsohn, M. Tilson, and J. R. Seeley. 1990. Dimensionality of coping and its relation to depression. *Journal of Personality and Social Psychology* 58: 499–511.

Rosenhan, R. 1973. On being sane in insane places. *Science* 179: 251–258.

Rutter, M., A. Pickles, R. Murray, and L. Eaves. 2001. Testing hypotheses on specific environmental causal effects on behavior. *Psychological Bulletin* 127: 291–324.

Sachs, O. 1995. *An anthropologist on Mars: Seven paradoxical tales.* New York: Knopf.

Sadler, J. Z., and G. J. Agich. 1995. Diseases, functions, values, and psychiatric classification. *Philosophy, Psychiatry, and Psychology* 2: 219–231.

Sarkar, S. 1998. *Genetics and reductionism.* Cambridge: Cambridge University Press.

Sass, L. 1992. *Madness and modernism: Insanity in the light of modern art, literature, and thought.* Cambridge, MA: Harvard University Press.

Sass, L. 1994. *The paradoxes of delusion: Wittgenstein, Schreber, and the schizophrenic mind.* Ithaca, NY: Cornell University Press.

Schaffner, K. F. 1993. *Discovery and explanation in biology and medicine.* Chicago: University of Chicago Press.

Schaffner, K. F. 1994. Reductionistic approaches to schizophrenia. In *Philosophical perspectives on psychiatric diagnostic classification*, ed. J. Sadler, O. Wiggins, and M. Schwartz, 279–294. Baltimore, MD: Johns Hopkins University Press.

Schaffner, K. F. 1998. Genes, behavior, and developmental emergentism: One process, indivisible? *Philosophy of Science* 65: 209–252.

Schaffner, K. F. 1999. Complexity and research strategies in behavioral genetics. In *Behavioral genetics: The clash of culture and biology*, ed. R. A. Carson and M. A. Rothstein, 1–88. Baltimore, MD: Johns Hopkins University Press.

Schaffner, K. F. 2002. Clinical and etiological psychiatric diagnoses: Do causes count? In *Descriptions and prescriptions: Values, mental Disorders, and the DSMs*, ed. J. Z. Sadler, 271–290. Baltimore, MD: Johns Hopkins University Press.

Scheff, T. J. 1999. *Being mentally ill: A sociological theory*. 3rd ed. New York: Aldine de Gruyter.

Schelling, T. 1984. The intimate struggle for self-command. In *Choice and consequence*, T. Schelling, 57–82. Cambridge, MA: Harvard University Press.

Schiffer, S. 1991. Ceteris paribus laws. *Mind* 100: 1–17.

Schneider, K. 1958. *Clinical psychopathology*. 5th ed. Trans. M. Hamilton. New York: Grune & Stratton.

Schreber, D. P. 1903. *Memoirs of my nervous illness*. Reprinted 2000, New York: New York Review Books.

Schwartz, J. M., P. W. Stoessel, L. R. Baxter, K. M. Martin, and M. E. Phelps. 1996. Systematic changes in cerebral glucose metabolic rate after successful treatment of obsessive-compulsive disorder. *Archives of General Psychiatry* 53: 109–113.

Sedgwick, P. 1982. *Psycho politics: Laing, Foucault, Goffman, Szasz, and the future of mass psychiatry*. New York: Harper & Row.

Segal, G. 1996. The modularity of theory of mind. In *Theories of theories of mind*, ed. P. Carruthers and P. K. Smith, 141–157. Cambridge: Cambridge University Press.

Sellars, W. 1963a. Philosophy and the scientific image of man. In *Science, perception and reality*, W. Sellars, 1–40. Reissued 1991, Atascadero, CA: Ridgeview.

Sellars, W. 1963b. Empiricism and the philosophy of mind. In *Science, perception and reality*, W. Sellars, 127–196. Reissued 1991, Atascadero, CA: Ridgeview.

Sen, A. 2003. *Rationality and freedom*. Cambridge, MA: Harvard University Press.

Shallice, T. 1988. *From neuropsychology to mental structure*. Cambridge: Cambridge University Press.

Shephard, B. 2001. *A war of nerves: Soldiers and psychiatrists in the twentieth century*. Cambridge, MA: Harvard University Press.

Shi, L., S. H. Fatemi, R. W. Sidwell, and P. H. Patterson. 2003. Maternal influenza infection causes marked behavioral and pharmacological changes in the offspring. *Journal of Neuroscience* 23: 297–302.

Shorter, E. 1997. *A brief history of psychiatry*. New York: Wiley.

Showalter, E. 1997. *Hystories*. New York: Columbia University Press.

Siekmeyer, C., and R. Hoffman. 2002. Enhanced semantic priming in schizophrenia: A computer model based on excessive pruning of local connections in association cortex. *British Journal of Psychiatry* 180: 345–350.

Simons, R. C. 1996. *Boo: Culture, experience, and the startle reflex*. New York: Oxford University Press.

Skyrms, B. 1996. *Evolution of the social contract*. Cambridge: Cambridge University Press.

Skyrms, B. 2000. Stability and explanatory significance of some evolutionary models. *Philosophy of Science* 67: 94–113.

Slater, L. 2004. *Opening Skinner's box: Great psychological experiments of the twentieth century*. New York: W.W. Norton.

Smith, M. 1994. *The moral problem*. Oxford: Blackwell.

Smith, V. L. 1991. Rational choice: The contrast between economics and psychology. *Journal of Political Economy* 99: 877–897.

Snowden, P. 1990. The objects of perceptual experience-I. *Proceedings of the Aristotelian Society* 64 (suppl.): 121–150.

Snyder, S. H., S. P. Banerjee, H. I. Yamamura, and D. Greenberg. 1974. Drugs, neurotransmitters, and schizophrenia. *Science* 184: 1243–1253.

Sober, E. 1988. *Reconstructing the past*. Cambridge, MA: The MIT Press.

Sober, E., and D. Sloan Wilson. 1998. *Unto others: The evolution and psychology of unselfish behavior*. Cambridge, MA: Harvard University Press.

Sorenson, R. 2000. Faking Munchausen's syndrome. *Analysis* 60: 202–208.

Spaulding, W., M. Sullivan, and J. Poland. 2003. *Treatment and rehabilitation of severe mental illness*. New York: Guilford Press.

Sperber, D. 1994. The modularity of thought and the epidemiology of representations. In *Mapping the mind: Domain specificity in cognition and culture*, L. A. Hirschfeld and S. Gelman, 39–67. Cambridge: Cambridge University Press.

Spitzer, R. L. 1975. On pseudoscience in science, logic in remission, and psychiatric diagnosis: A critique of Rosenhan's "On being sane in insane places." *Journal of Abnormal Psychology* 84: 442–452.

Spitzer, R. L., and J. Endicott. 1978. Medical and mental disorder: Proposed definition and criteria. In *Critical issues in psychiatric diagnosis*, ed. R. L. Spitzer and D. F. Klein, 15–39. New York: Raven Press.

Spitzer, R. L., M. Gibbon, A. Skodol, J. B. Williams, and M. B. First. 1994. *DSM-IV casebook*. Washington, DC: American Psychiatric Press.

Spitzer, R. L., and J. C. Wakefield. 1999. DSM-IV diagnostic criterion for clinical significance: Does it help solve the false positives problem? *American Journal of Psychiatry* 156: 1856–1864.

Steiger, H., L. Gauvin, M. Israel, N. Koerner, N. M. K. Ng Ying Kin, J. Paris, and S. N. Young. 2001. Association of serotonin and cortisol indices with childhood abuse in bulimia nervosa. *Archives of General Psychiatry* 58: 837–843.

Stein, E. 1998. Essentialism and constructionism about sexual orientation. In *The philosophy of biology*, ed. D. Hull and M. Ruse, 427–442. Oxford: Oxford University Press.

Stephens, L., and G. Graham. 2000. *When self-consciousness breaks*. Cambridge, MA: The MIT Press.

Stephens, G. L. and G. Graham, G. 2006. The delusional stance. Forthcoming in *Reconceiving schizophrenia*, ed. M. Chung, K. W. M. Fulford, and G. Graham. Oxford: Oxford University Press.

Sterelny, K. 2003. *Thought in a hostile world: The evolution of human cognition*. Oxford: Blackwell.

Stevens, A., and J. Price. 1996. *Evolutionary psychiatry: A new beginning*. London: Routledge.

Stich, S. 1998. *Deconstructing the mind*. New York: Oxford University Press.

Stone, T., and A. Young. 1997. Delusions and brain injury: The philosophy and psychology of belief. *Mind and Language* 12: 327–364.

Sulloway, F. *Freud: Biologist of the mind*. New York: Basic Books.

Sullum, J. 2003. *Saying yes: In defense of drug use*. New York: Tarcher.

Swedo, S. E., H. L. Leonard, M. Garvey, B. Mittleman, A. J. Allean, S. Perlmutter, L. Lougee, S. Dow, J. Zamkoff, and B. K. Dubbert. 1998. Pediatric autoimunne neuropsychiatric disorders associated with streptococcal infection: Clinical decription of the first 50 cases. *American Journal of Psychiatry* 155: 264–271.

Swerdlow, N., and G. Koob. 1987. Dopamine, schizophrenia, mania, and depression: Toward a unified hypothesis of cortico-striato-pallido-thalamic function. *Behavioral and Brain Sciences* 10: 197–245.

Symons, D. 1979. *The evolution of human sexuality*. New York: Oxford University Press.

Szasz, T. 1987. *Insanity*. New York: Wiley.

Taylor, K. A. 2000. What in nature is the compulsion of reason? *Synthese* 122: 209–244.

Thase, M. E., and R. H. Howland. 1995. Biological processes in depression. In *Handbook of Depression*, ed. E. E. Beckham and W. R. Leber. 2nd ed. New York: Guilford Press.

Thorne, J. 1996. *Simple cooking*. New York: North Point Press.

Tooby, J., and L. Cosmides. 1992. The psychological foundations of culture. In *The adapted mind: Evolutionary psychology and the generation of culture*, ed. J. Barkow, L. Cosmides, and J. Tooby, 19–136. Oxford: Oxford University Press.

Tooby, J., and L. Cosmides. 1995. Foreword. In *Mindblindness*, S. Baron-Cohen, xi–xviii. Cambridge, MA: The MIT Press.

Tourette Syndrome Classification Study Group. 1993. Definition and classification of tic disorders. *Archives of Neurology* 50: 1013–1016.

Troisi, A., and M. T. McGuire. 2000. Psychotherapy and Darwinian psychiatry. In *Genes on the couch: Explorations in evolutionary psychotherapy*, ed. P. Gilbert and K. G. Bailey, 28–41. Hove, UK: Brunner-Routledge.

Trivers, R. 1971. The evolution of reciprocal altruism. *Quarterly Review of Biology* 46: 35–57.

Trout, J. D. 1998. *Measuring the intentional world: Realism, naturalism, and quantitative methods in the behavioral sciences*. New York: Oxford University Press.

Turner, M. A. 2003. Psychiatry and the human sciences. *British Journal of Psychiatry* 182: 472–474.

Turner, S. 2002. *Brains/practices/relativism: Social theory after cognitive science*. Chicago: University of Chicago Press.

Ung, E. K. 2003. Eating disorders in Singapore: A review. *Annals of the Academy Medicine of Singapore* 32: 19–24.

Vaillant, G. 1995. *The natural history of alcoholism revisited*. Cambridge, MA: Harvard University Press.

Vaillant, G. 2003. Mental health. *American Journal of Psychiatry* 160: 1373–1384.

Valentine, C. W. 1930. The innate bases of fear. *Journal of Genetic Psychology* 37: 394–419.

Van Deth, R., and W. Vandereycken. 1995. Eating disorders: Clinical section. In *A history of clinical psychiatry*, ed. G. Berrios and R. Porter, 593–601. New York: New York University Press.

Verhoeven, W. M. A., and S. Tuinier. 1999. Neuropsychiatry or biological psychiatry? *Acta Neuropsychiatrica* 11: 80–84.

Wade, M. 1977. An experimental study of group selection. *Evolution* 31: 134–153.

Wakefield, J. 1992a. The concept of mental disorder. *American Psychologist* 47: 373–388.

Wakefield, J. 1992b. Disorder as harmful dysfunction: A conceptual critique of DSM-III-R's definition of mental disorder. *Psychological Review* 99: 232–247.

Wakefield, J. 1993. Limits of operationalization: A critique of spitzer and Endicott's (1978) proposed operational criteria of mental disorder. *Psychological Review* 99: 232–247.

Wakefield, J. 1996. Dysfunction as a value free concept. *Philosophy, Psychology, and Psychiatry* 2: 233–246.

Wakefield, J. 1997a. Diagnosing DSM-IV, part 1: DSM-IV and the concept of disorder. *Behavior Research and Therapy* 35: 633–649.

Wakefield, J. 1997b. Normal inability versus pathological inability: Why Ossorio's definition of mental disorder is not sufficient. *Clinical Psychology: Science and Practice* 4: 249–258.

Wakefield, J. 1999a. Philosophy of science and the progressiveness of the DSM's theory-neutral nosology: Response to Follette and Houts, part 1. *Behavior Research and Therapy* 37: 963–999.

Wakefield, J. 1999b. Philosophy of science and the progressiveness of the DSM's theory-neutral nosology: Response to Follette and Houts, part 2. *Behavior Research and Therapy* 37: 1001–1027.

Wakefield, J. 2000. Spandrels, vestigial organs, and such: A reply to Murphy and Woolfolk's "The Harmful Dysfunction Analysis of Mental Disorder." *Philosophy, Psychiatry, Psychology* 7 (4): 253–269.

Walsh, D. 1996. Fitness and function. *British Journal for the Philosophy of Science* 47: 553–574.

Waters, C. K. 1990. Why the anti-reductionist consensus won't survive: The case of classical Mendelian genetics. *Proceedings of the Biennial Meetings of the Philosophy of Science Association*, 125–139.

Watson, J. B. 1924. *Behaviorism*. New York: W.W. Norton.

Watson, P. J., and P. W. Andrews. 2002. Towards a revised evolutionary adaptationist analysis of depression: The social navigation hypothesis. *Journal of Affective Disorders* 72: 1–14.

Whewell, W. 1840. *Philosophy of the inductive sciences*. 2 vols. London: Parker.

Whitlock, F. A. 1967. The Ganser syndrome. *British Journal of Psychiatry* 113: 19–29.

Widiger, T. A., and L. A. Clark. 2000. Towards DSM-V and the classification of psychopathology. *Psychological Bulletin* 126: 946–963.

Widiger, T. A, and L. M. Sankis. 2000. Adult psychopathology: Issues and controversies. *Annual Review of Psychology* 51: 377–404.

Wiggins, O., and M. Schwartz. 1994. The limits of psychiatric knowledge and the problem of classification. In *Philosophical perspectives on psychiatric diagnostic classification*, ed. J. Sadler, O. Wiggins, and M. Schwartz, 89–103. Baltimore, MD: Johns Hopkins University Press.

Wilkerson, T. 1993. Species, essences, and the names of natural kinds. *Philosophical Quarterly* 43: 1–19.

Wilkinson, S. 2000. Is "normal grief" a mental disorder? *Philosophical Quarterly* 50: 290–304.

Wilson, G. T., P. E. Nathan, K. D. O'Leary, and L. A. Clark. 1996. *Abnormal psychology: Integrating perspectives*. Needham Heights, MA: Allyn & Bacon.

Wimsatt, W. C. 1976. Reductionism, levels of organization, and the mind-body problem. In *Consciousness and the brain: A scientific and philosophical inquiry*, ed. G. G. Globus, G. Maxwell, and I. Savodnik, 205–267. London: Plenum.

Wimsatt, W. C. 1986. Forms of aggregativity. In *Human nature and natural knowledge*, ed. A. Donagan, A. N. Perovich, and M. V. Wedin, 259–291. Dordrecht, the Netherlands: Reidel.

Winick, B. J. 1995. Ambiguities in the legal meaning and significance of mental illness. *Philosophy, Public Policy, and Law* 1: 534–611.

Winzeler, R. L. 1995. *Latah in Southeast Asia: The history and ethnography of a culture-bound syndrome*. Cambridge: Cambridge University Press.

Wood, J. M., M. T. Nezworski, S. O. Lilienfeld, and H. N. Garb. 2003. *What's wrong with the Rorschach? Science confronts the controversial inkblot test.* San Francisco, CA: Jossey-Bass.

Wood, J. W., K. A. O'Connor, D. J. Holman, E. Brindle, S. E. Barsom, and M. A. Grimes. 1998. The evolution of menopause by antagonistic pleiotropy. Department of Anthropology, Penn State University.

Woodward, J. 2000. Explanation and invariance in the special sciences. *British Journal for the Philosophy of Science* 51: 197–254.

Woodward, J. 2002. There is no such thing as a ceteris paribus law. *Erkenntnis* 57: 303–328.

Woolfolk, R. L. 1998. *The cure of souls.* San Francisco, CA: Jossey-Bass.

Woolfolk, R. L. 1999. Malfunction and mental illness. *The Monist* 82: 658–670.

Woolfolk, R. L. 2001. The concept of mental disorder: An analysis of four pivotal issues. *Journal of Mind and Behavior* 22: 161–178.

Woolfolk, R. L., and D. Murphy. 2004. Axiological foundations of psychotherapy. *Journal of Psychotherapy Integration* 14: 168–191.

Woolfolk, R. L., J. Novalany, M. Gara, L. Allen, and M. Polino. 1995. Self-complexity, self-evaluation, and depression: An examination of form and content within the self-schema. *Journal of Personality and Social Psychology* 68: 1108–1120.

World Health Organization. 1978. *Mental disorders: Glossary and guide to their classification in accordance with 9th revision of International Classification of Diseases.* Geneva: World Health Organization.

World Health Organization. 1992a. *DCR-10: The ICD-10 classification of mental and behavioural disorders: Diagnostic criteria for research.* Geneva: World Health Organization.

World Health Organization. 1992b. *ICD-10: The international statistical classification of diseases and related health problems.* 10th rev. 2 vols. Geneva: World Health Organization.

World Health Organization. 1993. *CDDG-10: The ICD-10 classification of mental and behavioural disorders: Clinical descriptions and dignostic guidelines.* Geneva: World Health Organization.

Wright, L. 1973. Functions. *Philosophical Review* 82: 139–168.

Yaffe, G. 2002. Recent work on addiction and responsible agency. *Philosophy and Public Affairs* 30: 178–221.

Yapko, M. 1994. *Suggestions of abuse.* New York: Simon & Schuster.

Young, A. 1995. *The harmony of illusions: Inventing post-traumatic stress disorder.* Princeton, NJ: Princeton University Press.

Young, A., and K. Leafhead. 1996. Betwixt life and death: Case studies of the Cotard delusion. In *Method in madness: Case studies in cognitive neuropsychology*, ed. P. Halligan and J. Marshall, 147–171. Hove, UK: Psychology Press.

Zachar, P. 2000. Psychiatric disorders are not natural kinds. *Philosophy, Psychiatry, Psychology* 7: 168–182.

Zawidski, T., and B. Bechtel. 2004. Gall's legacy revisited: Decomposition and localization in cognitive neuroscience. In *Mind as a scientific object: Between brain and culture*, ed. C. E. Erneling and D. M. Johnson. Oxford: Oxford University Press.

Index

Printed in the United States
by Baker & Taylor Publisher Services